Springer Series in Electrophysics
Volume 4

Edited by Leopold B. Felsen

Springer Series in Electrophysics

Editors: Günter Ecker   Walter Engl   Leopold B. Felsen

Volume 1   **Structural Pattern Recognition**
By T. Pavlidis

Volume 2   **Noise in Physical Systems**
Editor: D. Wolf

Volume 3   **The Boundary-Layer Method in Diffraction Problems**
By V. M. Babič, N. Y. Kirpičnikova

Volume 4   **Cavitation and Inhomogeneities**
in Underwater Acoustics
Editor: W. Lauterborn

Cavitation and Inhomogeneities
in Underwater Acoustics

Proceedings of the First International
Conference
Göttingen, Fed. Rep. of Germany,
July 9–11, 1979

Editor W. Lauterborn

With 192 Figures

Springer-Verlag
Berlin Heidelberg New York 1980

Professor Dr. *Werner Lauterborn*

Drittes Physikalisches Institut der Universität Göttingen,
Bürgerstraße 42–44, D-3400 Göttingen, Fed. Rep. of Germany

Series Editors:

Professor Dr. Günter Ecker

Ruhr-Universität Bochum, Theoretische Physik, Lehrstuhl I,
Universitätsstrasse 150, D-4630 Bochum-Querenburg, Fed. Rep. of Germany

Professor Dr. Walter Engl

Institut für Theoretische Elektrotechnik, Rhein.-Westf. Technische Hochschule,
Templergraben 55, D-5100 Aachen, Fed. Rep. of Germany

Professor Leopold B. Felsen, PhD

Polytechnic Institute of New York, 333 Jay Street, Brooklyn, NY 11201, USA

ISBN 978-3-642-51072-4   ISBN 978-3-642-51070-0 (eBook)
DOI 10.1007/978-3-642-51070-0

To Nature
Tormented by Men

Preface

The idea of organizing a conference on cavitation was born at the 8th International Symposium on Nonlinear Acoustics in Paris, 1978. This idea took further shape during a visit of Dr. L. Bjørnø to Göttingen where the final decision concerning the scope of the conference was made. We both felt the need to pierce the walls of traditional areas and to bring together specialists from related fields for fruitful interaction. Cavitation bubbles in liquids may be looked at as a kind of disturbance in an otherwise homogeneous medium and subsumed under the heading of *inhomogeneities*. A quick discussion revealed that so much work was in progress all over the world as to make an exchange of ideas really worthwhile. Thus the first conference on *Cavitation and Inhomogeneities in Underwater Acoustics* was set forth and entered a state of strong and serious activity beginning in July 1978 and culminating in the Proceedings now at hand.

As expected the conference brought together a wide variety of topics and characters and a vivid exchange of ideas. At least five sections were found necessary to group the papers: *Cavitation, Sound Waves and Bubbles, Bubble Spectrometry, Particle Detection* and *Inhomogeneities in Ocean Acoustics*.

In *Cavitation* (Part I) the new method of producing cavitation by photons has led to renewed attack on problems which some time ago were thought to be beyond the reach of experimenters, e.g., the bubble collapse problem. This well-established problem curiously reappears in a very modern context in connection with the laser driven collapse of hollow targets in laser fusion studies. But this topic was considered too far removed from the scope of this conference and therefore not included.

When bubbles are present in a liquid they may strongly influence the passage of sound waves (Part II). Indeed, such effects as shock wave and soliton formation, self-induced transparency and, of course, frequency mixing effects are reported. This terminology strongly bears relationship to the newly expanding field of nonlinear optics.

Inhomogeneities in the form of bubbles strongly determine the nonlinear properties of the medium. Therefore quite an effort has been undertaken from different sides to measure their size distributions (Part III). Acoustical and optical methods and extensive computer processing are involved in accomplishing this task.

Project DUMAND (Deep Undersea Muon and Neutrino Detection) was included as a special topic to show of how underwater acoustics may become relevant to farther removed fields in physics. The acoustical detection of astrophysical neutrinos in the ocean (Part IV) is a challenge to every ocean acoustician. But in order not to expand the conference in too many directions bubble chamber problems were not included.

Inhomogeneities also appear in the ocean on a larger scale than ordinary bubbles. They may not be so easily seen as bubbles, but due to the long distances involved they have a marked influence on sound propagation. The mixing of water of different temperatures and salinities gives rise to local sound speed variations. Currents, tides, eddies, and internal waves do likewise. These topics are covered in Part V.

A Congress cannot be organized without the aid of many helping hands and minds. First of all I want to thank Dr. L. Bjørnø for his steady help in all questions concerning the organization. He shared much of my burden and kept a watchful eye on bottlenecks in the countdown towards the conference. Then there was the help of the International Advisors Committee

V.A. Akulichev, USSR	*H. Medwin,* USA
R.E. Apfel, USA	*E.A. Neppiras,* UK
P.A. Crowther, UK	*A. Prosperetti,* Italy
V.A. Krasil'nikov, USSR	

in suggesting topics and lecturers. My special thanks here go to Dr. H. Medwin for his activities.

Our cheerful Local Organizing Committee from the Third Physical Institute where the conference took place:

K.J. Ebeling	*R. Timm*
E. Cramer	*A. Vogel*
G. Haussmann	*B. Binnewies*
W. Hentschel	*H. Präkelt*

made things function smoothly without interruption during the days of the conference. They, with their whole-hearted input were the sponsors of the conference making other sponsors superfluous.

Last but no least I admire the patience of Mrs. B. Binnewies and Mrs. H. Präkelt in executing the editorial work that partly turned out to become real labour. In the editorial work I obtained the help of Dr. J. Burt which is greatly appreciated. Many figures were redrawn by Mrs. Liebe for the benefit of the reader. The photographic work was done by Mrs. Kirschmann-Schröder. All this necessary work caused some delay in publishing but I hope will be compensated for by the final appearance of the book.

Herewith it is rendered to the scientific community.

Göttingen, December 1979   *Werner Lauterborn*

Contents

Part I *Cavitation*

Cavitation and Coherent Optics. By W. Lauterborn (With 10 Figures)   3

On the Dynamics of Non-Spherical Bubbles
By A. Prosperetti (With 3 Figures) 13

Oscillation and Collapse of a Cavitation Bubble in the Vicinity of a
Two-Liquid Interface. By G.L. Chahine and A. Bovis (With 3 Figures)   23

Experimental Investigation of Bubble Collapse at Laser-Induced
Breakdown in Liquids. By V.S. Teslenko (With 3 Figures) 30

Application of High Speed Holocinematographical Methods in
Cavitation Research. By K.J. Ebeling (With 8 Figures) 35

Bubble Collapse Studies at a Million Frames per Second
By W. Lauterborn and R. Timm (With 6 Figures) 42

Holographic Generation of Multi-Bubble Systems
By W. Hentschel and W. Lauterborn (With 8 Figures) 47

The Dynamics and Acoustic Emission of Bubbles Driven by a Sound Field
By E. Cramer (With 5 Figures) 54

Free and Forced Oscillations of Spherical Gas Bubbles and Their
Translational Motion in a Compressible Fluid
By H.J. Rath (With 8 Figures) 64

Acoustic Cavitation and Bubble Dynamics Due to a Tension Wave
By R.A. Wentzell and G.J. Lastman (With 3 Figures) 72

Some New Results on Cavitation Threshold Prediction and Bubble
Dynamics. By R.E. Apfel (With 2 Figures) 79

Acoustic Cavitation Thresholds in Water. By L.A. Crum (With 8 Figures)   84

The Influence of Modest Overpressures on the Persistence of Air
Bubbles in Water. By A. Evans (With 1 Figure) 90

On the Collapse of Cavity Clusters in Flow Cavitation
By K.A. Mørch (With 5 Figures) 95

Effect of Polarization on Electric Pulses Produced by Cavitation
Bubbles. By G. Gimenez and F. Goby (With 8 Figures) 101

Cavitation Effects at Megahertz Frequencies. By P.W. Vaughan,
S. Leeman, M. Hedges, E. Graham, and P. Sutton (With 5 Figures) ... 108

Nonlinear Sound-Scattering by Small Bubbles. By P.M. Tilmann 113

Dynamics of a Cylindrical Cavity in a Boundless Compressible Liquid
By V.K. Kedrinskii and V.T. Kuzavov (With 2 Figures) 119

Part II *Sound Waves and Bubbles*

Sound and Shock Waves in Bubbly Liquids
By L. van Wijngaarden (With 6 Figures) 127

On the Amplification of Modulated Acoustic Waves in Gas-Liquid
Mixtures. By F.H. Fenlon and J.W. Wonn (With 4 Figures) 141

Self-Induced Transparency and Frequency Conversion Effects for Acoustic
Waves in Water Containing Gas Bubbles. By Yu.A. Kobelev,
L.A. Ostrovsky, and A.M. Sutin (With 3 Figures) 151

Pressure Waves in a Liquid with Gas or Vapour Bubbles
By V.E. Nakoryakov, B.G. Pokusaev, and I.R. Shreiber (With 4 Figures) 157

Dynamics of a Liquid with Gas Bubbles During Interaction with Short
Large-Amplitude Pulses. By N.V. Malykh and I.A. Ogorodnikov
(With 7 Figures) 164

Shock Wave Transformation in Bubbly Liquids
By V.K. Kedrinskii (With 6 Figures) 170

Relaxation Effects in the Propagation of Underwater Shock Waves
By T.N. Fedoseeva, F.E. Fridman, V.N. Goldberg, and I.G. Zarnitsina
(With 6 Figures) 177

Part III *Bubble Spectrometry*

Acoustical Bubble Spectrometry at Sea. By H. Medwin (With 4 Figures)   187

Acoustical Scattering from Near-Surface Bubble Layers
By P.A. Crowther (With 5 Figures) 194

Density of Air-Bubbles Below the Sea Surface, Theory and Experiments
By Ir.P. Schippers (With 2 Figures) 205

Acoustic Measurements of the Gas Bubble Spectrum in Water
By A. Løvik (With 8 Figures) 211

Determination of Bubble Size Spectra by Digital Processing of Holograms
By G. Haussmann (With 6 Figures) 219

Determination of Bubble Sizes by Far Field Diffraction of Photographic
Recordings. By R. Butt and K. Hinsch (With 1 Figure) 225

Complementing Discussion Contribution to the Papers of H. Medwin,
Ir.P. Schippers, and A. Løvik. By E.-A. Weitendorf (With 3 Figures)   230

Part IV *Particle Detection*

Acoustical Detection of Astrophysical Neutrinos in the Ocean
By A. Parvulescu and G.D. Curtis (With 15 Figures) 237

Part V *Inhomogeneities in Ocean Acoustics*

Inhomogeneities in Underwater Acoustics. By L. Bjørnø 261

Sound Propagation in an Inhomogeneous Ocean
By R.H. Mellen (With 11 Figures) 272

Acoustic Fluctuations in the Ocean. By Y.J.F. Desaubies 281

Mesoscale Inhomogeneities and Turbulence in Ocean Acoustics
By P. Scully-Power (With 4 Figures) 294

On the Influence of Stochastic Sound Speed Variations on Acoustic
Transmission Loss in Shallow Water. By H.G. Schneider
(With 9 Figures) 308

The Inverse Backscattering Problem - a Different Approach
By S. Leeman and P.W. Vaughan 315

Index of Contributors 319

Part I

Cavitation

Cavitation and Coherent Optics

W. Lauterborn

Drittes Physikalisches Institut, Universität Göttingen, Bürgerstr. 42-44
D-3400 Göttingen, Fed. Rep. of Germany

1. Introduction

Cavitation, the rupture of liquids, is one of the longstanding problems in hydrodynamics as well as acoustics. It is first mentioned, as far as is known today, by Leonhard Euler in 1754 in his work on the theory of turbines [1]. But it was not until 150 years later, at the turn of the century, that it really became a problem in connection with ship propellers. Nowadays, cavitation presents a problem in all kinds of hydraulic machinery, especially ship propellers, turbines, pumps, and hydrofoils. This type of cavitation is called *hydraulic cavitation* [2].

In acoustics cavitation showed up in connection with high intensity underwater sound projectors which first became available in the late 1920's. It presents a problem in sonar systems and has become known as *acoustic cavitation.* [3], [4].

Besides these two classical fields of cavitation new ones are gradually emerging. Soon after the invention of the laser cavitation has been observed to occur in a liquid when irradiated with light of high intensity [5,6]. But it took some time until this new possibility has been explored for the benefit of solving problems in cavitation physics, mainly in cavitation bubble dynamics [7-10]. This new form of cavitation is called *optic cavitation* [10].

In optic cavitation photons are used to rupture the liquid. But there is no cause why not any other sort of elementary particles, may it be protons or neutrinos, may give rise to a similar breakdown in the liquid. Indeed, this effect is used since the 1950's in the bubble chamber. From the viewpoint taken here, one may speak of *particle cavitation*. As in the other cases there will be a threshold for this type of cavitation to occur. Below the threshold sound waves may be generated via the thermoacoustic effect [11]. Indeed, thoughts have been put forward of how to detect muons and neutrinos from astrophysical sources via this effect in the ocean [12].

In Fig.1 a first systematic approach is tried to group the quite different kinds of cavitation. According to this scheme cavitation may be brought about by two mechanisms, either *tension* in the liquid or a *local deposit of energy*. These seem to be the only possible main categories to classify cavitation. Optic cavitation has been emphasized in Fig.1 because it is the central topic of this paper.

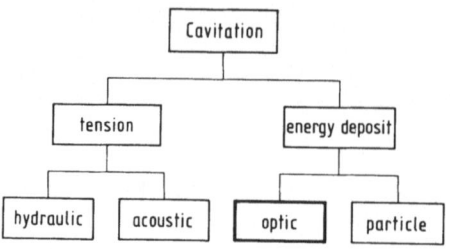

Fig.1 Classification scheme for the different kinds of cavitation

Besides giving rise to the new field of optic cavitation coherent optics has a further impact on cavitation physics. Cavitation bubbles usually appear statistically throughout a larger volume of the liquid and are small and fast moving objects, especially during collapse. Thus they are not easily photographed and, indeed, have revealed significant features of their dynamics only in sophisticated model systems [13], [14]. Holography due to its three-dimensional image storage capability has vastly increased the possibility to observe cavitation in liquids. It may also be used in conjunction with optic cavitation to produce multiple breakdown sites in the liquid for bubble interaction studies. Both applications of holography have been introduced into cavitation physics in our laboratory and will be discussed as a second topic of this paper.

2. Optic Cavitation

High light intensities are needed to produce cavities in liquids. Nowadays, many different laser systems exist being capable of delivering the necessary intensity and energy in a short enough time like the ruby, neodymium glass and of course any laser planned for nuclear fusion studies provided the liquid under study is transparent enough for the wavelength of the laser. In our experiments Q-switched ruby laser pulses are used to achieve breakdown and cavity formation in the liquid. The experimental set-up is given in Fig.2. The single-stage ruby is Q-switched by a Kerr-cell and delivers light pulses of up to one Joule total energy at a pulse duration of about 30 to 50 ns. The beam diameter is about 1 cm. The light is focused into the liquid (mostly water) with a single lens of short focal length (44 mm when submerged in water). The cavity or the cavities appear at, or in the vicinity of, the focal point. They are diffusely illuminated by a flash lamp through a ground glass plate and photographed by a Beckman-

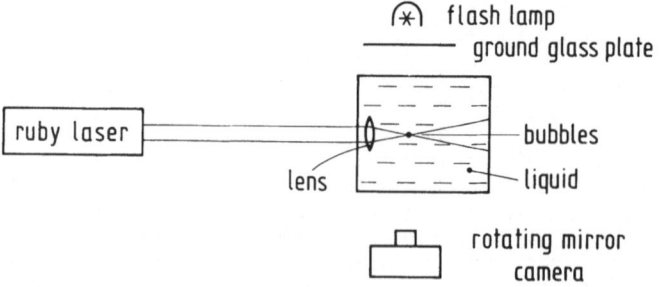

Fig.2 Experimental set-up used for high speed photographic studies of laser-induced cavities in liquids

4

Whitley Model 330 rotating mirror camera which allows framing rates of up
to a million frames/s. Under these backlighting conditions the cavities
look dark on a bright background because the light is deflected off at the
bubble-liquid interface. Smooth spherical bubbles show a bright central
spot because there the light can pass undeflected.

In a series of experiments cavity oscillations in the bulk of the liquid
(water and silicone oil), cavity dynamics near plane solid boundaries and
at air-liquid interfaces, and the dynamics of interacting cavities have
been investigated [7-10],[15], [16]. The main results of these high speed
photographic studies are also documented in a film available from the Institut für den Wissenschaftlichen Film [17].

The investigations show that one peculiar dynamical feature seems to
pervade all dynamics, i.e. jet formation. It has been predicted by KORNFELD
and SUVOROV [18] in 1944 to explain cavitation erosion, but jets were not
observed before the 1960's by NAUDE and ELLIS [13] and BENJAMIN and ELLIS
[14]. The first numerical calculations of jet formation from initially
spherical cavities were done by PLESSET and CHAPMAN [19] in 1971. They
could be largely confirmed experimentally with laser-produced cavities in
1975 by LAUTERBORN and BOLLE [10].

A typical sequence of pictures with jet formation upon collapse of a cavity in the neighborhood of a solid boundary is shown in Fig.3. The bubble
is produced at a distance of b = 4.5 mm from the solid boundary to be seen
in the lower part of each frame. It reaches a maximum radius of R_m = 1.1 mm.
Thus the ratio b/R_m, important for normalization, becomes 4.2. This is a
rather high value, nevertheless a pronounced jet towards the boundary is
formed upon collapse as can be seen in the third row of Fig.3. The sequence
has been taken at 75 000 frames/s. This is not sufficient to really resolve
the collapse. Much higher framing rates are needed, even exceeding a million
frames/s [20]. The liquid jet towards the boundary is formed by involution
of the top of the cavity leading to a torus-like cavity in the later stages

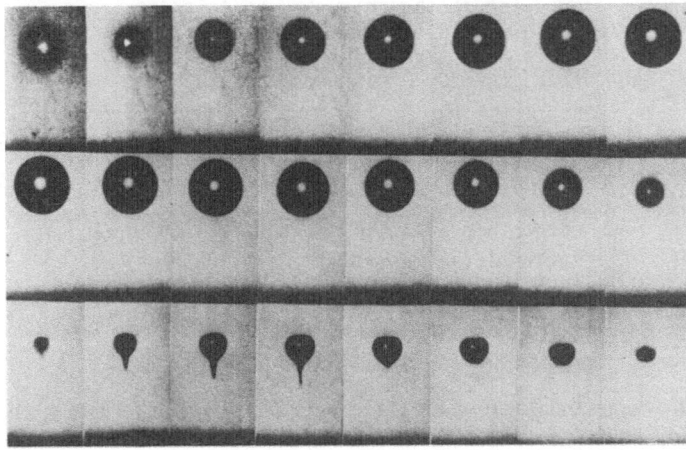

Fig.3 Jet formation upon collapse of a spherical cavity in the neighborhood
of a plane solid boundary in water. Maximum bubble radius: 1.1 mm, distance
of bubble center from the wall: 4.5 mm, framing rate: 75 000 frames/s. (The
first frame (upper left) is doubly exposed with another cavity.)

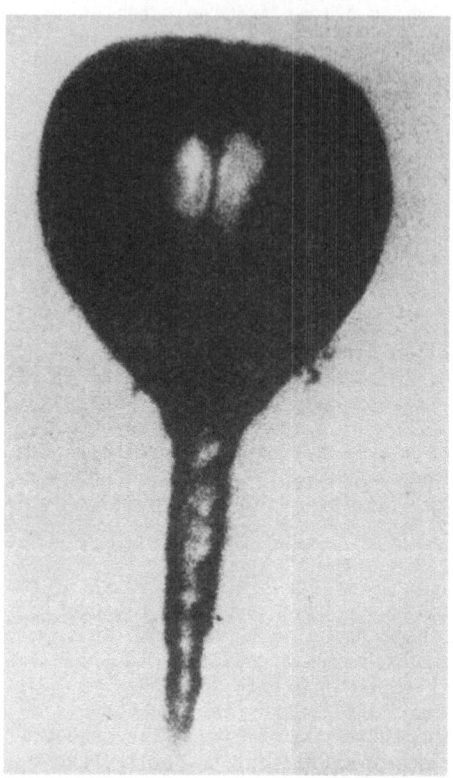

Fig.4 Cavity in water during its
first rebound after collapse near a
plane solid boundary below with a
typical, well developed jet

of the collapse [19,10]. The funnel-shaped protrusion which gives the re-
bounding bubble its remarkable appearance is the cavity-liquid boundary
produced by the impinging liquid jet in the interior. An enlarged frame of
a cavity after its first collapse with a typical, well developed jet is
given in Fig.4. The liquid jet is directly visible as a fine dark line only
in the bright central spot of the cavity. Of interest are the sizes and
velocities of such jets. Both quantities are difficult to measure and only
preliminary values can so far be given. Jet diameters seem to lie in the
range from 10 to 100 µm and maximum jet velocities in the range of 100 to
200 m/s [9], [10].

Cavities collapsing near each other also show strong jet formation. This
is evident in the case of two equally sized cavities collapsing simultaneously. The problem is equivalent to the collapse of a cavity in the neighborhood of a solid wall at half the distance between the cavities. Both
cavities should develop a jet towards each other. This is confirmed by
experiments [21], [9].

When the cavities are of different sizes strongly different phenomena
are observed depending upon the relative size of the cavities and their
distance. A very small cavity near a big one develops a jet away from the
big one [17], perhaps due to the shock waves emitted upon breakdown. A cavity of smaller but comparable size develops a jet towards the bigger one. An
example is shown in Fig.5. The sequence has been taken at 75 000 frames/s.
The smaller cavity is flattened at the side of the bigger one in the expan-

6

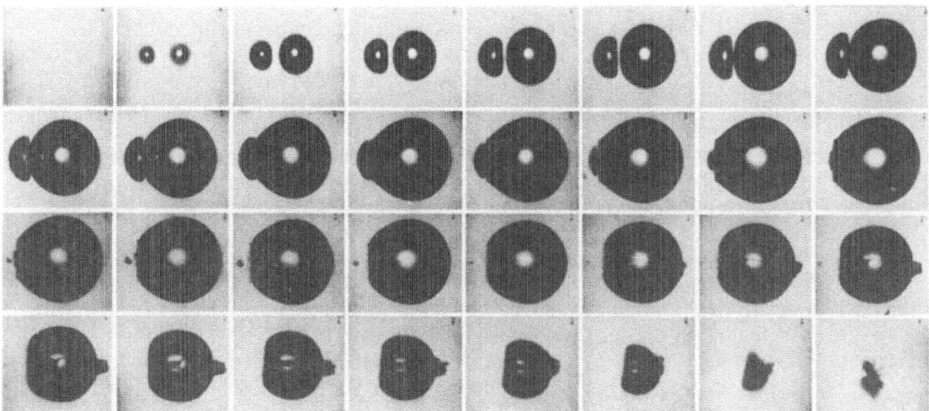

Fig.5 Jet formation upon collapse of a smaller cavity in the neighborhood of a bigger one. Framing rate: 75 000 frames/s, frame size: 5 mm x 6 mm

sion cycle and then collapses on the surface of the bigger one leaving behind a strong jet penetrating the remaining bigger cavity. The smaller cavity itself has totally disappeared except for a few tiny bubbles appearing attached to the now flattened side of the bigger cavity.

The intermediate case between jet formation away from the bigger cavity and towards it shows an interesting solution of nature. It decides just to produce two jets simultaneously, one away and one towards the bigger cavity as shown in Fig.6. The sequence has again been taken at 75 000 frames/s. The double jet formation cannot be explained by a simple involution of one side of the cavity. Calculations of CHAPMAN and PLESSET [22] on earthlike

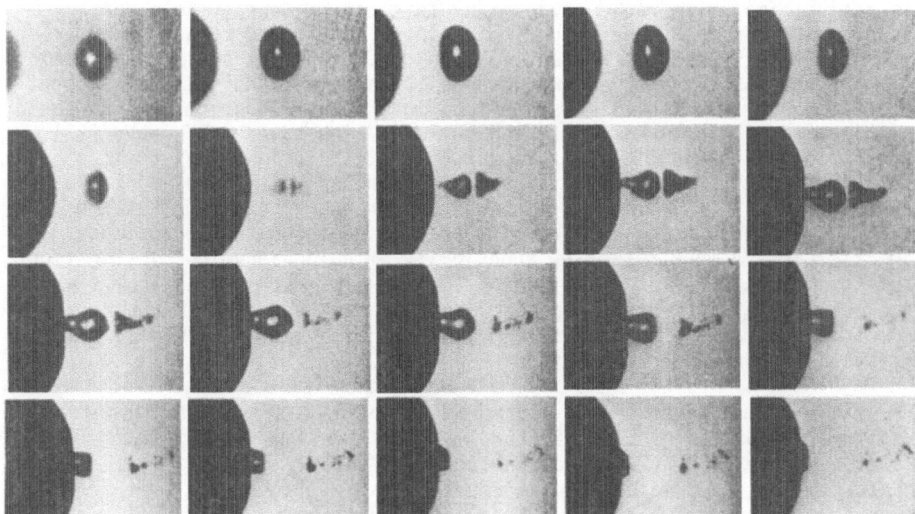

Fig.6 Double jet formation upon collapse of a smaller cavity in the neighborhood of a bigger one. Framing rate: 75 000 frames/s, frame size: 2.25 mm x 3.5 mm

flattened cavities (like that in Fig.6) show that the equator contracts faster than the polar regions leading to a dumbbell shape. The calculations cannot easily be conducted beyond that point. But obviously the liquid masses rushing inwards from the equator are squeezed out at the poles. This result is in agreement with a conjecture made to explain the abundance of jets in cavitation bubble dynamics [10], [15]. The conjecture is that at least in the simple cases of the experiments jet formation depends on the curvature of the cavity-liquid interface in the following manner: parts of higher curvature collapse faster than less curved parts.

An interesting question next to be posed is what happens when two cavities collapse in the neighborhood of a solid wall because this situation is approaching more real ones. It is found that the jet is now inclined both to the boundary and to the other cavity. The issue is that another cavity can deflect the jet away from the boundary. This is an important fact as jet formation is believed to be at least one source of damage to solids from cavitation. The observation that jets are deflected away from the boundary by other cavities suggests that dense clouds of cavities as often encountered in hydraulic as well as acoustic cavitation may have less damage capabilities than loosely distributed ones. Of course, this cannot be said for sure because of the expected totally different dynamics of bubble clouds due to cooperative effects. Such cooperative effects might be brought about by an acoustic coupling through the sound and shock waves emitted during the motion of the cavities.

3. Holography in Cavitation Physics

3.1 High-Speed Holocinematography to Observe Cavitation

Cavitation bubbles usually appear in clouds or clusters. To study their threedimensional motion and interaction a method is needed to store three-dimensional images in rapid succession. This can be done by holographic methods, but, in fact, the holographic equivalent of the rotating mirror or image converter camera is needed. No such devices are at present available. They are necessarily complex and not easily conceived and set up. We spent a considerable amount of time and effort into the development of high-speed holocinematography starting from conventional high-speed holography [23]. An example of an acoustic cavitation bubble field inside a cylindrical piezoelectric transducer as obtained holographically is shown in Fig.7, taken from BADER [23]. The two pictures are photographed from the same hologram by focusing onto different planes in depth of the real image. Fig. 7 clearly demonstrates the superiority of holography in investigating cavitation bubble fields. Just one of the two pictures of Fig.7 could have been obtained by conventional photography and many more planes can be taken from the hologram. By using double pulse holography velocity fields of cavitation bubbles have been evaluated [23], [24].

Essentially three different arrangements for high-speed holocinematography were reported by EBELING [25-31]. He uses either a multiply Q-switched ruby laser [25-29] or an argon ion laser externally modulated by an acousto-optic modulator [30]. A fourth method has recently been given by KUHNKE [32]. The set-up is shown in a condensed form in Fig.8. The cw-light beam of an argon ion laser pumped dye laser, operating at the wavelength of the ruby laser (694,3 nm), is chopped at up to 30 kHz to deliver an (almost) infinite train of light pulses. A short sequence of these pulses (about eight) is amplified in a ruby amplifier and used in a conventional holo-

8

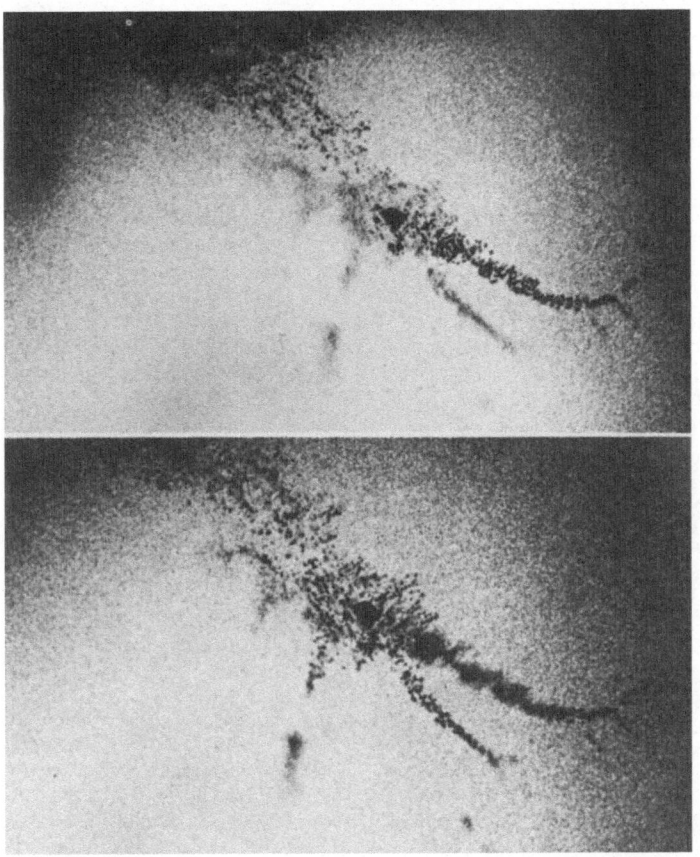

Fig.7 Two different planes in depth about 5 mm apart as obtained from a hologram of an acoustic cavitation field inside a hollow cylindrical piezoelectric transducer

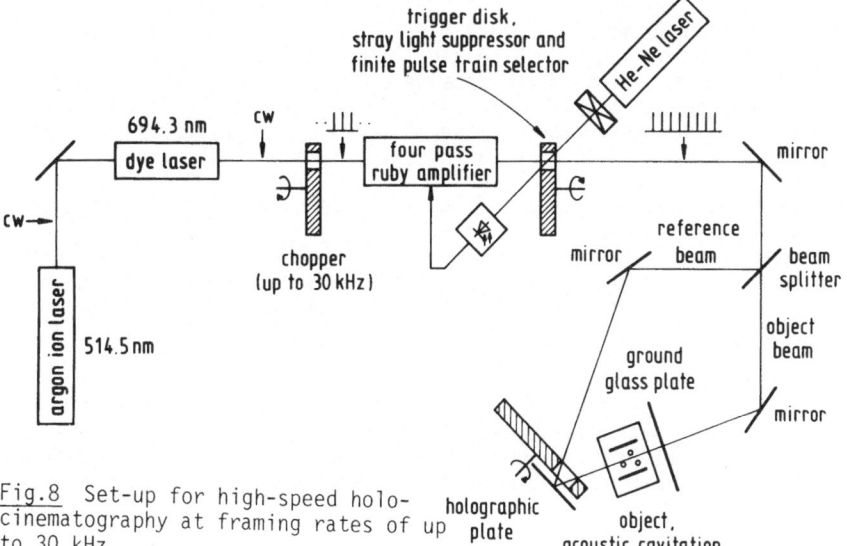

Fig.8 Set-up for high-speed holocinematography at framing rates of up to 30 kHz

9

graphic arrangement to illuminate the holographic plate. To illuminate different areas of the plate with different pulses a disk with apertures rotates in front of the plate at the appropriate speed. Thus a series of holograms is obtained with separately reconstructable images. The trigger disk in the output beam of the ruby amplifier also acts as a shutter for the non-amplified dye laser pulses which would give rise to a preexposure of the holographic plate.

The collapse of a cavity in an acoustic cavitation bubble field as reconstructed from a hologram series is shown in Fig.9. The holographic framing rate is 30 000 holograms per second. The shock wave radiated upon collapse is clearly visible and seems to be spherical. This indicates a spherical collapse down to a very small size of the cavity. Aspherical cavities may collapse by parts giving rise to several shock waves at different times upon collapse [20]. More examples illustrating the holographic methods are given by EBELING in this volume [31].

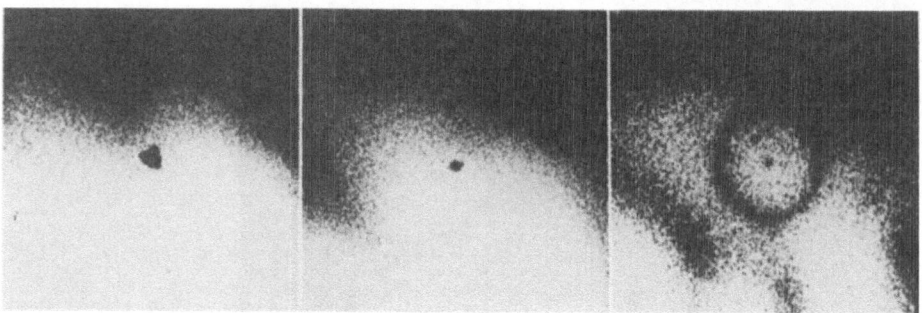

Fig.9 Example of the collapse of an acoustic cavitation bubble. Reconstruction from three frames of a holographic series taken at 30 000 holograms per second. Frame size 27 mm x 27 mm [33]

3.2 Holography in Producing Cavitation

Holography can also be applied to achieve multiple breakdown in a liquid for cavity interaction studies. Such studies are of great importance in connection with noise emission and erosion problems in both hydraulic and acoustic cavitation. To clarify the cavity interaction dynamics reproducible cavity configurations are needed. It seems that optic cavitation is so far the only method which allows complex investigations of that kind.

The holographic approach as shown in Fig.10 is especially appealing as the experimental set-up remains conceptually as simple as before. Only the focusing lens (see Fig.2) needs to be replaced by the appropriate hologram which in this case is also called a holographic lens. Such a lens can be regarded as a generalization of a usual lens as the incoming light beam is focused into different points in space.

Strong difficulties are encountered when trying to realize this idea. Only phase holograms can be used due to the high light intensities. Then, calculated holograms would be desirable, otherwise the flexibility of the method would be lost. This presents strong problems for the copying process, as very small structures of a few micrometer are involved. We nevertheless

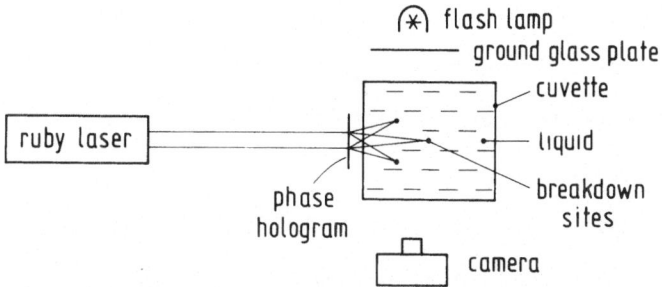

Fig.10 Set-up for multiple cavity generation in liquids by holographic focusing

succeeded in realizing this approach [33]. Once the technical difficulties are overcome it will become a powerful method for sophisticated measurements on cavitation bubble dynamics.

4. Acknowledgment

The author wants to express his thanks to both the Fraunhofer-Gesellschaft, München, and the Deutsche Forschungsgemeinschaft, Bonn, which only slightly hesitated to support the ideas which at the time of proposal were likely to fail. I also thank all my numerous coworkers who enthusiastically and with great ability conducted much of the work reported here.

5. References

1   L. Euler: Histoire ae l'Académie Royale des Sciences et Belles Lettres, Mem. R. 10, 1754. Berlin 1756. Classe de Philosophie expérimentale, p. 227-295; the remarks on the rupture of the liquid from the walls are made in chapter 81, p. 266-267 (in French)

2   R.T. Knapp, J.W. Daily, F.G. Hammitt: *Cavitation* (McGraw-Hill, New York 1970)

3   H.G. Flynn: "Physics of acoustic cavitation in liquids", in: *Physical Acoustics*, Vol 1B, W.P. Mason (ed.), (Academic Press, New York 1964) pp. 57-172

4   R.E. Apfel: "Acoustic cavitation", in: *Ultrasonics*, P. Edmonds (ed.), Academic Press, New York, to appear

5   G.A. Askar'yan, A.M. Prokhorov, G.F. Chanturiya, G.P. Shipulo: Sov. Phys. - JETP 17, 1463 (1963)

6   R.G. Brewer, K.E. Rieckhoff: Phys. Rev. Lett. 13, 334a (1964)

7   W. Lauterborn, in: *Non-steady flow of water at high speed*, Proc. IUTAM-Symposium, Leningrad June 22-26, 1971, L.I. Sedov and G.Yu. Stepanov (eds.), (Nauka Publ. House, Moscow 1973) pp. 267-275

8   W. Lauterborn: Appl. Phys. Lett. 21, 27 (1972)

9 W. Lauterborn: Acustica <u>31</u>, 51 (1974)

10 W. Lauterborn, H. Bolle: J. Fluid Mech. 72, 391 (1975)

11 P. Westervelt: J. Acoust. Soc. Am. <u>64</u>, S105 (1978)

12 A. Parvulescu, G.D. Curtis: this volume

13 C.F. Naudé, A.T. Ellis: J. Basic Eng., Trans. ASME D <u>83</u>, 648 (1961)

14 T.B. Benjamin, A.T. Ellis: Phil. Trans. London A <u>260</u>, 221 (1966).

15 W. Lauterborn: Phys. Bl. <u>32</u>, 553 (1976)

16 W. Lauterborn: Laser + electro-optik <u>9</u>, 26 (1977)

17 W. Lauterborn, H. Bolle, Inst. Wiss. Film: Film E 2353 (1977). Available from: Institut für den Wissenschaftlichen Film, Nonnenstieg 72, D-3400 Göttingen, Germany

18 M. Kornfeld, L. Suvorov: J. Appl. Phys. <u>15</u>, 495 (1944)

19 M.S. Plesset, R.B. Chapman: J. Fluid Mech. <u>47</u>, 283 (1971)

20 W. Lauterborn, R. Timm: this volume

21 E.E. Timm, F.G. Hammitt: *Cavitation Forum* (ASME, 1971) pp. 18-20

22 R.B. Chapman, M.S. Plesset: J. Basic Eng., Trans. ASME D <u>94</u>, 142 (1972)

23 F. Bader: Ph.D. dissertation (University of Göttingen 1973)

24 K. Hinsch, F. Bader, W. Lauterborn, in: *Finite-amplitude wave effects in fluids*, L. Bjørnø (ed.), (IPC Press, Guildford 1974) pp. 240-244

25 K.J. Ebeling: Ph.D. dissertation (University of Göttingen 1976)

26 K.J. Ebeling: Optik <u>48</u>, 383-397 and 481-490 (1977)

27 K.J. Ebeling, W. Lauterborn: Opt. Comm. <u>21</u>, 67 (1977)

28 W. Lauterborn, K.J. Ebeling: Appl. Phys. Lett. <u>31</u>, 663 (1977)

29 K.J. Ebeling, W. Lauterborn: Appl. Opt. <u>17</u>, 2071 (1978)

30 K.J. Ebeling, in: Proc. First European Conference on Optics Applied to Metrology, M. Grosman and P. Meyrneis (eds.), SPIE Vol. <u>136</u>, 348 (1977)

31 K.J. Ebeling: this volume

32 K. Kuhnke: Ph.D. dissertation (University of Göttingen 1979)

33 W. Hentschel, W. Lauterborn: this volume

On the Dynamics of Non-Spherical Bubbles

A. Prosperetti

Istituto di Fisica, Università di Milano
I-20133 Milano, Italy

To-day one of the most challenging questions that the theoretical modelling of the behavior of bubbles in liquids must face is that of the dynamics of non-spherical bubbles. Indeed, while the study of situations endowed with spherical symmetry has reached a high degree of sophistication (see [1,2] for recent reviews), we may say that, on both absolute and relative terms, our current theoretical understanding of the general nonsymmetric case is still in its infancy. In essence three approaches have been developed so far to deal with this difficult problem, namely perturbation methods, numerical techniques, and variational formulations. In this study we shall describe applications of these approaches, and we shall present a new technique which appears superior to the variational method in a number of cases. The emphasis of this study is on technique rather than on results. We feel that this feature is amply motivated by the present state of our knowledge in this field.

1. The Instability of a Collapsing Bubble

A number of studies which make use of linearized perturbation theory to deal with slightly non-spherical bubbles is available [1-3] and, in a few others, the approximation has been carried one step farther to the weakly non-linear approximation [3-5]. It appears that this approach is at present the only one capable of yielding results with resolution of detail comparable to the spherically symmetric case. As an application of this method we shall investigate the effect of viscosity on the instability of the spherical shape for a collapsing bubble.

In [6] it was shown that, if the bubble surface is described by

$$r = R(t) + a_n(t) Y_{nm}(\theta,\varphi) \tag{1}$$

where R is the average radius, $|a_n| \ll R$, and Y_{nm} is a spherical harmonic, then R satisfies the Rayleigh-Plesset equation [1,2] while a_n is the solution of

$$\ddot{a}_n + [3\dot{R}/R - 2(n-1)(n+1)(n+2)\nu/R^2]\dot{a}_n + (n-1)[-\ddot{R}/R + 2(n+1)(n+2)\nu\dot{R}/R^3$$

$$+ (n+1)(n+2)\sigma/\varrho R^3]a_n + n(n+1)(n+2)(\nu/R^2)T(R,t)$$

$$+ n(n+1)\dot{R}/R^2 \int_R^\infty [(R/s)^3 - 1] (R/s)^n T(s,t) ds = 0 \quad , \tag{2}$$

with dots denoting time differentiation. The quantity $T(r,t)$ appearing here is related to the toroidal component of the vorticity and satisfies the equation

$$\nu \frac{\partial^2 T}{\partial r^2} - \frac{\partial T}{\partial t} - \frac{\partial}{\partial r}\left[(R/r)^2 \dot{R}\, T\right] - n(n+1)\frac{\nu}{r^2} T = 0 \qquad , \tag{3}$$

subject to the boundary condition

$$2 R^{n-1} \int_R^\infty s^{-n}\, T(s,t)\, ds + T(R,t) = \frac{2}{n+1}\left[(n+2)\dot{a}_n - (n-1)(R/R)a_n\right], \tag{4}$$

at the bubble (unperturbed) boundary $r = R(t)$. The system (2-4) is still too complex to be treated analytically. Some progress can however be made if it is assumed that vorticity is confined to a thin layer of thickness $\delta \ll R$ adjacent to the bubble sur-face. Clearly this assumption cannot remain valid throughout the collapse of the bub-ble, but it can still be useful to gain some insight into the effect of viscosity on the behavior of the distortion of the spherical shape. An order-of-magnitude discussion of its validity can be found in [7]. Let us set

$$T(r,t) = f(t)\, U(r,t) / r \quad , \tag{5}$$

where $\quad f(t) = (R/R_o)^3 \exp\left\{-[n(n+1)-2]\nu \int_o^t R^{-2}\, dt\right\} \quad$,

with $R_o = R(0)$, and let us introduce the Lagrangian coordinate $h = (r^3 - R^3)/3$. Then from (3) we obtain the following equation for U

$$\nu(R^3+3h)^{4/3} \frac{\partial^2 U}{\partial h^2} + \left\{3R^2 \dot{R}(r^{-3} - R^{-3}) - \nu[n(n+1)-2](r^{-2}-R^{-2})\right\}U = \frac{\partial U}{\partial t} \quad .$$

If we now set $\quad \tau = R_o^{-4}\int_o^t R^4\, dt\quad$ and neglect terms of order 1 or greater in h this

equation becomes, with $\bar{\nu} = \nu R_o^4$

$$\bar{\nu} \frac{\partial^2 U}{\partial h^2} = \frac{\partial U}{\partial \tau} \quad .$$

The solution is readily found to be

$$U(h,\tau) = \frac{1}{2}(\pi\bar{\nu})^{-1/2} h \int_o^\tau U_o(\tau-\theta)\, \theta^{-3/2} \exp(-h^2/4\bar{\nu}\theta)\, d\theta \quad ,$$

where $U_o(\tau) = U(0,\tau)$ must be determined from the boundary condition (4) which, again to zero order in h, is

$$U_o(\tau) + 2 R^{-3}(\bar{\nu}/\pi)^{1/2} \int_o^\tau U_o(\tau-\theta)\, \theta^{-1/2}\, d\theta =$$

$$\frac{2}{n+1} \frac{R}{f}\left[(n+2)\dot{a}_n - (n-1)(\dot{R}/R)a_n\right] \quad . \tag{6}$$

14

An approximate solution of this Volterra integral equation can be obtained truncating the corresponding Neumann series at its second term,

$$U_0(\tau) \cong \frac{2}{n+1} \frac{R}{f} \left[(n+2)\dot{a} - (n-1)(\dot{R}/R)a \right]$$

$$- \frac{2}{n+1} (\bar{\nu}/\pi)^{1/2} (2/R^3) \int_0^\tau \frac{R}{f} \left[(n+2)\dot{a}_n - (n-1)(\dot{R}/R)a_n \right] \theta^{-1/2} d\theta \quad . \tag{7}$$

Notice that successive terms of the complete Neumann series differ by powers of $\nu^{1/2}$. Thus we expect this approximate solution to be valid for small viscosity, which is consistent with the approximation made in neglecting terms of order h and higher. To zero order in h the integral in (2) is negligible. Making use of (5) and (7) to evaluate $T(R,t)$ we then obtain

$$\ddot{a}_n + \left[3\dot{R}/R + 2(n+2)(2n+1)\nu/R^2 \right]\dot{a}_n + (n-1)\left[(n+1)(n+2)\sigma/\varrho R^3 - \ddot{R}/R \right.$$

$$\left. + 2(n+2)\nu/R^3 \, \dot{R} \right] a_n \tag{8}$$

$$- 4n(n+2)\nu(\bar{\nu}/\pi)^{1/2} \int_0^\tau K(\tau,\theta) \left[(n+2)\dot{a}_n - (n-1)(\dot{R}/R)a_n \right] d\theta = 0 \quad ,$$

where $\tau = \tau(t)$ and the kernel K is given by

$$K(\tau,\theta) = \left[\theta^{1/2} R^3(\tau) R^2(\theta) \right]^{-1} \exp\left\{ -[n(n+1)-2]\bar{\nu} \int_\theta^\tau R^{-6} d\eta \right\} \quad .$$

To the present order of approximation in ν the exponential in this formula may be taken as 1. Eq.(8) is an integro-differential equation for a_n for the solution of which efficient numerical techniques exist [8]. An analytical approximate result for $R \to 0$ can however be obtained neglecting the integral term to find, as shown in [7],

$$a_n \sim R^{-1/4} \exp\left\{ \pm i \left[6(n-1) - \frac{1}{4} \right]^{1/2} \log(R/R_0) - (n+2)(2n+1)\nu \int_0^t R^{-2} dt \right\} \quad . \tag{9}$$

It follows from the Rayleigh-Plesset equation that $\dot{R} \propto - R^{-3/2}$ for $R \to 0$, so that the integral appearing here is bounded. It is therefore clear that the effect of a small viscosity is not sufficient to inhibit substantially the instability of the spherical shape for moderate n. However the viscous contribution may be significant in another respect, the breakup of a collapsing bubble. We may picture the bubble breaking up into a number of fragments of order n when $a_n \sim R$, so that the liquid on different sides of the interface comes into contact. It can be seen from (9) that the frequency with which a_n oscillates as $R \to 0$ increases as $n^{1/2}$. Hence, in an interval of time Δt it is much more probable for a_n to go through a maximum for large than for small n. Thus one would expect that breakup into a very large number of fragments is much more likely than breakup into a limited number. The fact that experimentally this number is found to be

15

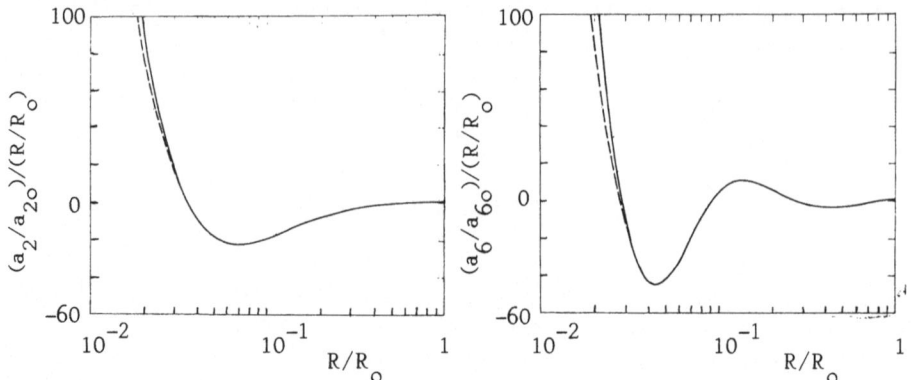

Fig.1 The solutions of (8) in which the integral is neglected for n=2 (left) and n=6 (right) as functions of R/R_o. The solid lines are for N=0 and $N=10^{-4}$, the dashed lines for $N=10^{-3}$. Here $N = \nu/R_o[(p_\infty-p_i)/\varrho]^{1/2}$ (p_∞=ambient pressure, p_i=bubble internal pressure) is the dimensionless viscosity.

of order 10 appears to be a consequence of the viscous contribution in (9) the importance of which increases as n^2. In Fig. 1 we show numerical results obtained from (8) without the last term for $n=2$ and n=6 [7].

2. Numerical Methods for the Dynamics of Non–Spherical Bubbles

Recourse to numerical methods appears to be strongly motivated by the complexity of the problem that we are considering and a limited number of studies applying numerical techniques is in fact available[9–11]. These studies, as well as the numerical approaches to be described in this section, are important for the achievement of specific results (see e.g.[9] on the mechanism of cavitation damage) but, from a methodological viewpoint, they are even more important as a guide towards the development of approximate analytical or semi–analytical techniques.

The problem is considerably simplified if the approximation is made that viscous and thermal effects are negligible. It is not difficult to convince oneself that these assumptions are justified for a "cold" liquid [1] of small viscosity such as water at ambient temperature. We can therefore introduce a velocity potential φ which satisfies Laplace's equation in the liquid and determines the pressure through Bernoulli's equation. The appropriate form of this equation for a point just outside the bubble is

$$\frac{\partial \varphi}{\partial t} + \frac{1}{2}|\nabla\varphi|^2 = \frac{p_\infty - p_i + \sigma K}{\varrho} \quad , \tag{10}$$

since the condition on the normal stresses stipulates that there the pressure p is given by $p = p_i - \sigma K$, where p_i is the internal cavity pressure, σ is the surface tension, and $K = \underline{\nabla}\cdot\underline{n}$ is the curvature of the interface. The negligibility of thermal effects allows one to take p_i constant during the process and equal to the equilibrium vapor pressure of the liquid. The kinematic boundary condition requires that the elements of the free surface displace with the local liquid velocity . Thus the time evolution of the free

16

surface is determined once that φ is known. The boundary conditions associated with Laplace's equation for φ are of the Neumann type at all rigid boundaries, $\partial \varphi/\partial n = 0$, and of the Dirichelet type on the free surface. The velocity potential here can be obtained integrating the equation

$$\frac{d\varphi}{dt} = \frac{\partial \varphi}{\partial t} + \underline{u} \cdot \nabla \varphi = \frac{\partial \varphi}{\partial t} + |\nabla \varphi|^2 = \frac{p_\infty - p_i + \sigma K}{\rho} + \frac{1}{2}|\nabla \varphi|^2 \quad , \tag{11}$$

where d/dt denotes the convective derivative following the motion of the interface and the last step follows from (10).

The final boundary condition for φ stipulates that

$$\lim_{|\underline{x}| \to \infty} \varphi(\underline{x},t) = 0 \quad .$$

It is clear that this condition engenders some difficulty if a finite–difference solution of the problem is attempted, because its application as it stands would require an excessively large integration domain. In [9,10] this problem was avoided with the aid of an iterative procedure which allowed to determine approximately at each time step the potential on a surface at a sufficient distance from the bubble. We have followed here a different route which appears to have the advantages that (i) the outer boundary of the integration domain can be taken nearer to the bubble, and (ii) although in principle an iterative procedure would be required, in practice already the first estimate is sufficiently accurate. This procedure is applicable when the domain is unbounded in all directions or when the boundary is an impermeable plane (so that the problem can be extended by reflection to all space) as would be the case, for instance, for bubble collapse or growth in the neighborhood of a rigid wall. We shall outline the procedure for the case in which the system has also an axis of symmetry (z axis) normal to the plane of symmetry. The solution of Laplace's equation in the liquid may be expanded in a series of Legendre polynomials as follows

$$\varphi(\underline{x},t) = \sum_{k=0}^{\infty} A_{2k} |\underline{x}|^{-(2k+1)} P_{2k}(\cos \theta) . \tag{12}$$

Only terms of even order appear if there is a plane of symmetry. An alternative representation follows from Green's theorem as [12] (the time dependence of φ is temporarily omitted)

$$\varphi(\underline{x}) = \frac{1}{4\pi} \int_S \left(|\underline{x} - \underline{x}'|^{-1} \frac{\partial \varphi}{\partial n'}(\underline{x}') - \varphi(\underline{x}') \frac{\partial}{\partial n'} |\underline{x} - \underline{x}'|^{-1} \right) dS' \quad , \tag{13}$$

where the integration is on the bubble surface S. The relation between the two representations (12) and (13) can be established taking the scalar product of the two equations with P_{2k} and equating the results. In this way, using the well known expansion of $|\underline{x} - \underline{x}'|^{-1}$ in terms of Legendre polynomials and the orthogonality relations between polynomials of different orders, one readily finds $A_{2k} = B_{2k} - C_{2k}$ where

$$B_{2k} = \int_0^{\pi/2} \frac{\partial \varphi}{\partial n}(\underline{x},t) |\underline{x}|^{2k+2} P_{2k}(\cos \theta) \sin \theta \, d\theta \quad , \tag{14}$$

17

$$C_{2k} = \int_0^{\pi/2} \varphi(\underline{x},t) \, |\underline{x}|^{2k+1} \left(2k \, P_{2k} \, n_r + \frac{dP}{d\theta} 2k \, n_\theta \right) \sin\theta \; d\theta \quad . \tag{15}$$

In writing this equation advantage has been taken of the existence of a plane of symmetry and (n_r, n_θ) are the components of the unit normal along the vector $\underline{x}$. If (12) is used at some distance from the bubble (e.g. an order of magnitude greater than its linear dimensions) the series can be truncated at its third or fourth term and the result used as a boundary condition for the finite-difference solution of Laplace's equation. Conceptually, the computational procedure is then the following: suppose φ known at time t; (a) compute φ at time t+dt on the surface from (11) and determine the coefficients C from (15);(b) compute the coefficients B from (14) at time t;(c) solve the Laplace equation; compute new values of the coefficients B from this solution;(e)compare with the previous values: if their difference is sufficiently small proceed to the next time step, if not repeat (c-e) until convergence is achieved. As mentioned earlier, the iterative procedure can be avoided using as boundary condition at "infinity" for the potential at time t+dt (12) with the coefficients B evaluated from φ at time t. The procedure just described has been used to set up a numerical code which is near completion. A preliminary result obtained from this code is shown in Fig. 2 for   the collapse of an initially ellipsoidal bubble. The results are very similar to those reported in [10]. In the future this code will be used to study several processes such as the interaction of two collapsing or growing bubbles near a rigid wall, the coalescence of two equal bubbles, and others.

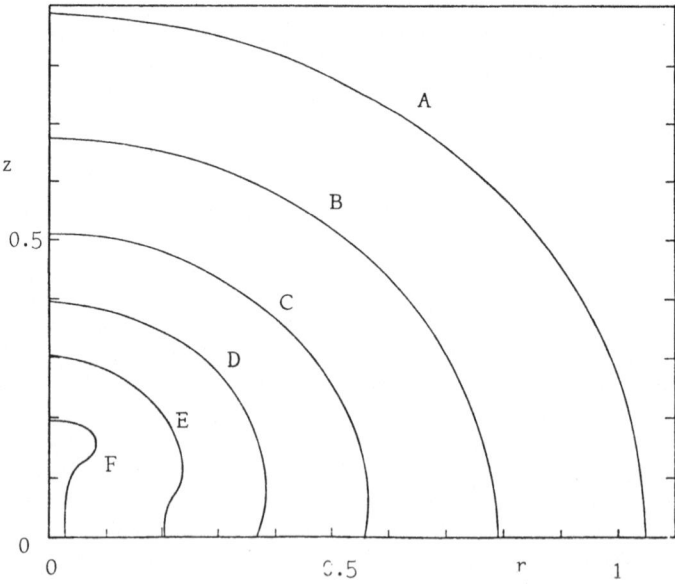

Fig.2   Results of the finite-difference code for the collapse of an initially ellipsoidal bubble. The shapes shown correspond to the following values of the dimensionless time: A, t=0; B, t=0.575; C, t=0.725; D, t=0.79375; E, t=0.82656; F, t=0.84258.

3. Variational Method

It can be shown that the solution of the potential problem posed in the previous section minimizes the functional [3]

$$J = \int_{t_1}^{t_2} dt \left\{ \int_{V_i} (p_i - p_\infty) \, dV_i - \rho \int_{V_o} \left(\frac{\partial \varphi}{\partial t} + \frac{1}{2} |\nabla \varphi|^2 \right) dV_o \right\} \quad , \tag{16}$$

where V_i and V_o are the volumes internal and external to the bubble respectively.(For simplicity surface tension effects have been neglected in this equation, but they can readily be included[3]). This fact has been used to obtain numerical results by the so-called "spectral method"[14]. The principle is to assume a representation of the potential in the form (12) and of the free surface in the form of a superposition of terms of the type appearing in (1). After substitution of these expressions into (16) and evaluation of the spacial integrals the functional J depends only on the unknown coefficients A_k and a_k, which can be determined integrating the Euler-Lagrange (ordinary) differential equations to which the minimization of J leads. This method has the attractive feature of a certain simplicity, both conceptual and computational.On the other hand, it does not appear capable of handling very severe distortions of the spherical shape unless many terms of the series are retained, and even then only for those shapes for which the relation between r and (θ, φ) on the bubble surface is single-valued.

An alternative way to use the minimum property of the functional J would be to assume a certain functional form for the potential (e.g. the superposition of sources, see (17) below) and then obtain the Euler-Lagrange equations for the unknown functions of time appearing in this form. In this way no specific assumption need be made on the shape of the free surface, which is to be determined as that particular surface which satisfies (10) when the assumed form for the potential is substituted for φ. An application of this idea already to the simplest possible case, that of a spherical bubble, is sufficient to convince oneself that this procedure is bound to result in very awkward computations which soon become unmanageable if a system of coordinates other than the spherical one is used. A variant of this idea which considerably simplifies the algebra is described in the next section.

4. A Singularity Method for the Dynamics of Non-Spherical Bubbles

The solution of fluid mechanical problems by the superposition of flows produced by suitable singularities is a well-established procedure for potential flow and Stokes flow problems but, to our knowledge, it has never been applied to bubble dynamics. We shall describe here some preliminary work done in this direction.

Consider the potential produced by N sources of strength $q_j(t)$ located at positions $\underline{X}_j(t)$,

$$\varphi(\underline{x}, t) = \sum_{j=1}^{N} q_j(t) / |\underline{x} - \underline{X}_j(t)| \quad . \tag{17}$$

Inserting this expression into the Bernoulli integral (10) (where we neglect the surface tension contribution) we obtain

19

$$F(\underline{x},t) \equiv \sum_{j=1}^{N} \left[\dot{q}_j(t) \, |\underline{x} - \underline{X}_j(t)|^{-1} - q_j(t) \, \dot{\underline{X}}_j \cdot \underline{\nabla} \left(|\underline{x} - \underline{X}_j(t)|^{-1} \right) \right]$$

$$+ \frac{1}{2} \sum_{j,k=1}^{N} \left[q_j \, q_k \, \underline{\nabla} \left(|\underline{x} - \underline{X}_j| \right)^{-1} \cdot \underline{\nabla} \left(|\underline{x} - \underline{X}_k| \right)^{-1} \right] = \frac{p_\infty - p_i}{\rho} \tag{18}$$

For each specified value of $p_i - p_\infty/\rho$ this equation determines a surface $F(\underline{x},t)$ which may be taken as the surface of a cavity of internal pressure p_i in a liquid of density ρ and ambient pressure p_∞. For $|\underline{X}_i| \ll |\underline{x}|$ this equation gives $\dot{Q}/|\underline{x}| \simeq (p_\infty - p_i)/\rho$ (where $Q = q_1 + q_2 + \dots + q_N$ is the total source strength), which shows that the surface described by (18) is bounded. Select now $4N$ arbitrary points $\underline{\xi}_k$ satisfying (18) and consider the equations

$$F(\underline{\xi}_k, t) = p_\infty - p_i/\rho \quad , \qquad k = 1, 2, \dots, 4N \quad . \tag{19}$$

Viewed as a linear system for $\dot{a}_j$, $\dot{\underline{X}}_j$ these equations can be solved to give

$$\dot{a}_j = \dot{a}_j(a, \underline{\xi}) \quad , \qquad \dot{\underline{X}}_j = \dot{\underline{X}}_j(a, \underline{\xi}) \quad . \tag{20a,b}$$

On the other hand, since particles on the free surface remain on the free surface, we can follow the motion of the points $\underline{\xi}_k(t)$ through the equations

$$\dot{\underline{\xi}}_k(t) = \underline{\nabla} \, \varphi(\underline{\xi}_k, t) \quad . \tag{20c}$$

Eqs.(20) are a set of 16N scalar equations in the 16N unknowns $(a_j, \underline{X}_j, \underline{\xi}_k)$. Once these quantities have been calculated, the shape of the bubble at any instant will be given by (18).

The simplest application of this idea consists of two sources of equal strength $q(t)$ located at positions $(0,0,\pm Z(t))$. (Notice that $Z(t)$ can be pure imaginary still leaving (17) and (18) real). Clearly in this case the free surface has an axis and a plane of symmetry and it is possible to consider the situation in one quadrant of a single meridian plane. It is convenient to take $\underline{\xi}_1 = (0, z_o)$ as the point in this quadrant at which the bubble intersects the axis of symmetry and $\underline{\xi}_2 = (r_o, 0)$ as the point where it crosses the plane of symmetry. Due to the high level of symmetry of this configuration it is found that only four equations are needed. The results of a test case with $q(0)=0$, $Z(0)=0.1\,i$, $r_o(0) = 1.048$, $z_o(0) = 0.8871$ are shown in Fig.3. These values of r_o and z_o are the same as for the case treated by the finite-difference technique and shown in Fig.2. On comparing these two figures it can be seen that the simple source distribution used here does not reproduce exactly the peculiar shape of the finite-difference solution near the end of the collapse, but still gives an acceptable picture of the history of the process with only a small fraction of the effort required by the complete numerical treatment. Incidentally we may add that a very similar case has been studied in [14] by the spectral method with 8 terms in the series. The results are surprisingly similar to those of our simple singularity approach. Thus we think that this idea deserves to be examined in more detail considering more complex singularity distributions. Obviously, in assessing the degree of approximation of this technique, we shall rely on the results obtained from the numerical codes.

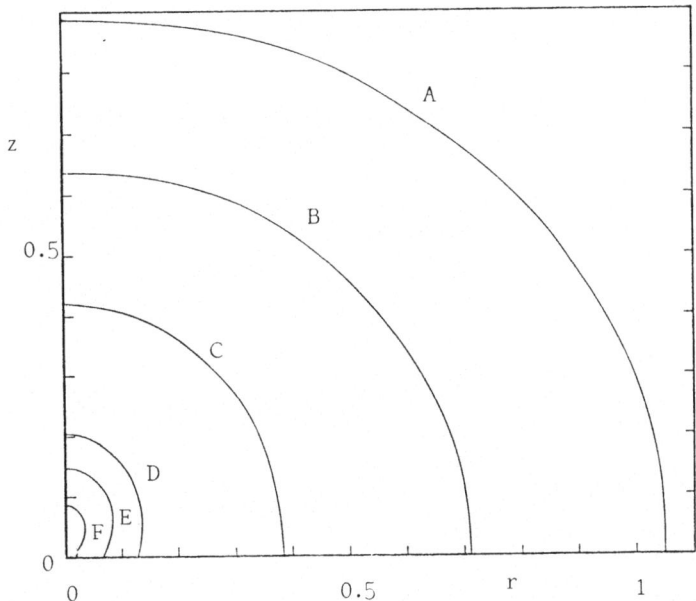

Fig.3   Results of the singularity method with two sources of equal strength for the collapse of an initially ellipsoidal bubble. The shapes shown are for the following values of the dimensionless time: A, t=0; B, t=0.7; C, t=0.86; D, t=0.9125; E, t=0.915625; F, t=0.916875. As in the previous figure the time is made dimensionless by multiplying by $(p_\infty - p_i / \rho)^{1/2}$ and by dividing by the unit of length.

Finally, we wish to remark that, in the limit $|Z| \ll r_0, z_0$ the equations obtained from the singularity method are identical with the Rayleigh–Plesset equation and the appropriate form of (8) (i.e., with $\nu = 0$, $\sigma = 0$, $n = 2$) provided that q be identified with $-(1/2)R^2\dot{R}$ and $Z^2 = (1/3)R^2(\dot{a}_2 / \dot{R} + 2a_2 / R)$.

Acknowledgments

It is a pleasure to acknowledge the determining contribution of Dr. M. Gibilterra to the numerical work described in section 2. This study has been supported in part by Gruppo Nazionale di Fisica Matematica of C.N.R.

Appendix

We would like to describe very briefly another numerical procedure for the solution of the potential problem described in Section 2 which we are currently developing, the so-called boundary-integral method[13]. The starting point is again (13) which we now evaluate for the point $\underline{x}$ on the boundary replacing 4π by 2π. It is clear that the equation can be considered as an integral equation of the Fredholm type for the potential, which can be solved dividing the boundary into segments (or strips) on which

φ and $\partial\varphi/\partial n$ are taken as constants. The integral equation is thus reduced to an algebraic system the right-hand side of which is related to φ on those portions of the boundary where Dirichelet data are given and to $\partial\varphi/\partial n$ on those portions where Neumann data are prescribed. Clearly the integral in (13) is singular on that element of the boundary to which $\underline{x}$ belongs, but satisfactory techniques have been developed to handle this problem [13]. This method appears attractive because it yields directly $\partial\varphi/\partial n$ on the bubble boundary, thus leading immediately to the configuration of the free surface at time t+dt without the need to determine the value of the potential in the liquid.

<u>References</u>

1. M. S. Plesset and A. Prosperetti, Ann.Rev. Fluid Mech. <u>9</u>, 145–185 (1977).
2. A. Prosperetti, "Current topics in the dynamics of gas and vapor bubbles", Meccanica, in press.
3. D. Y. Hsieh, J. Basic Eng. <u>94</u>, 655–665 (1972).
4. C. F. Naudé and A. T. Ellis, J. Basic Eng. <u>83</u>, 648–656 (1961).
5. P. Hall and G. Seminara, in preparation.
6. A. Prosperetti, Quart. Appl. Math. <u>35</u>, 339–352 (1977).
7. A. Prosperetti and G. Seminara, Phys. Fluids <u>21</u>, 1465–1470 (1978).
8. A. Prosperetti, Int. J. Numer. Methods in Eng. <u>11</u>, 431–438 (1977).
9. M. S. Plesset and R. B. Chapman, J. Fluid Mech. <u>47</u>, 283–290 (1971).
10. R. B. Chapman and M. S. Plesset, J. Basic Eng. <u>94</u>, 142–145 (1972).
11. C. L. Kling and F. G. Hammitt, J. Basic Eng. <u>94</u>, 825–833 (1972).
12. J. D. Jackson, <u>Classical Electrodynamics</u> (Wiley, New York, 1977).
13. M. A. Jaswon and G. T. Symm, <u>Integral Equation Methods in Potential Theory and Elastostatics</u> (Academic Press, New York, 1977).
14. A. Shima and K. Nakajima, J. Fluid Mech. <u>80</u>, 369–391 (1978).

Oscillation and Collapse of a Cavitation Bubble in the Vicinity of a Two-Liquid Interface

Georges L. Chahine and Alain Bovis

Groupe G.P.I., Ecole Nationale Supérieure de Techniques Avancées, F-Paris, 32 Bd Victor 75015, France

1. Introduction

The problem of the interaction between an oscillating bubble and a free surface was first investigated in the case of underwater explosions, the main concern being the influence of the free surface on the bubble migration and period of oscillations [1-4]. The extension to the case of cavitation bubbles appeared to be of interest since flexible membranes and elastomeric coatings were seen to have a similar effect on a collapsing bubble as that of a free surface [5,6]. The cavitation damage resistance capability of these materials seems to be due to their repulsing effect on the bubble. High speed photographic evidence showed that, while in the case of a solid boundary the re-entering jet formed during the bubble collapse is directed towards the wall [7-9], this same jet is oriented away from a free surface or a flexible boundary [5,6,10]. In some practical applications, injection of bubbles [11] and formation of a preserving liquid layer [12] on solid walls to be protected were reported to be effective on cavitation damage reduction. Even if in the latter case the effect of a magnetic field in the layer is the main interest of the authors, in all cases the presence of a free surface (or an elastic boundary) seems to be fundamental. Thus the investigation of the interaction between an isolated bubble and these different boundaries is a first and basic step in the understanding of their cavitation damage resistance capability. Moreover, since the observation of the deformation of an interface is much easier than the measurement of a solid deformation, it provides a simple and more effective way to investigate the validity of theoretical models on cavitation bubble dynamics.

In a previous study [10], the case of an air-water interface was investigated theoretically and experimentally. This study is currently being continued through the investigation of a two non-miscible liquid interface.

2. Experimental Study

The simultaneous behavior of the bubble collapse and the liquid-liquid interface is observed using high-speed photography. The variation of the bubble size and its distance from the interface allows for the determination of the limiting distance of interaction. Below this distance non-spherical deformations of the bubble and the interface can be observed.

Spark-generated bubbles were produced by discharging a capacitor across a pair of platinum electrodes for a very brief period of time. The electrodes are mounted in a large vessel hermetically sealed and connected to a vacuum pump capable of a minimum pressure of 0.06 atm. Lowering the ambient pressure has two favorable effects : water degasing is facilitated and the period of bubble collapse is significantly increased. This slowing down of the phenomenon is directly responsible for the obtaining of valuable information with a HYCAM high speed camera with a limiting framing rate of 20 000 frames per second. Focus lighting is provided by eight 750 watts flash bulbs placed in front of the tank and on both sides of the camera. A L-W photo Optical Analyser is used to study fast evolutions by projecting frame by frame the part of the film which is of interest.

Liquid-liquid interfaces were obtained by using two non-miscible liquids. Limitations in the choice of the liquids are due to the fact that they must satisfy simultaneously a certain number of conditions : They must be neither flammable nor explosive. The two liquids should be able to be colored differently. Some precautions have to be taken concerning their noxiousness and their harmful effects on the environment.

A series of photographic experimental runs were made to examine the effects upon bubble behavior and free surface disturbance of the initial distance ℓ_o from the liquid-liquid interface to the electrode-gap. The pressure P_o was maintained at approximately 0.1 atm. Four different combinations of liquids have been tested : water-tetrabromo ethane, water-aniline, water-oil, water-white spirit. The two last configurations, being the more convenient, were investigated extensively. We present here the qualitative results and compare them to those obtained in the case of an air-liquid interface [10].(Fig.1).

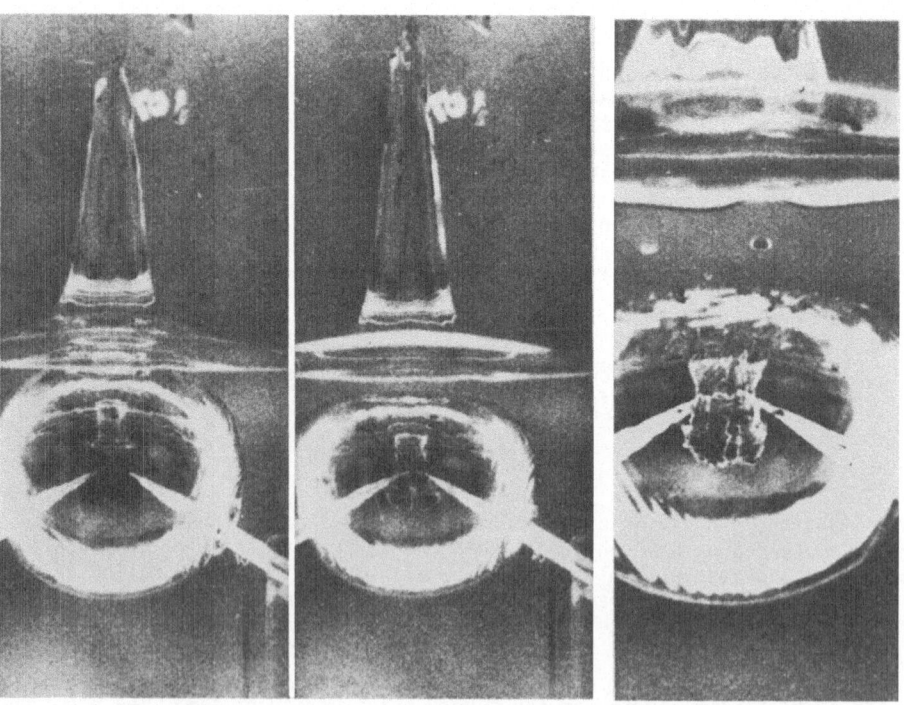

Fig.1 Case of an air-water interface $\beta \approx 1.25$

24

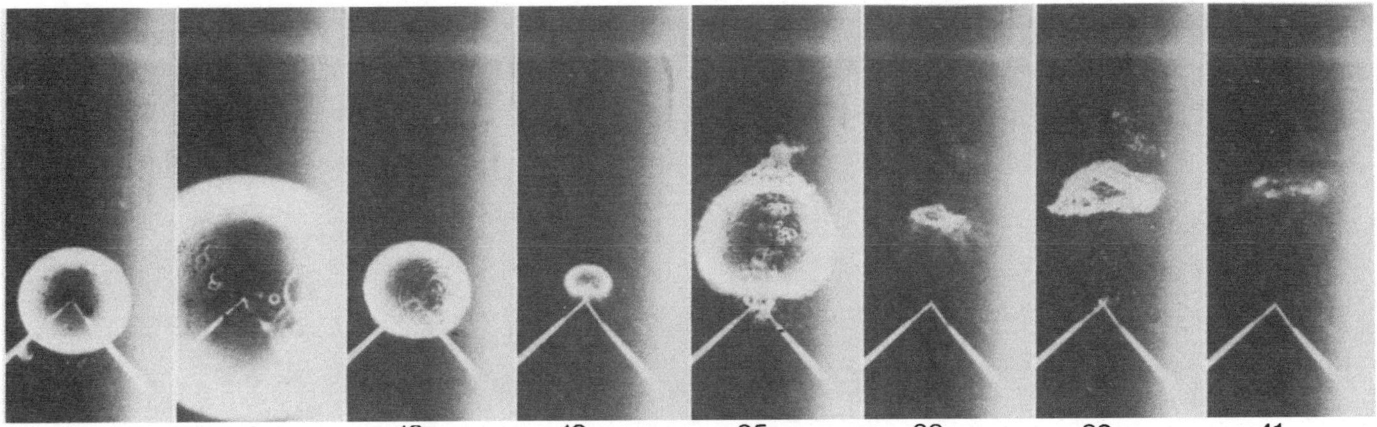

Fig.2

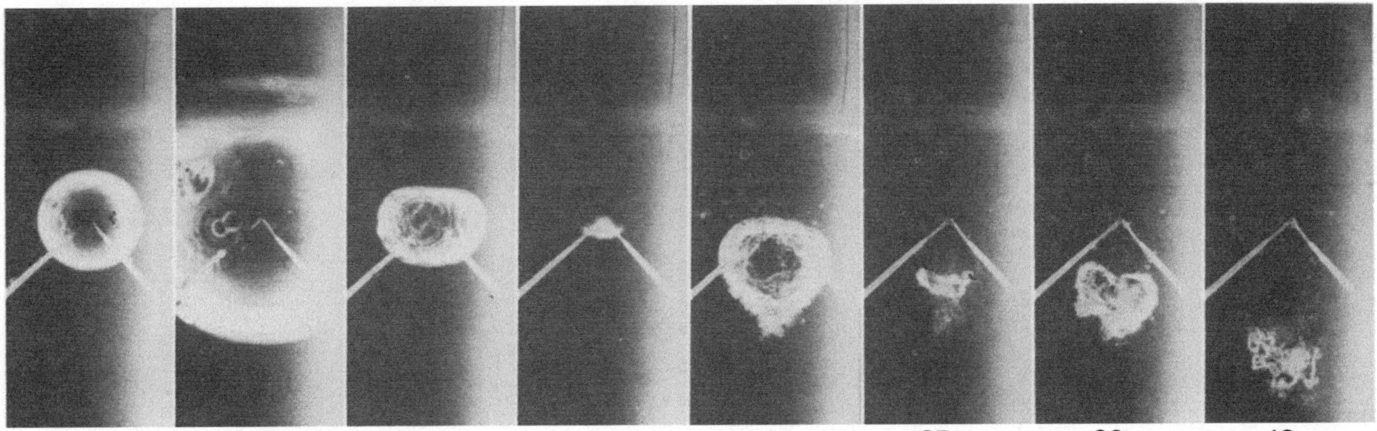

Fig.3

Fig.2 Case of a water-white spirit interface: ATTRACTION $\beta \approx 0.45$ t=0.06, 0.6, 1.05, 1.1, 1.5, 1.9, 2.3, 2.4

Fig.3 Case of a water-white spirit interface: REPULSION $\beta \approx 1.15$ t=0.06, 0.5, 0.9, 1.0, 1.3, 1.6, 1.75, 2.3

A dimensional analysis shows that the normalized deformation of the interface η, depends on four adimensional parameters : a Froude number τ^{-1}, the growth rate of the bubble K, the ratio β of a characteristic length of the bubble a_o to the initial interface to bubble center distance ℓ_o, and the ratio of the two liquid densities. In the case of cavitation bubbles gravity has no time to take effect, τ is negligible and, as shown below in the theoretical study, the limiting distance of interaction ℓ^*, as in the case of an air-water interface, depends only on the grouping $K\beta^3$. Observations show that when $K\beta^3 \ll 1$ the interaction between the bubble and the interface is negligible. In this case the spark-generated bubbles remain practically spherical and the values of the growth rate K, for the same energy of discharge and the same ambient pressure, are comprised between 35 and 65. This leads to a limiting value of $\beta^* \simeq 0.3$ and thus to a limiting distance $\ell^* \simeq 3a_o$.

Below this distance, high-speed photography shows that the behavior of the bubble in the presence of a liquid-liquid interface and in the presence of an air-liquid interface is not the same, even if in the two cases gravity has no time to take effect. An oscillating bubble is always seen to be repelled from the air-liquid interface. A reentering jet penetrates it during the collapse and is oriented away from the free surface on which a liquid jet is formed simultaneously (Fig.1). In the liquid-liquid interface repulsion or attraction of the collapsing bubble occurs depending on whether its initial distance from the interface is less than another limiting distance ℓ^{**}, or comprised between ℓ^* and ℓ^{**}. Figures 2 and 3 show these two cases for the same two-liquid interface (water-white spirit). In Fig.2, $\beta \simeq 0.45$ and the interface acts as a solid wall . The bubble, practically spherical during its first period of oscillation, is violently attracted towards the surface during the final stage of the collapse and during the rebound. The re-entering jet is oriented towards the interface while the latter remains undisturbed. In the second case (Fig.3) $\beta \simeq 1.15$, and a behavior comparable to that in the presence of an air-water is observed. However, the deformation of the interface is much weaker than that of a free surface. This statement is always valid for a two liquid case. The deformation is significantly delayed and the interface jet formed weaker.

3. Theoretical Study

Let us consider two inviscid incompressible liquids of densities ρ_1 and ρ_2. To study the axisymmetric collapse of a bubble near an interface formed by these two liquids, one has to solve:

$$\Delta \phi^{(i)} = 0 \qquad ; \qquad i = 1,2 \qquad\qquad (1)$$

$\phi^{(i)}$ being the velocity potential of the flow in the fluid (i). Eq. (1) has to be subjected to the kinematic and dynamical conditions on the interface between the two liquids and the bubble surface. The free boundary problem in its general form is rather complex due to the unknown shapes of the interfaces, and an analytical solution is unlikely. However, solutions using matched asymptotic expansions can be considered if the bubble is assumed to be small, compared to its initial distance to the boundary ($\beta = a_o/\ell_o \ll 1$). In this case there are two characteristic lengths, and depending on whether the choice is a_o or ℓ , the solution applies to the vicinity of the bubble (inner problem) or of the interface (outer problem). We will proceed here in the following way : we will first look at the outer problem to determine the order of magnitude of the perturbation on the interface. The inner problem will then be considered with more precision in order to justify the outer approach and to set up the equations giving the deformation and the motion of the bubble.

3.1 Outer Problem

In the Outer Problem the bubble , whose radius vanishes in the first approximation, is a singular perturbation of the flow. The problem to be solved is, therefore, the determination of the velocity potential $\overline{\Phi}$ which satisfies simultaneously the conditions on the two-liquid interface and a matching condition. with the Inner Problem when $\overline{r} = r/\ell$ tends to zero. As can be shown, to the first order of approximation, the bubble remains spherical and its radius variations a(t) are given by the RAYLEIGH-PLESSET equation [13]. The condition to be satisfied at the origin of coordinates is,therefore,the presence of an oscillating source with a known variable intensity with time ($q(t) = 4\pi a^2 \dot{a}$).

Consequently, if lengths are reported to ℓ_o, time to the period of oscillation of the bubble T, source intensity to $Q_o = 4\pi a_m^2 \dot{a}_m = \max(q(t))$ and velocity potential to $\Phi_o = Q_o/\ell_o$, the condition to be satisfied by $\Phi^{(i)}$ can be written as follows in cylindrical coordinates :

$$\overline{\eta_t} + \zeta[\overline{\Phi}_r^{(i)}\,\overline{\eta_r} - \overline{\Phi}_z^{(i)}] = 0 \qquad ; \quad \text{on } \overline{z} = 1 + \overline{\eta}(r,t) \; ; \; i = 1,2 \quad (2)$$

$$\sum_{i=1,2} (-1)^i \rho_i[\overline{T\eta} + \zeta\overline{\Phi}_t^{(i)} + \frac{1}{2}\zeta^2(\overline{\Phi}_r^2 + \overline{\Phi}_z^2)^{(i)}] = 0 \; ; \; \text{on } \overline{z} = 1 + \overline{\eta}(r,t) \quad (3)$$

$$\overline{\Phi}_r^{(i)}, \overline{\Phi}_z^{(i)} \to 0 \qquad ; \quad |r| \to \infty \,,\, z \to \infty \qquad\qquad i = 1,2 \quad (4)$$

$$\lim_{\overline{r},z \to 0} \overline{\Phi}^{(2)} = -\overline{q}(t)\,(\overline{r}^2 + \overline{z}^2)^{-1/2} \qquad\qquad (5)$$

with the following definitions :

$$\zeta = K\beta^3 \quad ; \quad K = 4\pi\dot{a}_m T/a_m \quad ; \quad \beta = a_o/\ell_o \qquad\qquad (6)$$

$$T = \overset{+}{-}\,gT^2/\ell_o \quad ; \quad (+) \text{ if } \rho_2 > \rho_1 \quad ; \quad (-) \text{ if } \rho_1 > \rho_2 \qquad (7)$$

For cavitation bubbles $T \ll 1$, so that the importance of the different terms in (2), (3) depends only upon the order of magnitude of ζ. The linearization of the problem is seen to be possible only if $\zeta \ll 1$. In this case the less degeneracy is obtained with $\zeta = O(T)$. At order ζ^o, the interface is not disturbed, and the problem is reduced to the determination of the harmonic velocity potentials $\overline{\Phi}_o^{(1)}$, $\overline{\Phi}_o^{(2)}$ satisfying (4) and (5) in addition to :

$$\rho_1 \overline{\Phi}_t^{o(1)} - \rho_2 \overline{\Phi}_t^{o(2)} = 0 \qquad\qquad ; \qquad \text{on } z = 1 \qquad (8)$$

$$\overline{\Phi}_z^{o(1)} = \overline{\Phi}_z^{o(2)} \qquad\qquad ; \qquad \text{on } z = 1 \qquad (9)$$

The solution is given by :

$$\overline{\Phi}^{o(1)} = \frac{\rho_2}{\rho_1}\,\Psi(\overline{r},\overline{z},t) = \frac{1}{2\pi} \cdot \iint \frac{\rho\,\overline{\eta}_t^1\,d\rho\,d\phi}{[(\overline{z}-1)^2 + \overline{r}^2 + \overline{\rho}^2 - 2\overline{r}\rho\cos\phi]^{1/2}} \qquad (10)$$

$$\overline{\Phi}^{o(2)} = -q(t)\,[\frac{1}{(\overline{r}^2 + z^2)^{1/2}} - \frac{1}{(\overline{r}^2 + (\overline{z}-2)^2)^{1/2}}] + \Psi(\overline{r},\overline{z},t) \qquad (11)$$

$$\overline{\eta}^1 = \frac{2\rho_2}{\rho_2 - \rho_1} \cdot \frac{\int_o^t q(\tau)d\tau}{(r^2 + 1)^{3/2}} \qquad\qquad (12)$$

27

To continue the study at higher orders one has to take into account the motion and the departure from sphericity of the bubble. In the case of a collapsing spherical bubble near a wall the motion was taken into account by adding a dipole of moment $a^3 v(t)$ [14]. The velocity of the sphere $v(t)$ is determined by writing that the net pressure thrust on the bubble vanishes. The same method and approximations applied here give :

$$\dot{v}(t) = -q^2/4 + \rho_1 q/2(\rho_2-\rho_1) \qquad (13)$$

This indicates that the tendency of the bubble is to move away from an air-water interface ($\rho_1 \ll \rho_2$), while this tendency depends upon $q(t)$ for the two-liquid case.

3.2 Inner Problem

In the Inner Problem the interface is at infinity so that in the first approximation ($\beta=0$) a variation of the ambient pressure induces a spherical radial motion $a(t)$ of the bubble, without migration. Thus we can write in spherical coordinates originated at the center of the bubble :

$$R(t) = a_o(\tilde{a}(t) + \mu_1(\beta)\tilde{R}_1(\theta,t) + ...) \qquad (14)$$

$$\dot{\ell}(t) = a_o/T(0 + \nu_1(\beta)\tilde{\dot{\ell}}_1(t) + ...) \qquad (15)$$

$$\Phi(\tilde{r},\theta,\tilde{t}) = a_o^2/T(\frac{\tilde{a}^2\tilde{\dot{a}}}{\tilde{r}} + \xi_1(\beta)\tilde{\Phi}_1(\tilde{r},\theta,\tilde{t}) + ...) \qquad (16)$$

The principle of less degeneracy, and the matching conditions with the outer problem in which the first perturbation of the β^o-order potential, is at order β when $\tilde{r} \to 0$, lead to the choice :
$$\mu_1 = \nu_1 = \xi_1 = \beta^2$$

As the deformation of the two-liquid interface is at order β^3 the potential Ψ (10) does not contribute to the writing of (17). In this case the following equations to be solved apply to an interface ($s = -1$) as well as to a solid wall ($s = 1$).

$$\tilde{\Delta}\tilde{\Phi}_1 = 0 \qquad (18)$$

$$\frac{\partial\tilde{\Phi}_1}{\partial\tilde{r}}\Big|_{\tilde{r}=\tilde{a}} = 2\frac{\tilde{\dot{a}}_o}{\tilde{a}_o} \cdot \tilde{R}_1 + \tilde{\dot{R}}_1 - \tilde{\dot{\ell}}_1\cos\theta \qquad (19)$$

$$-\tilde{\dot{\Phi}}_1\Big|_{\tilde{r}=\tilde{a}} = \tilde{a}_o\tilde{\dot{R}}_1+(2\frac{\tilde{\dot{a}}_o}{\tilde{a}_o}+\tilde{a}_o)\tilde{R}_1+\tilde{a}_o\tilde{\dot{\ell}}_1\cos\theta+W_e^{-1}\tilde{a}_o^{-2}(2R_1+\tilde{R}_1'\cot g\theta+\tilde{R}_1'') \qquad (20)$$

$$\lim_{\tilde{r}\to\infty}\tilde{\Phi}_1 = \beta^2\lim_{\tilde{r}\to 0}\tilde{\Phi}_1 - s\frac{a^2\dot{a}}{4}\tilde{r}\cos\theta \qquad (21)$$

$\tilde{\Phi}_1$ can be expanded in a series of spherical harmonics. This allows for the expansion of R in a series of LEGENDRE polynomials. Eqs. (19) to (21) then show that : (with the initial conditions $R_1=\dot{R}_1=0$)

$$\tilde{\Phi}_1 = \frac{h(t)\cos\theta}{\tilde{r}^2} - s\frac{\tilde{a}^2\tilde{\dot{a}}}{4}\tilde{r}\cos\theta \qquad (22)$$

Substituting in (18), (20)

$$\tilde{R}_1 = f_1(t)\cos\theta \qquad (23)$$

We obtain :

$$\widetilde{a}\ddot{f}_1 + 3\dot{\widetilde{a}}\dot{f}_1 = \ddot{\widetilde{\ell a}} - \ddot{\ell}\dot{\widetilde{a}} - s\,\frac{9\widetilde{a}^2\dot{\widetilde{a}}^2 + 3\widetilde{a}^3\ddot{\widetilde{a}}}{4} \tag{24}$$

$$\widetilde{a}^3\dot{f}_1 + 2\,\widetilde{a}^2\dot{\widetilde{a}}f_1 = \dot{\widetilde{\ell a}}^3 - h - s\,\frac{\widetilde{a}^2\dot{\widetilde{a}}}{2} \tag{25}$$

The third equation, necessary to determine the three unknowns f,h and ℓ_2 is obtained by writing that the net pressure thrust on a sphere of radius aβ vanishes when β→ 0. This gives

$$h = -s\,\frac{\widetilde{a}^5\dot{\widetilde{a}}}{8} \tag{26}$$

Eqs. (24) to (26) have to be solved numerically. Similar equations are obtained without difficulty at the β^3-order of expansion and give a deformation $\widetilde{R}_2$

$$\widetilde{R}_2 = f_2(t)\,[3\cos^2\theta - 1] \tag{27}$$

To sum up we can say that when β is small the bubble oscillates spherically at order β^0. Migration and shape deformation then intervene simultaneously at order β^2. Even if numerical solutions have not already been obtained for this order, we notice that the bubble flattens on one of its sides depending upon the nature of the boundary (solid or free surface). In either case the bubble is attracted or repelled. At order β^3 singular points appear on its axis of symmetry.

References

[1] C.Herring, Columbia Univ. NDRC rep. C-4-sr 20-010 (1941).
[2] R.H.Cole, Underwater Explosions, Dover Publications (1948).
[3] J.W.Pritchett, Information Research Associates, Berkeley, IRATR-Z-71 (1971)
[4] M.Holt, Ann. Rev. Fluid Mech. 9, 187 (1977).
[5] E.E.Timm and F.G.Hammitt, ASME Cavitation Forum, 18 (1971).
[6] D.C.Gibson, Proc. of the 3rd Australasian Conference on Hydraulics and Fluid Mechanics, Sydney, 210 (1968).
[7] C.F.Naude and A.T.Ellis, J.Basic Eng. 83,648 (1961).
[8] M.S.Plesset and R.B.Chapman, J.Fluid Mech. 47,283 (1971).
[9] W.Lauterborn and H.Bolle, J.Fluid Mech. 72,391 (1975).
[10] G.L.Chahine, J.Fluids Eng. 99,709 (1977).
[11] R.H.Smith and R.B.Mesler, J.Basic Eng., 933 (1972).
[12] K.K.Shalnev, I.A.Shalobasov and V.G.Degtar, Symposium Grenoble IARH-SHF, 189 (1976).
[13] M.S.Plesset, ASME, J.Appl. Mech. 16, 228 (1949).
[14] G.Birkhoff and E.H.Zarantonello, "Jets, Wakes and Cavities", Academic Press (1957).

This work was sponsored by the Direction des Recherches, Etudes et Techniques Contract 1116/78.

Experimental Investigation of Bubble Collapse at Laser-Induced Breakdown in Liquids

V.S. Teslenko

Institute of Hydrodynamics, Siberian Branch of the USSR Academy of Sciences
Novosibirsk 630090, USSR

1. Introduction

Currently, the problem of bubble collapse has not been completely investigated experimentally since producing spherical monobubbles is technically
difficult. The method of creating monobubbles in a liquid using an electric
spark is known in the literature. However, as is shown by experiments [1],
this method does not allow a sufficiently spherical bubble to be obtained.

Recently, lasers have found wide application in the investigation of
cavitation problems. For example, focusing Q-switched laser radiation into a
liquid makes it possible to produce a sufficiently spherical monobubble, to
obtain any number bubbles with prescribed phases of their pulsation periods
and different maximum sizes and to realize the conditions of interaction of
pulsating bubbles with a rigid or free surface [2,3,10].

This paper presents the investigations of acoustic and light radiation
at the collapse of monobubbles in correlation with cinematographic investigations of the bubble pulsation kinetics.

2. Experiment

Cavitation bubbles were produced by focusing Q-switched ruby laser radiation in vessels containing various liquids.

The pressure pulses were measured using barium titanate pressure transducers having a resolution of 0.1 μs.

The light radiation was recorded at the bubble collapse by the methods
described in [4,5] using photomultipliers-types FEU-18 and FEU-22 as well
as the type OK-17M oscilloscope.

The bubble pulsation kinetics was studied by the Schlieren method [2].

3. Experimental Results

For the case of non-surfacing bubbles, the ratio of the bubble energy of
the subsequent pulsation to the foregoing one is a function of the relative consumption of energy stored by a bubble. The experimental results are
presented in Table 1 for different cases.

Table 1

	$E_r^{(2)}/E_r^{(1)}$	$E_r^{(3)}/E_r^{(2)}$
Explosion of chemical charges in water	0.34	0.54
Electrical discharge in water	0.05÷0.01	-
Laser breakdown in water	0.07÷0.01	0.5÷0
Laser breakdown in glycerine	0.26	0.5
Laser breakdown in vaseline	0.35÷0.32	0.51÷0.59

In practice it is assumed that .the energy difference between two subsequent pulsations corresponds to the shock acoustic radiation energy of the collapsing bubble. However, as noted in [2], the energy is lost not only due to the shock-acoustic radiation of the bubble but also due to non-symmetric bubble collapse.

Presented in Fig. 1a,b,c are oscillograms of pressure pulses from a pulsating bubble in water. Oscillogram 1a corresponds to the case of shock-acoustic radiation of the collapsing bubble for which the collapse is very nearly symmetric. Oscillogram 1b corresponds to the case when a non-symmetry of the bubble appears at the minimum bubble size, and oscillogram 1c to the case when it appears at earlier stages before collapsing.

Analysis of the oscillograms shows that in the case of a non-symmetric bubble collapse the pressure transducer records the "precursors", i.e. the pressure pulses preceding the basic pulse radiated at the bubble collapse. In the cases when the precursors are observed, the steepness of the pressure growth of the basic pulse is decreased, and the maximum pressure amplitude is decreased by more than 50 % as well.

Fig.2 shows the results measured for maximum pressure amplitudes of different collapsing bubbles recorded at a distance of 1.5 cm.

The maximum pressure amplitude of bubbles with the same maximum radii has a considerable spread. This is caused by the bubble collapse kinetics, its sphericity and the stability of the spherical shape at the moment of collapse. Therefore, comparison of acoustic radiation parameters of the collapsing bubble with theoretical data [7-9] should be carried out from the maximum amplitude values of measured pulses without precursors.

Calculation of the acoustic radiation flux of the collapsing bubble from oscillograms without precursors shows that the bubble radiates about 75 % of stored bubble energy into sound.

The maximum velocity of a symmetrically collapsing bubble is 530 m/s ($M = u/c = 0.35$).

Comparison of theoretical and experimental results [8] shows that, as follows from the estimated results of efficiency of converting bubble effi-

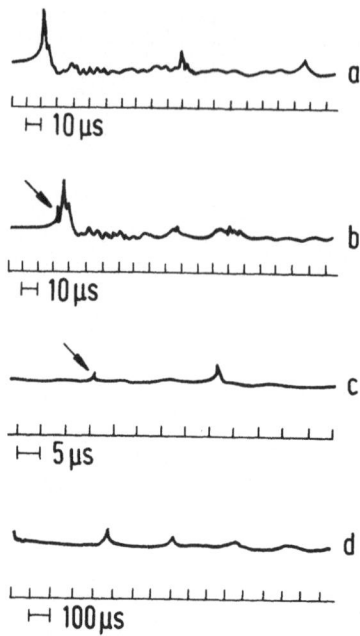

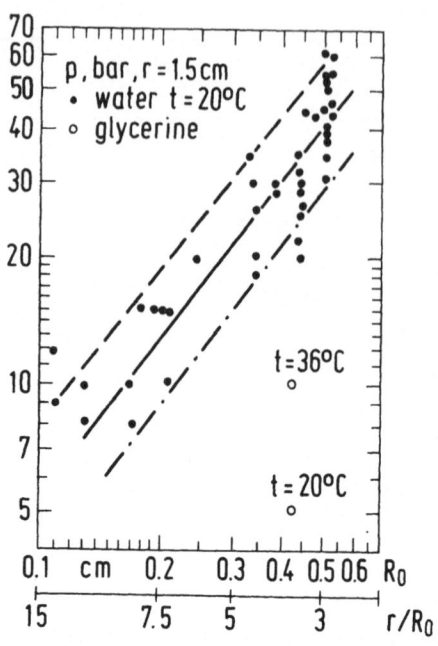

Fig.1 Oscillograms of pressure
pulses of pulsating bubbles
a - collapse of an almost
spherical bubble
b,c - non-spherical collapse of a
bubble in water
d - pressure pulsations of a bubble in vaseline
'Precursors' are marked by
arrows.

Fig.2 Experimental values of
pressures at bubble collapse at a
distance of 1.5 cm from the breakdown centre.

ciency of converting bubble energy into acoustic radiation energy
($\eta = E_p/E_r$, where E_p is the acoustic radiation energy) and from the Mach
number M, the region of initial pressures varies from 10^{-2} to 10^{-3} atm.

For the bubbles obtained by laser breakdown in vaseline or glycerine,
the gas pressure P_g exceeds 10^{-2} atm (see Table 1). When heating glycerine,
the viscosity decreases by three orders, but it does not lead to an appreciable increase in the pulse amplitude nor the parameter η (Fig.2).

A qualitative pulse structure in viscous liquids is similar to that of
a symmetrically collapsing bubble in water. Presented in Fig.1d
is the oscillogram of the pressure pulses of a pulsating bubble in vaseline.
It is clear that pulsations in viscous liquids are stable and repeated.
The artificial asymmetry of the bubble disappears at subsequent stages of
pulsation. "Precursors" are not observed probably due to the significantly
smaller velocities for the cumulative jets. This is due to the fact that
the "precursors" arise from hydraulic impact of cumulative jets with the
opposite wall of the bubble.

Investigation of the acoustic radiation spectrum of the bubble correlates with the sonoluminescence of the bubble produced by laser breakdown in the liquid.

Fig.3 shows the oscillograms of light radiation pulses (a) from bubble pulsations induced by the laser breakdown in water. Similar oscillograms have been obtained in carbon tetrachloride, acetone, and benzene.

We have shown experimentally that the sonoluminescence amplitude increases with increasing acoustic radiation amplitude. This shows the degree of adiabatic gas compression in the bubble is increased for those bubbles whose form is almost spherical at collapse. Estimates show that the gas temperature inside the bubble at the moment of collapse is as great as $6000^{\circ}K$. However, it should be noted that sonoluminescence in glycerine and vaseline was not observed.

Thus, a symmetric bubble collapse in liquid can radiate up to 90 % of the energy contained in the bubble as sound. Asymmetry influences the stability of succeeding bubble pulsations and decreases acoustic and light radiation pulses.

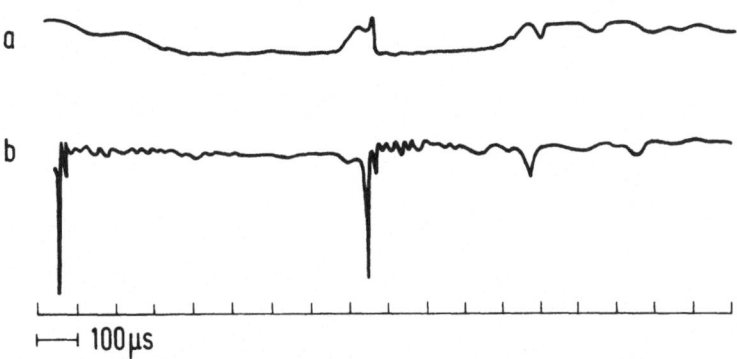

a

b

⊢⊣ 100µs

Fig.3 Typical oscillograms of light radiation pulses (a) and pressure pulses (b) during the bubble pulsations

4. References

1 Gibson, D.K., O prirode kolebanii pri paroobrazovanii pod deistviem iskrovogo zaryada. Trudy Amerikanskogo obschestva inzhenerov mekhanikov. Teoreticheskie osnovy inzhenernykh raschetov. (Russian translation), 94, No 1, 273 (1970).
2 Buzukov, A.A., Popov, Yu.A., Teslenko, V.S., Eksperimental'noe issledovanie vzryvnogo protsessa vyzvannogo fokusirovkoi monoimpul'snogo izlucheniia lazera v vodu, PMTF, No 5, 17 (1969) (in Russian).
3 Lauterborn, W., Kavitation durch Laserlicht, Acustica 31, 51 (1974).
4 Buzukov, A.A., Teslenko, V.S., Sonoljuminestsentsiia pri fokusirovke lazernogo izlucheniia v zhidkost. Pis'ma v ZhETF 14, 286 (1971) [JETP-Lett. 14, 189 (1971)].
5 Akmanov, A.G., Ben'kovsky, V.G., Golubichii, P.I., Maslenikov, S.I., Shamanin, V.G., Issledovaniie lazernoi sonoluministsentsii v zhidkosti. Akusticheskii zhurnal 19, 649 (1973) [Sov. Phys.-Acoustics 19, 417 (1973)].
6 Coul, P., Podvodnye vzryvy. Izd-vo Inostrannaya literatura, M., 1950.

7 Hickling, R., Plesset, M., Collapse and Rebound of Spherical Bubble in Water. Phys. Fluids 7, 7 (1964).

8 Morozov, V.P., Chislennyi analiz islucheniia zvuka sfericheskoi kavernoi. Trudy akusticheskogo instituta vyp. 7, 115 (1969).

9 Kedrinskii, V.K., Osobennosti dinamiki sfericheskogo gazovogo puzyr'ka v zhidkosti. Zhurnal PMTF, No 3, 120 (1967).

10 Teslenko, V.S., Eksperimental'nye issledovaniia kinetikoenergeticheskikh osobennostei kollapsiruiuschego puzyr'ka ot lazernogo proboia v vyazkih zhidkostiakh. PMTF, No 4, 109 (1976).

Application of High Speed Holocinematographical Methods in Cavitation Research

K.J. Ebeling

Drittes Physikalisches Institut, Universität Göttingen, Bürgerstr. 42-44
D-3400 Göttingen, Fed. Rep. of Germany

1. Introduction

The dynamics of cavitation bubbles are usually studied by means of high
speed cinematographical methods. Several devices like rotating drum, rotating mirror, or image converter cameras are commercially available for this
purpose. But all photographic methods suffer from the fact that the process
under study has to take place in a thin region restricted to the depth of
focus. Thus, objects distributed in a deep volume like, for example, acoustically produced cavitation bubbles cannot be recorded satisfactorily.

As is well-known, the problems connected with the limited depth of observation can be overcome using holography. We adopted this new method and
developed high speed holocinematographical recording techniques for investigating cavitation bubble dynamics. In this paper we present two basic
holographic devices and show typical series of reconstructed holograms in
order to demonstrate the performance of the systems applied. Moreover, we
compare some experimental results obtained for spherical laser produced
bubbles with theory.

2. Holocinematography with the multiply Q-switched ruby laser

As has been shown [1-3], several holographic devices are possible for recording hologram series with framing rates up to 20 kHz. Here, we concen-
trate on a standard system for studying laser produced bubbles. The setup
is shown in Fig. 1. The ruby laser for holography produces up to eight
Q-switch light pulses for hologram exposure. Each pulse has a duration of
30 ns. All pulses go the same path. A beam splitter divides the light into
object wave and reference wave. The reflected beam is expanded and collimated to cover the holographic plate uniformly serving as reference beam.
The transmitted light diffusely illuminates a water-filled container
through a ground glass plate and constitutes the object beam. Cavitation
bubbles to be investigated holographically are produced by focusing a
giant pulse from a second ruby laser into the water. The bubbles are pre-
magnified in order to increase resolution. The holographic plate is placed
exactly into the exit pupil of the imaging system. As the key element of
the whole device a rotating disk with apertures is placed directly in
front of the holographic plate and selects that small portion of the plate
which is to be exposed by a single pulse of the sequence. The disk rotates
so fast that each pulse illuminates another portion of the plate.

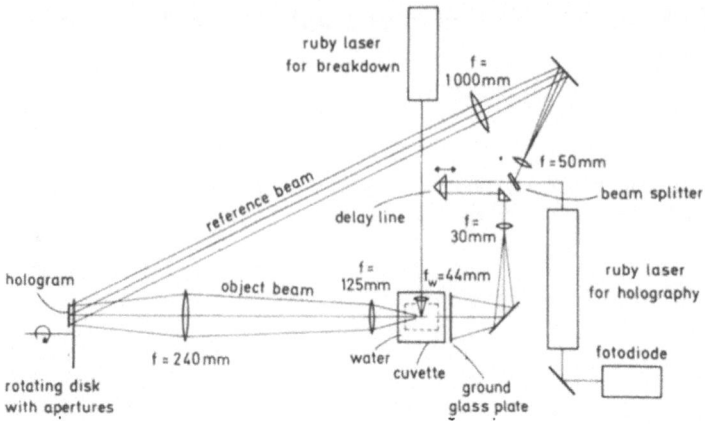

Fig.1 Standard experimental setup for high-speed holography of laser-induced breakdown in liquids

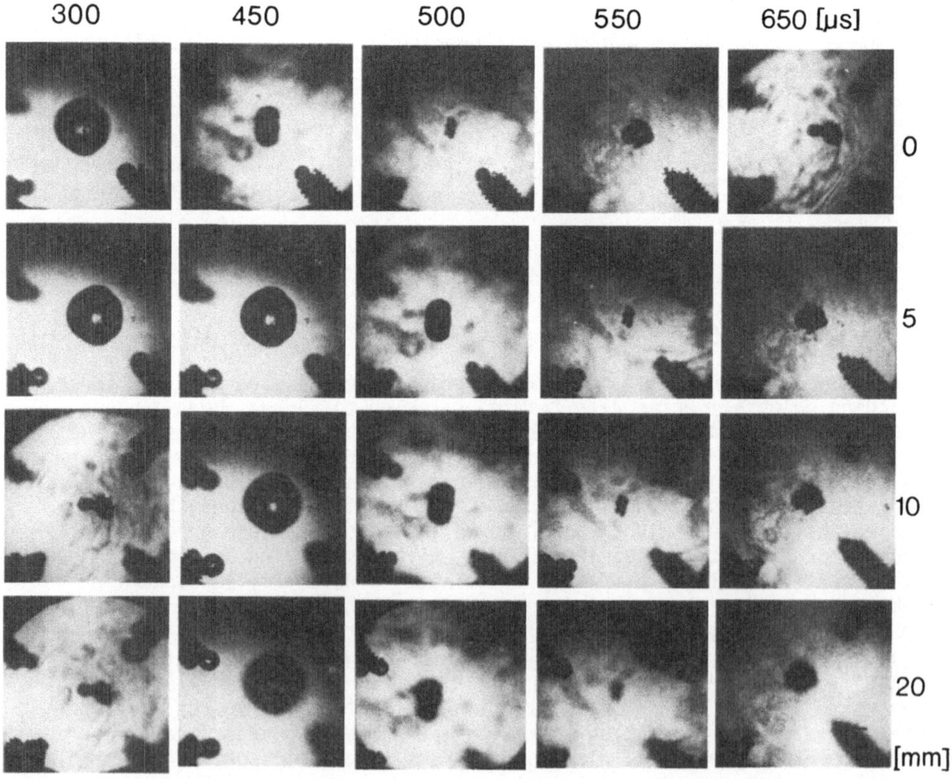

0

5

10

20

[mm]

Fig.2 Asymmetric collapse of laser produced bubble and interaction with adjacent air bubbles

For reconstruction a continuous wave He-Ne-laser is used. To reproduce a single hologram of the sequence simply that small area containing the information is illuminated by the reconstruction beam. Fig. 2 shows an example of a reconstructed hologram series. The interaction of laser produced bubbles and adjacent air bubbles is studied. Five holograms out of the whole series of eight are displayed. Each hologram is reconstructed in another column. The frames in one column show different planes in depth at the same time. The depth coordinate is given at the right of each row. The recording time measured from the instant of breakdown is written on top of each row. The laser produced bubble is visible in the middle of the frames. The air bubbles attached to screws of 3 mm diameter are in focus in different depths. For instance, the air bubble attached at the upper left screw is in focus in the lower row, the laser produced bubble and the air bubble attached at the upper right are in focus in the second row from the top. The spherical collapse of the laser produced bubble and its rebound accompanied by jet formation and shock wave radiation can be seen. Shock waves exhibit as dark clouds in front of the bright background. Deformations of the air bubbles excited by the motion of the laser produced bubble can be followed in different rows.

The laser induced breakdown in the liquid usually occurs at several points [4]. In the initial stadium of bubble growth strong shock waves are radiated from each breakdown point. The shock waves detach in less than 1 μs from the emerging bubbles which can be seen from the hologram reconstructions in Fig. 3. The shock waves surrounding each breakdown center are displayed as dark circles. It should be mentioned that a photographic investigation of the shock wave detachment suffers a great deal from the scattered laser light and the intense almost white light that is radiated from the bubble interior. For holographic recording this additional light is no significant problem because it constitutes only an incoherent background which does not disturb the interference pattern on the holographic plate.

150 ns after   **1.75 μs after**

breakdown   **breakdown**

1mm ⊢——⊣   ⊢ **1mm**

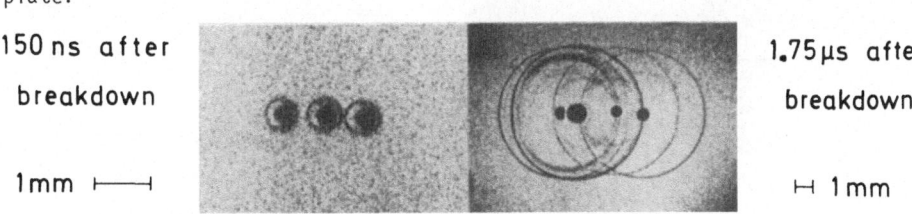

Fig.3 Shock wave detachment in the initial stage after breakdown. Shock waves exhibit as dark circles

As can be seen from Fig. 4 the shock wave profiles built up in the liquid in the surrounding of a single breakdown center after the breakdown can be computed according to Gilmore's theory. More details can be found in [5]. The computed half-widths of the shock wave thickness are in good agreement with the widths of the shock wave circles observed in Fig. 3.

Sometimes when working near above breakdown threshold one succeeds in producing breakdown mainly at one single point in the focal region of the giant pulse. Such a situation is shown in Fig. 5. A big bubble has formed which is almost fully spherical during its phase of maximum expansion, as indicated in the upper left frame. The maximum bubble diameter is about 2.4 mm. The first bubble collapse lies between the second and third frame.

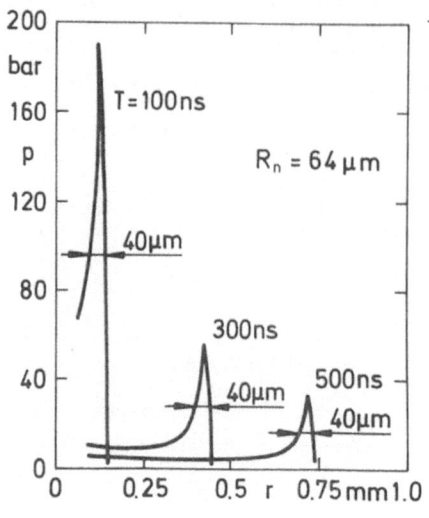

◄ Fig.4 Shock wave pressure profiles in the liquid at three times T = 100 ns, 300 ns, 500 ns after breakdown. R_n is the euqilibrium radius of the bubble

Fig.5 Oscillation of a spherical laser produced bubble in water. Maximum bubble diameter 2.35 mm

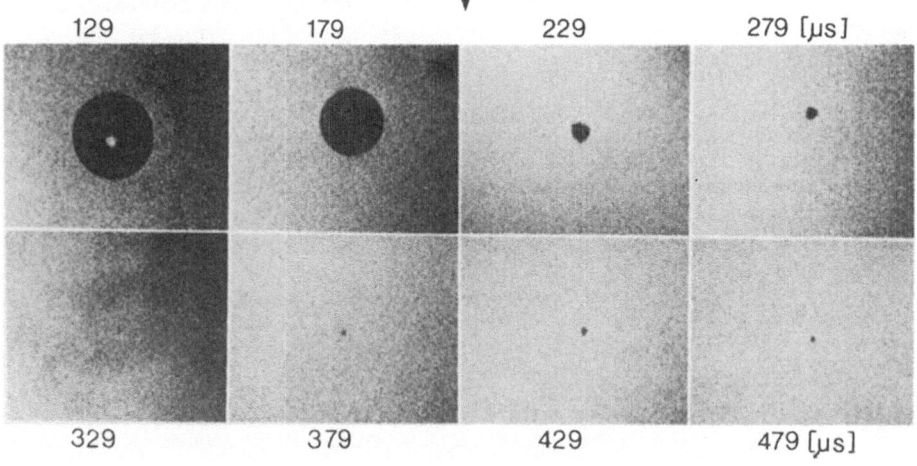

The bubble rebounds with an irregular surface structure. A shock wave radiated at another collapse is faintly visible as dark circular cloud in the fifth frame in the lower left. After first collapse the bubble never reaches its original maximum size. This strong damping is typical for spherical laser produced bubbles in water.

The strong damping cannot be explained by Rayleigh's theory of bubble wall motion even when gas content, surface tension, and viscosity are taken into account. A rather good fit of theoretical and experimental data is only obtained when the liquid is assumed to be compressible [5]. This is demonstrated by Fig. 6 where the radius-time curve of bubble wall motion is computed according to Gilmore's model. The crosses and open squares are experimental. The strong damping of the oscillation after the first collapse is caused by sound radiation.

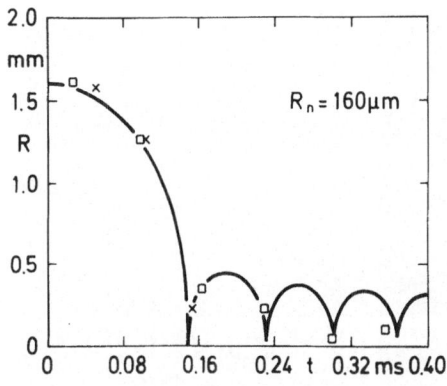

Fig.6 Bubble wall motion as a function of time computed according to Gilmore's model. Crosses and squares are experimental

3. Holocinematography with the multiply cavity-dumped argon ion laser

In this section we describe first experiments done with a multiply cavity-dumped argon ion laser as illuminating light source [6]. The setup is shown in Fig. 7. The argon ion laser produces 60 coherent light pulses of 30 ns pulse length at a repetition rate of 1 kHz for exposing the holographic plate. The three elements, He-Ne-laser, mechanical shutter, and photodiode are used in order to prevent a preexposure of the holographic plate. The geometry for hologram recording is conventional. The object wave illuminates cavitation bubbles which are produced in a water filled cuvette by a Mason horn oscillating at 20 kHz. The light scattered from the bubbles interferes on the holographic plate with the plane wave reference beam. The area of the plate exposed by one single pulse of the series is only 3 by 4 mm. The holographic plate itself rotates so fast that each pulse exposes another portion of the plate, thus separating successively recorded holograms.

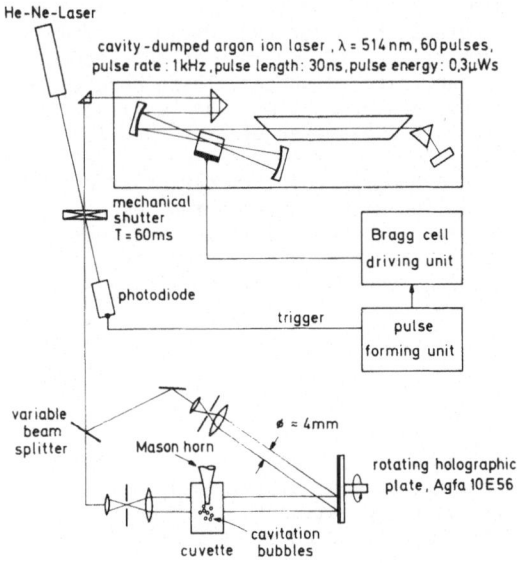

Fig.7 Setup for recording hologram series with the multiply cavity-dumped argon ion laser

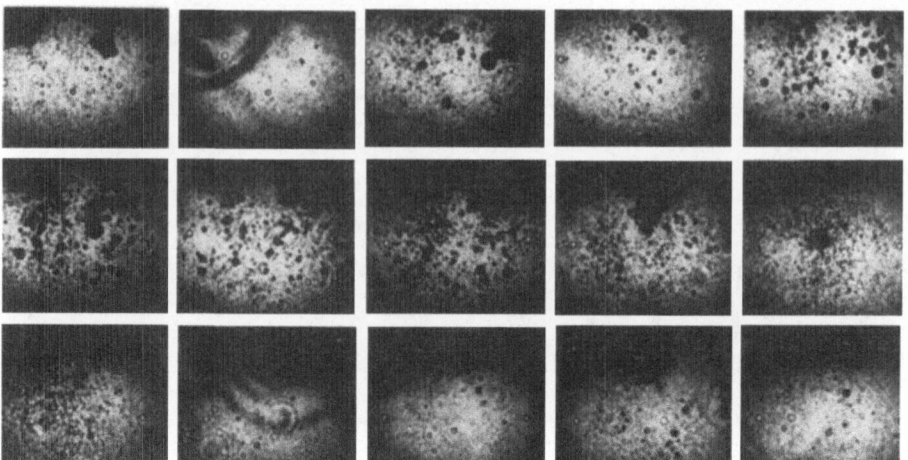

Fig.8 Reconstruction of a series of 15 holograms showing the cavitation bubble field below an oscillating Mason horn. All frames show the same plane in depth. The time interval between successive frames is 1 ms. The sequence runs from the left to right and from the top to the bottom

Images of fifteen successively recorded holograms from the whole of 60 are reconstructed in Fig. 8. The series shows the development of a cavitation bubble field below the Mason horn in time steps of one millisecond. The sequence runs from the left to the right and from the top to the bottom. The frame area is 3.0 by 2.8 mm. The bubbles are imaged as dark spots in front of a bright background. In all frames it is focussed on the same plane in depth. One finds great changes in the arrangement of the bubbles from one hologram to the next. Periods in which many small bubbles are present in the field of view follow periods of rather low bubble density. This is a typical behaviour found in many other series. Single bubbles cannot be identified in successive frames. This is due to the relatively long time interval between successive holograms which corresponds to 20 periods of the oscillating Mason horn. In some frames, for example in the second and in the twelfth, somewhat blurred dark circles appear. These have to be attributed to shock waves emitted by collapsing bubbles. The smallest bubbles which can be resolved clearly have a diameter of less than 10 μm.

Currently, our work concentrates on producing hologram series at higher repetition rates and at improved resolution with the argon ion laser. The application of an argon ion laser is very desirable as it is much more handy than the pulsed ruby laser. So far, the only disadvantage is that because of the relatively low light energy per pulse only small scenes can be illuminated.

4. Conclusion

Summarizing it can be stated that high speed holocinematography is a powerful tool for studying cavitation bubble dynamics. Two special features should be stressed in particular. Firstly, bubbles distributed in a deep volume can easily be recorded because of the great depth of observation, and, secondly, shock wave propagation can be studied successfully due to the short exposure times.

40

5. References

1 K.J. Ebeling, Optik 58 (1977) 383 and 481.
2 K.J. Ebeling and W. Lauterborn, Appl. Opt. 17 (1978) 2071.
3 K.J. Ebeling and W. Lauterborn, Opt. Commun. 21 (1977) 67.
4 W. Lauterborn and K.J. Ebeling, Appl. Phys. Letters 31 (1977) 663.
5 K.J. Ebeling, Acustica 40 (1978) 229.
6 K.J. Ebeling, Proc. First European Conference on Optics Applied to Metrology, M. Grosman and P. Meyrneis (Eds.), Strasbourg 1977, SPIE Volume 136, p. 348.

Bubble Collapse Studies at a Million Frames per Second

W. Lauterborn and R. Timm

Drittes Physikalisches Institut, Universität Göttingen, Bürgerstr. 42-44, D-3400 Göttingen, Fed. Rep. of Germany

1. Introduction

The phenomenon of laser-induced breakdown in liquids, for a long time just investigated for its own sake [1-4] has been applied for several years to the longstanding problem of cavitation bubble dynamics [5-9]. High speed photographic [5-8] and holographic methods [9] have been employed. However, since the cavities occurring upon breakdown proved to be never exactly repeatable in size the dynamics of the cavity a long time after breakdown, as e.g. its first collapse, could not be studied systematically with high time resolution. This can now be achieved by a sophistication in the experimental set-up which we report here. The essential idea is that the laser-induced cavity itself is used to trigger the high speed photographic equipment via the modulation of a He-Ne laser light beam.

2. The Experimental Set-up

Figure 1 shows the experimental set-up. Breakdown and cavity formation is achieved with giant pulses from a ruby laser (30-50 ns duration, 0.1 - 1 Joule energy per pulse) which are focused into the liquid under study with

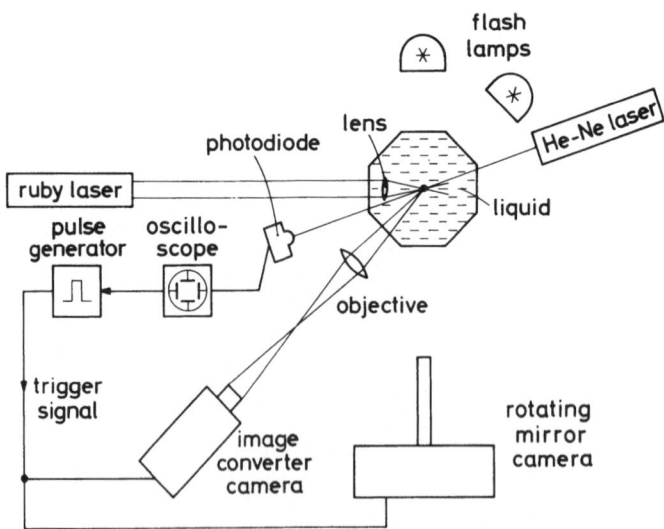

Fig.1 Block diagram of the experimental set-up for high speed photography of laser-induced cavities with triggering from the cavity itself

a single lens of short focal length. The lens is submerged in the liquid (water) and has a focal length of 44 mm in water. The cavity or cavities appearing after breakdown in the focal region are photographed by a rotating mirror camera (Beckman and Whitley Model 330) and an image converter camera (Hadland Imacon 790). The set-up allows simultaneous studies of the same cavity with both cameras. They then run at different framing rates to cover the whole life cycle of the cavity as well as selected parts at higher framing rates and thus better time resolution. The cameras can be triggered by the cavity itself. This is accomplished by letting the cavity modulate the intensity of a He-Ne laser light beam (15 mW) which is picked up by a fast photodiode. For convenience we used the trigger of an oscilloscope to derive the trigger signal for the cameras.

3. The Trigger Signal

A typical signal picked up by the photodiode when a cavity is formed in the path of the He-Ne laser light beam is shown in Fig.2. The curve has been redrawn from a photograph taken from a storage oscilloscope. It starts at a time when the laser-induced cavity has already reached such a size as to fill the total cross section of the beam. Then almost no light reaches the diode and a small signal results. After the expansion the bubble begins to shrink with increasing velocity. When it becomes smaller than the cross section of the laser beam the intensity at the diode rises until the cavity reaches its minimum volume (or, more carefully stated, its minimum cross section in the direction of the He-Ne laser beam). The point of maximum light intensity can thus be identified with the bubble collapse. Immediately after collapse the light intensity again drops drastically to some minimum value. This minimum arises from the emission of a strong shock wave (or a whole series of shock waves) as best understood intuitively from the third picture in Fig.3. There the black spot is a bubble shortly after collapse surrounded by the emitted shock wave. Obviously this shock wave deflects the He-Ne laser light out of its path so that the diode gets less light. The shock wave quickly travels out of the laser beam and once more there in high light intensity at the diode. After collapse and shock wave emission the bubble

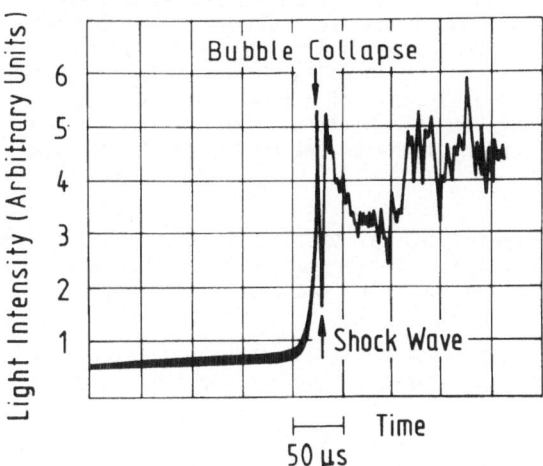

Fig.2 Typical bubble modulated He-Ne light intensity as picked up by the photodiode of Fig.1

rebounds as is seen from the lowering of the light intensity. In fact, the bubble executes a few minor oscillations after the first collapse. The irregularities of the curve during rebound and the subsequent oscillations of the bubble indicate strong irregularities in the shape of the bubble after its first collapse. This is well known from photographic studies (e.g. [6, 7]) but can more easily and continuously be monitored by the experiment described here.

4. Bubble Collapse and Shock Wave emission

The sudden rise in intensity of the He-Ne laser light at the photodiode before bubble collapse has been used to trigger the high speed cameras (see Fig.1). In this way bubble collapse studies can be done rather easily even at high framing rates. So far a million frames per second has been reached, and the method seems to be extendable to five or even twenty million frames per second.

Figure 3, taken at a million frames per second, gives an example of the collapse of a single laser-induced cavitation bubble which proceeds very nearly spherically. In the third frame the radiated shock wave is seen. This indicates that the frame has been taken shortly after collapse. In the fourth frame the bubble shows deviations from sphericity. The shock wave is just visible in the outer parts of the frame. When there is not just one single point of breakdown in the liquid but several nearby, a single big cavity may nevertheless result upon growth of the individually created bubbles. Such cavities usually collapse with large distortion and radiate a multiplicity of shock waves as exemplified in Fig.4. In the case of Fig.4 two shock waves are radiated upon collapse of the aspherical bubble and a grossly distored shape develops after collapse. As frame two shows no sign of a shock wave it must have been taken just before collapse.

Of strong interest in cavitation physics is the collapse of cavitation bubbles in the neighborhood of solid boundaries. A knowledge of how bubbles behave near solid boundaries is the key to an understanding of how erosion and destruction of mechanical parts in cavitating liquids come about. Only a few investigations exist [6,7,11-14] due to the difficulties both experimentally and theoretically. The trigger method works also well with laser-induced bubbles produced in the vicinity of solid walls. Fig.5 shows an example. The bubble has been produced at a distance of 3.5 mm away from the boundary and its collapse has been taken at a million frames per second with the image converter camera. The boundary is below the bottom of the frames.

Fig.3 Collapse of a spherical laser-induced cavitation bubble taken at a million frames per second with the image converter camera. The shock wave radiated upon collapse is to be seen on the third and fourth frame. The width of the frames is 2.7 mm.

Fig.4 Collapse of an aspherical laser-induced cavitation bubble taken at a million frames per second with the image converter camera. In this case two shock waves are radiated. The width of the frames is 2.7 mm.

Fig.5 Collapse of an initially spherical cavity in the neighborhood of a plane solid boundary (below the bottom of the frames) taken at a million frames per second with the image converter camera. The distance of breakdown from the boundary is 3.5 mm, the width of the frames 2.7 mm.

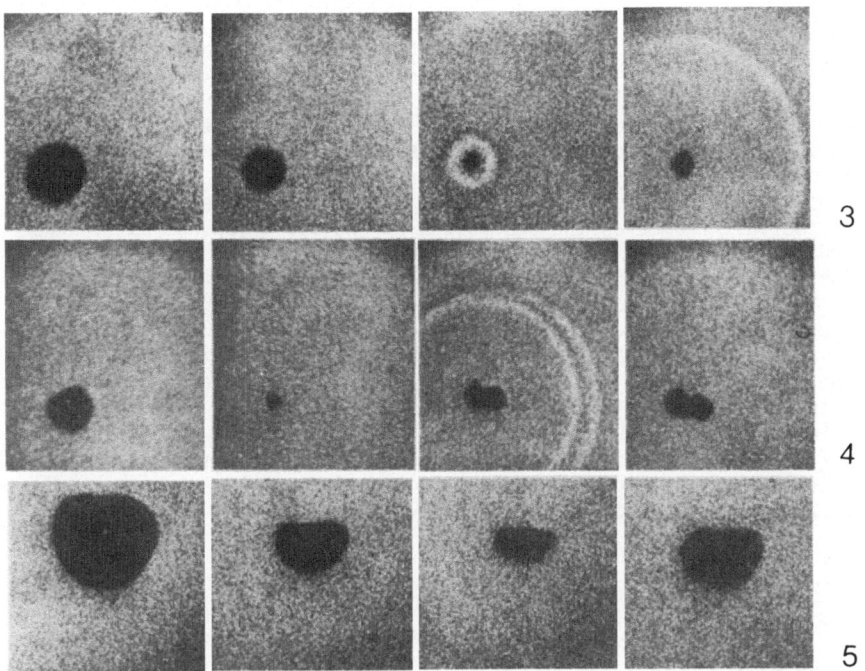

Figs. 3-5. Caption see opposite page

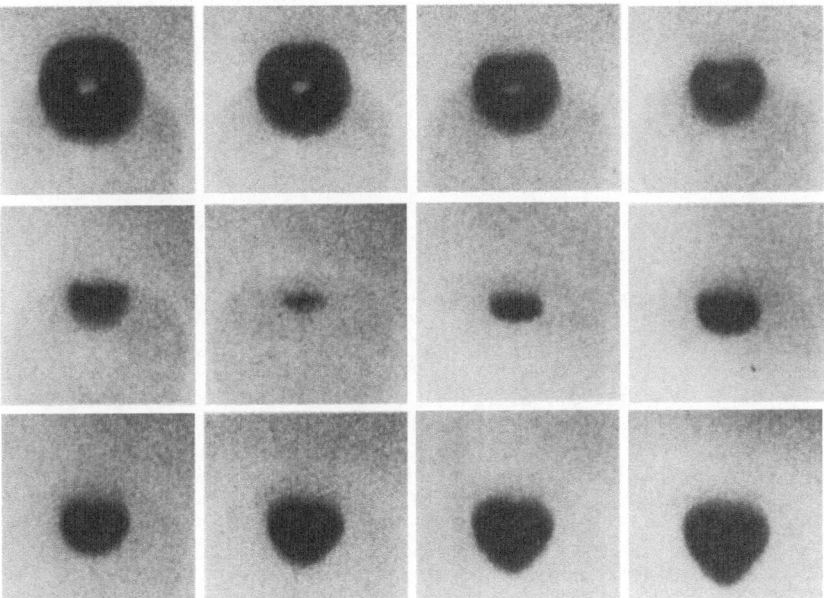

Fig.6 Collapse of an initially spherical cavity in the neighborhood of a plane solid boundary (below the bottom of the frames) taken at 940 000 frames per second with the rotating mirror camera. The distance of breakdown from the boundary is 5.25 mm, the width of the frames 2.7 mm.

The bubble shows the characteristic flat top [7,13,14]. The collapse seems to be not as strong as in the free liquid case. The shock wave is only faintly to be seen in the outer parts of frame 3. It is known [6,7,11-14] that a liquid jet is developed towards the boundary. The investigations made here show that this jet takes a relatively long time to develop (Fig.5 and especially Fig.6). Upon collapse and even some time (microseconds) afterwards no sign of a jet may be detected whereas in the later stages of rebound (see [6,7]) a pronounced liquid jet is observed. It is indicated in Fig.6 on the last frames as the protrusion on the bubble downwards towards the boundary. We observed that the collapsed state of the cavity seems to be of the form of a flat disk (third frame of Fig.5, sixth frame of Fig.6) contrary to the theoretical calculations which predict a more spherical appearance on collapse and an already well developed jet [13]. The calculations in [13] were done for an incompressible liquid without surface tension. One might argue that an inclusion of surface tension and compressibility of the liquid may alter the calculated collapsed state of the cavity in the direction of the observed shape. More details can be found in the work of Timm [10], in particular a comparison of the experimentally measured radius-time curves of single spherically collapsing bubbles with theory including the compressibility of the liquid.

5. Acknowledgements

This work was sponsored by the Fraunhofer-Gesellschaft. Much of the work has been done at the Institut für den Wissenschaftlichen Film, Göttingen. We thank Dipl.-Ing. R. Tilke for his help with the rotating mirror camera and Dr. K. J. Ebeling and Dr. E. Cramer for many discussions on the collapse of cavitation bubbles.

6. References

1 G.A. Askar'yan, A.M. Prokhorov, G.F. Chanturiya, and F.P. Shipulo: Sov. Phys.-JETP 17, 1463 (1963).
2 R.G. Brewer and K.E. Rieckhoff: Phys. Rev. Lett. 13, 334a (1964).
3 E.F. Carome, E.M. Carreira, and C.J. Prochaska: Appl. Phys. Lett. 11, 64 (1967).
4 M.P. Felix and A.T. Ellis: Appl. Phys. Lett. 19, 484 (1971).
5 W. Lauterborn: Appl. Phys. Lett. 21, 27 (1972).
6 W. Lauterborn: Acustica 31, 51 (1974).
7 W. Lauterborn and H. Bolle: J. Fluid Mech. 72, 391 (1975).
8 W. Lauterborn: Phys. Bl. 32, 553 (1976).
9 W. Lauterborn and K.J. Ebeling: Appl. Phys. Lett. 31, 663 (1977).
10 R. Timm: Masters Thesis, Göttingen, 1979 (University of Göttingen, Germany).
11 C.F. Naudé and A.T. Ellis: J. Basic Engng, Trans. ASME D83, 648 (1961).
12 T.B. Benjamin and A.T. Ellis: Phil. Trans. Royal. Soc. London A 260, 221 (1966).
13 M.S. Plesset and R.B. Chapman: J. Fluid Mech. 47, 283 (1971).
14 W. Lauterborn, H. Bolle, and Inst. Wiss. Film: Film E 2353 (1977). Available from: Institut für den Wissenschaftlichen Film, Nonnenstieg 72, D-3400 Göttingen, Germany.

Holographic Generation of Multi-Bubble Systems

W. Hentschel and W. Lauterborn

Drittes Physikalisches Institut, Universität Göttingen, Bürgerstr. 42-44
D-3400 Göttingen, Fed. Rep. of Germany

Abstract

A new holographic method for the simultaneous production of several bubbles in a liquid is presented. Computer generated Fresnel and Fourier transform binary phase holograms formed in photoresist are used to split a ruby laser pulse. Holograms yielding systems of two or five bubbles each in silicone oil have been successfully employed in the laboratory. Preliminary results are given.

1. Introduction

The dynamics of single cavitation bubbles has been investigated experimentally in our laboratory for many years [1,2,3] by the method of optical cavitation. More details about these investigations are given by EBELING, LAUTERBORN and TIMM in this volume.

In hydrodynamics and acoustics the dynamics of cavitation bubble *fields* is of great interest. To study the interaction between individual bubbles multi-bubble systems must be considered. But there is a gap between generating a single bubble by optic cavitation and generating large bubble fields acoustically or mechanically. It is the aim of this paper to present a method for the simultaneous and systematic production of several bubbles.

2. Computer Generated Holograms

The obvious way to generate several bubbles by light is to use a combination of conventional beam splitters, mirrors and lenses to split and focus the giant pulse of a ruby laser. The optical breakdown at the focal points leads to a cavitation bubble system in a liquid. To change the bubble configuration the total optical setup has to be modified.

A more attractive method is to use computer generated holograms to split the giant pulse. Now only the hologram has to be changed to get another bubble system. The computer generation of holograms has proven useful in optical spatial filtering and testing of optical surfaces.

For our application there are two restrictions on this type of holograms. High power laser light pulses are used, and no absorbing materials (e.g. an amplitude hologram) are allowed in the laser beam. Only transparent structures will not evaporate, hence our hologram has to be a *phase* hologram.

47

Secondly, we need large and distortion free plots [5]. Most plotters are only able to plot or not plot a point at a given position. This leads to a *binary* hologram. Several types of computer generated holograms are described in the literature [e.g. 4]. Two of these, the Fresnel and the Fourier transform hologram will be discussed in this paper.

2.1 Fresnel Hologram

In normal holography a plane reference wave E_R is added to the wavefront coming from the object. The intensity of the interference pattern is stored on the holographic film. By reconstructing the hologram with the same reference wave we get back the object wavefront. In digital holography this is done with the help of a digital computer.

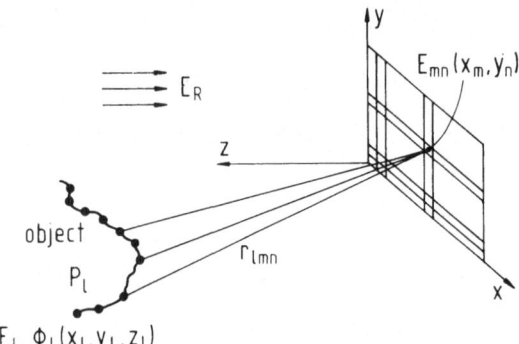

Fig.1 Computer generation of a Fresnel hologram. Object and hologram are sampled in space (details see text).

In a first step the object is sampled at a number of points P_ℓ with the coordinates (x_ℓ, y_ℓ, z_ℓ) (see Fig. 1). Each point has an amplitude E_ℓ and a starting phase ϕ_ℓ and acts as a point source from which a spherical wave is radiated. The second step is to sample the hologram in space and to quantize the amplitude of each sample to a finite number of levels. The complex amplitude E_{mn} of a point (x_m, y_n) in the hologram is given by

$$E_{mn} = E_R + \sum_\ell \frac{E_\ell}{r_{\ell mn}} \exp \left\{ i \left(\phi_\ell + \frac{2\pi}{\lambda} r_{\ell mn} \right) \right\}$$

where

$$r_{\ell mn} = \sqrt{(x_\ell - x_m)^2 + (y_\ell - y_n)^2 + z_\ell^2}$$

and λ is the wavelength of light. A point is plotted if the intensity I_{mn} of a hologram element is

$$I_{mn} = |E_{mn}|^2 > \bar{I} \ .$$

The intensity threshold $\bar{I}$ is chosen so that about half of the hologram points are plotted whereas the other half is left blank.

2.2 Fourier Transform Hologram

A twodimensional Fourier transform can be constructed with the help of a spherical lens (refer to Fig. 2). We get the Fourier transform $G(u,v)$ of an initial transparency in the object plane $g(x,y)$ in the back focal plane of the lens L_1. Adding a plane reference wave, E_R, the Fourier transform may be stored as an intensity distribution on a holographic film. Reconstructing the hologram with the same reference wave E_R and using a second lens L_2 for Fourier transforming we will get back the initial image in the image plane $\bar{g}$, but with changed coordinates.

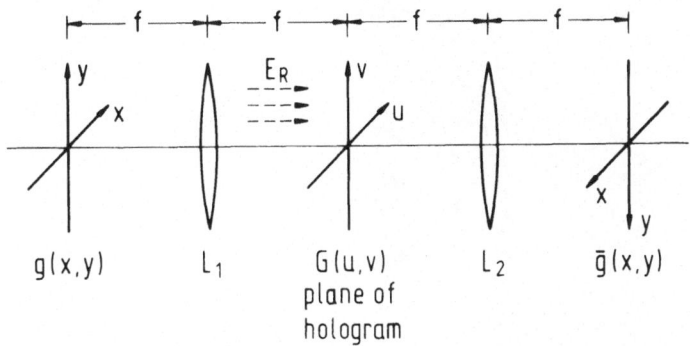

Fig.2 Setup for twodimensional Fourier transform with lenses L_1 and L_2 with focal length f, E_R reference wave, $g(x,y)$ in the object plane, $G(u,v)$ in the Fourier plane and $\bar{g}(x,y)$ in the image plane.

The mathematical description of the twodimensional Fourier transform is given by

$$G(u,v) = \int\int_{-\infty}^{+\infty} g(x,y) \cdot \exp\left\{-2\pi i (x\nu_x + y\nu_y)\right\} dxdy = \mathcal{F}\left\{g(x,y)\right\}$$

with

$$\nu_x = \frac{u}{\lambda f} \quad \text{and} \quad \nu_y = \frac{v}{\lambda f}$$

the spatial frequencies, λ the wavelength of light and f the focal length of the lens used. In the case of a computer generated Fourier transform hologram the first part of the optical arrangement is replaced by a digital computer. For the computation special algorithms are used like the Fast Fourier Transform.

2.3 Production of the Binary Phase Hologram

There are several techniques for producing phase holograms. For our application photoresist has been chosen since it is easy to handle and is highly insensitive to ruby laser light. After computation of the Fresnel or Fourier transform hologram (even a big computer needs several hours to compute just one hologram) an OPTRONICS plotter is used to plot the 4096 x 4096 picture elements point by point. Fig. 3 shows an example of such a plot with 16 million picture elements. Photographic reduction onto high resolution and high contrast film (KODAK High Resolution Plates) leads to the mask used in the technique. The mask is 10 mm x 10 mm and the smallest single picture element is only 2.5 μm wide. Photoresist has been spin coated to a thickness of approximately half a μm on to a carefully prepared glass substrate. The photoresist layer is brought into contact with the mask and exposed by UV light. In the developing process the resist is removed where it has been exposed a phase relief being left.

Fig.3 Computer generated two point Fresnel hologram. It consists of 4096 x 4096 picture elements (by courtesy of HATTENBACH).

2.4 The Phase Grating as an Example of a Binary Phase Fourier Transform Hologram

The most simple example of a binary phase Fourier transform hologram is a phase grating. If the height h of the phase relief grating is given by

$$h = \frac{\lambda}{2} \frac{1}{n - n_a}$$

this leads to a phase shift π for the transmitted light, where n_a is the index of refraction of air, n the index of refraction of the photoresist and λ the wavelength of light. Fig. 4 shows a histogram of the diffraction efficiency of the first three diffraction orders of a relief phase grating with a ratio s:d = 0.5 where s is the wall width and d the grating period (compare [6]). The left column in each diffraction order shows the diffraction efficiency measured using cw dye laser light at the wavelength of the ruby laser. In the middle column the theoretical values are shown. The reasons for the differences between the theoretical and the experimental results are deficiencies in the manufacturing process. A calculation for a real binary phase grating with imperfections in the grating height and width leads to the diffraction efficiency shown in the right column. In the case s:d = 0.41 and a phase shift of 0.85 π we get approximately the measured values.

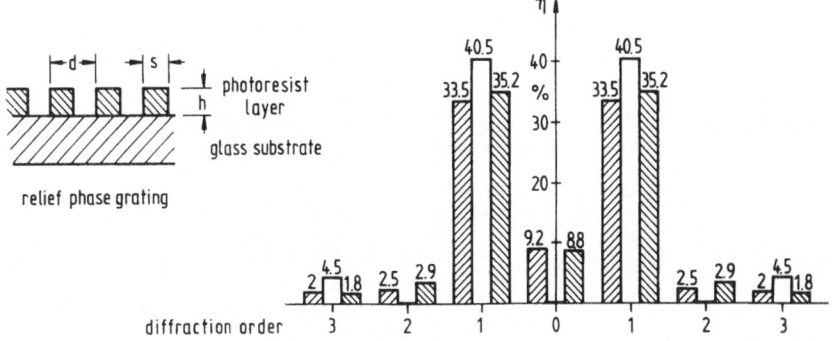

relief phase grating

diffraction order

Fig.4 Diffraction efficiency η of the first three diffraction orders of a relief phase grating in a photoresist layer.
Left column: experimental results at 694 nm wavelength of light.
Middle column: theoretical values under best conditions.
Right column: calculated values for a real binary phase grating with imperfections in grating height and width.

3. Experiments

In the experimental setup (see Fig. 5) a giant pulse from a passively Q-switched ruby laser passes through a phase grating (or, in general, a Fourier transform hologram) and then is focused into a cuvette filled with silicone oil. The optical breakdown at the focal points leads to a cavitation bubble system in the oil. Highly viscous silicone oil is used to store the bubbles for a short time. Thus a conventional camera can be used to photograph the bubble arrangement illuminated from the back. An example of a two bubble system in silicone oil formed by a Q-switched ruby laser pulse is shown in Fig. 6. The bubbles have a spatial separation of about 4 mm and reach 1 to 2 mm in diameter.

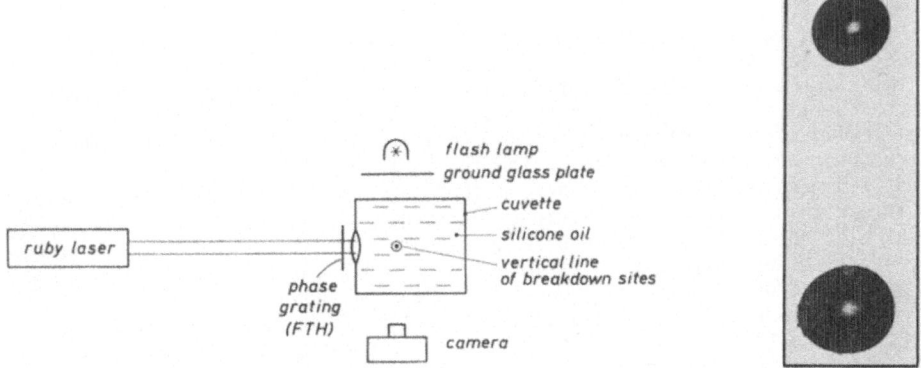

Fig.5 Experimental setup for the production and photographing of multi-bubble systems with a Q-switched ruby laser.

Fig.6 Two-bubble system in silicone oil formed by a Q-switched ruby laser pulse passing a phase grating.

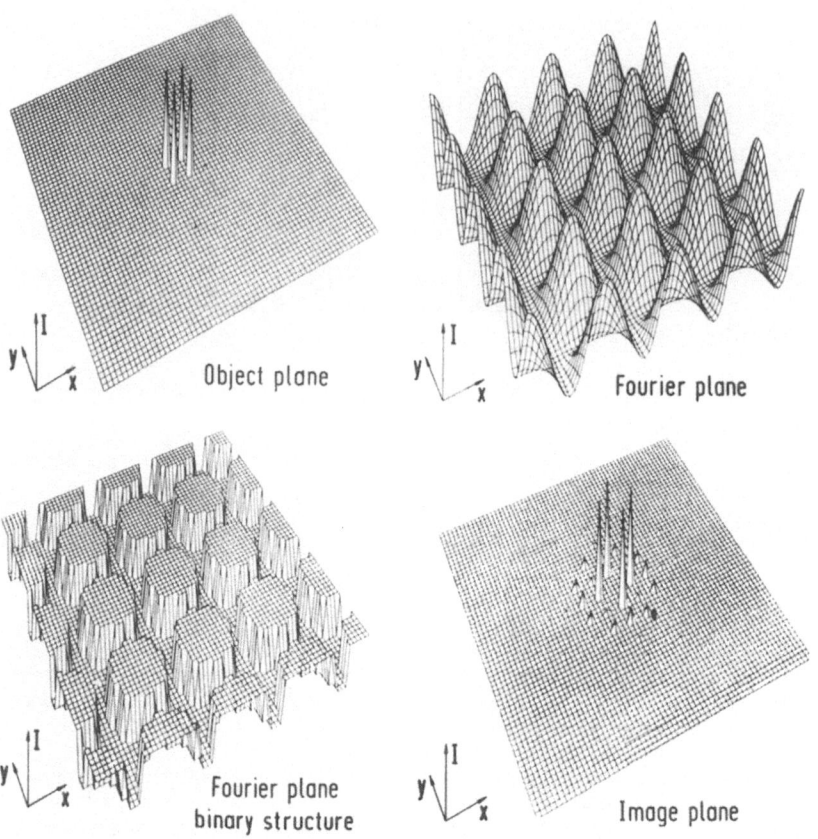

Fig.7 Model calculation of the diffraction efficiency of a twodimensional binary phase Fourier transform hologram.

To get a twodimensional bubble configuration a model calculation of a special twodimensional binary phase Fourier transform hologram has been made (Fig. 7). The calculation starts with the initial distribution in the upper left. The Fourier transform of the data in the upper right is quantized into two levels to get a binary structure. Interpreting this as a binary phase structure the intensity distribution in the image plane is given by a second twodimensional Fourier transform shown at the lower right. Four intense points with a diffraction efficiency of about 15% each may be seen in the centre of the plot surrounded by a great number of higher order points with about 2% or less diffraction efficiency each. A relief binary phase Fourier transform hologram has been produced of the type shown in Fig. 7 at the lower left. When the ratio s:d does not equals 0.5 we get an additional zero order diffraction point in the centre. Two examples of five-bubble systems are presented in Fig. 8. An interesting point may be noted in the second example where the centre bubble is surrounded by a great number of very small bubbles. Important bubble interaction must have taken place before this photograph was taken.

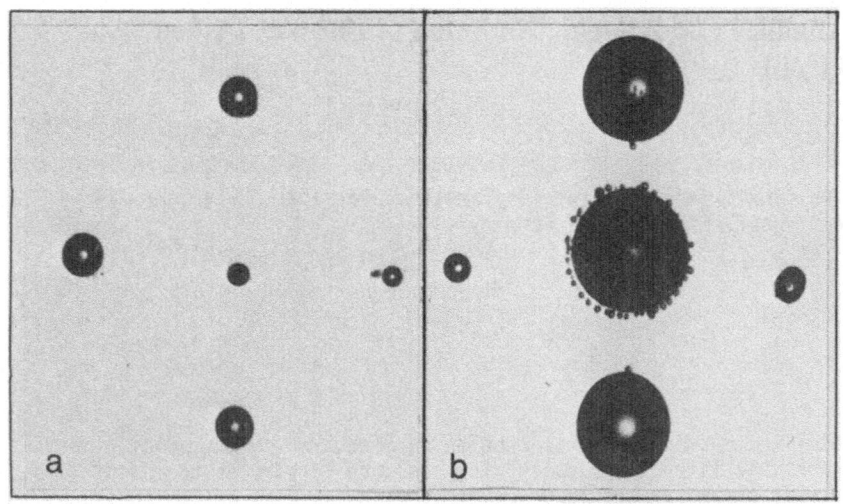

Fig.8 Five-bubble systems in silicone oil formed by Q-switched ruby laser
pulses passing a binary phase Fourier transform hologram

4. Discussion

A new method has been presented for simultaneous production of several
bubbles. Computer generated Fresnel and Fourier transform holograms are
able to produce multi-bubble systems. Comparison between Fresnel and Fourier
transform holograms shows that Fresnel holograms permit three dimensional
distributions, while Fourier transform holograms will give only two dimensional distributions but double the total diffraction efficiency. High speed
holocinematographic studies will help ascertain the mechanism for cavitation
bubble interaction [7,8].

5. Acknowledgement

We want to thank A. Lohmann and R. Hauck (Erlangen) for making available
their plotting facilities to us and H. Dammann from the Philips Research
Laboratories (Hamburg) for demagnifying our Fresnel hologram plots.

6. References

1 Lauterborn, W., Kavitation durch Laserlicht, Acustica 31, 51 (1974).
2 Ebeling, K.J., Zum Verhalten kugelförmiger, lasererzeugter Kavitationsblasen in Wasser, Acustica 40, 229 (1978).
3 Timm, R., Masters Thesis, Göttingen 1979.
4 Huang, T.S., Digital Holography, Proc. IEEE 59, 1335 (1971).
5 Biedermann, K., Holgren, O., Large-size distortion-free computer generated holograms, Appl. Opt. 16,2014 (1977).
6 Sirohi, R.S., Blume, H., On the diffraction efficiency of synthetic
  binary holograms, Optica Acta 22, 943 (1975).
7 Ebeling, K.J., Lauterborn, W., Acustooptic beam deflection for spatial
  frequency multiplexing in high speed holocinematography, Appl. Opt. 17,
  2071 (1978).
8 Ebeling, K.J., Lauterborn, W., High speed holocinematography using
  spatial multiplexing for image separation, Opt. Comm. 21, 67 (1977).

The Dynamics and Acoustic Emission of Bubbles Driven by a Sound Field

E. Cramer

Drittes Physikalisches Institut, Universität Göttingen, Bürgerstr. 42-44
D-3400 Göttingen, Fed. Rep. of Germany

Abstract

Forced oscillations of spherical bubbles in a compressible viscous liquid (water) are calculated numerically. The response of bubbles to sound fields for a special parameter set is given along with examples of the pressure distribution around a single bubble during oscillation.

1. Introduction

To understand the emission of noise in liquids irradiated by sound with an intensity beyond the cavitation threshold knowledge is required of the response of bubbles to sound fields of different frequencies and different pressure amplitudes. Response curves have previously been given for bubbles oscillating in an incompressible liquid [1]. In this paper numerical calculations are given for bubbles in a compressible liquid. The compressibility becomes important as soon as bubble wall velocities become comparable with the speed of sound in the liquid.

2. The Bubble Model

The GILMORE model [2] is based on the KIRKWOOD-BETHE hypothesis [3] which consists in specifying an invariant of the motion. For a compressible liquid two families of characteristic curves, $x_\alpha(t)$ and $x_\beta(t)$, may be defined covering the r-t plane. The motion is then completely specified by two quantities, X_α and X_β, which are such that X_α is an invariant of $x_\alpha(t)$ and X_β is an invariant of $x_\beta(t)$. A fruitful approach to the solution of such problems has been the suggestion of KIRKWOOD and BETHE, that the function $G = r \, (h + u^2/2)$ is an adequate approximation to the invariant X_α, when the other invariant X_β is everywhere constant. Here h is the specific enthalpy of the liquid, u is the particle speed and r is the distance from the middle of the bubble.

A value of G is propagated along a curve $x_\alpha(t)$ with a speed $u + c$ in such a manner that it remains unchanged. Here c is the local speed of sound. From this condition we may derive GILMORE's equation of motion of the interface:

$$R\left(1 - \frac{U}{C}\right) \frac{d^2R}{dt^2} + \frac{3}{2}\left(1 - \frac{U}{3C}\right)\left(\frac{dR}{dt}\right)^2 - \left(1 + \frac{U}{C}\right)H - \frac{U}{C}\left(1 - \frac{U}{C}\right) R\frac{dH}{dR} = 0$$

(1)

$$\frac{dR}{dt} = U$$

Capital letters denote quantities at the interface of the bubble. R is the radius of the bubble, U is the speed of the bubble wall, C is the speed of sound in the liquid and H is the specific enthalpy. The basic restriction on the KIRKWOOD-BETHE approximation is that the hydrodynamic field must contain only outgoing waves.

Both the enthalpy H and the speed of sound C are functions of the pressure p(R) at the bubble wall. The pressure is given by

$$p(R) = \left(p_o + \frac{2\sigma}{R_n}\right)\left(\frac{R_n}{R}\right)^{3\gamma} - \frac{2\sigma}{R} - \frac{4\mu}{R} U$$

(2)

and the pressure $p_\infty(t)$ at infinity is given by

$$p_\infty = p_o - p_m \sin \omega t$$

(3)

With the equation of state

$$p = A\left(\frac{\rho}{\rho_o}\right)^n - B$$

(4)

the enthalphy H becomes

$$H = \int_{p_\infty}^{p(R)} \frac{dp}{\rho} = \frac{n}{n-1} \frac{A^{1/n}}{\rho_o} \left\{\left(p(R) + B\right)^{\frac{n-1}{n}} - \left(p_\infty + B\right)^{\frac{n-1}{n}}\right\}$$

(5)

and the speed of sound at the interface is given by

$$C = \left[c_o^2 + (n-1) \ H\right]^{1/2}$$

(6)

As water was chosen as the liquid, the constants are:
$n = 7$, $A = 3001$ atm, $B = 3000$ atm, $\rho_o = 0.998$ g·cm^{-3}, $\sigma = 72.5$ dyn·cm^{-1}, $\mu = 0.01$ poise, $p_o = 1$ bar, $\gamma = 1.33$, $c_o = 1.482 \ 10^5$ cm·sec^{-1}.

AKULICHEV [4] has given an approximation for the pressure distribution in the liquid surrounding the bubble. From the approximation of KIRKWOOD and BETHE it follows that the function G(r,t) remains constant along the outgoing characteristic $dr/dt = \tilde{c} = c+u$ (c = speed of sound, u = speed of particles). If this quantity G is known at the interface it becomes at every distance r:

$$G(r,t) = G(R,t_R) = R\left(H + \frac{u^2}{2}\right) \tag{7}$$

with

$$t = t_R + \int_R^r \frac{dr}{\tilde{c}} \tag{8}$$

We can combine (7) with the equation of state (4) as in ref. [5]:

$$G = rc_o u \left(1 + \frac{n+1}{4c_o} \cdot u\right) \tag{9}$$

and

$$\tilde{c} = c_o\left(1 + \frac{n+1}{2c_o} \cdot u\right) \tag{10}$$

Consequently we get for the integral in (8):

$$\int_R^r \frac{dr}{\tilde{c}} = -\frac{G}{c_o^2} \int_U^u \frac{du}{u^2(1 + \frac{n+1}{4c_o})^2} \tag{11}$$

The result is

$$t = t_R + \frac{G}{c_o^2} \beta \left[\frac{1+2\beta u}{\beta u(1+\beta u)} - \frac{1+2\beta U}{\beta U(1+\beta U)} - 2 \ln \frac{(1+\beta u)\beta U}{\beta u(1+\beta U)}\right] \tag{12}$$

With

$$\beta u = \left[\left(1 + \frac{n+1}{rc_o^2} G\right)^{1/2} - 1\right] \tag{13}$$

$$\beta U = \left[\left(1 + \frac{n+1}{Rc_o^2} G\right)^{1/2} - 1\right] \tag{14}$$

If for small disturbances u is small compared to unity, this means $G \ll r \cdot c_o^2$ and one gets the well-known result from linear acoustics:

$$t - t_R = \frac{r-R}{c_o} \tag{15}$$

Using

$$c^2 = \left(\frac{\partial p}{\partial \rho}\right)_S \tag{16}$$

(S is the entropy) and (4) we get an expression for the pressure in the liquid

$$p = A\left[\frac{2}{n+1} + \frac{n-1}{n+1}\left(1 + \frac{n+1}{rc_o^2}\,G\right)^{1/2}\right]^{2n/n-1} - B \,. \tag{17}$$

4. The Calculation Procedure

Eq. (1) is solved by a RUNGE-KUTTA method on a digital computer evaluating principally the steady-state solution of the radius-time curve $R(t)$. A solution is considered periodic, if in three successive periods the differences between corresponding minima and maxima are lower than one percent. After 150 periods of the excitation sound field the calculations are stopped.

Most of the calculations were done starting with the initial conditions

$$R(t = 0) = R_n \,, \quad \dot{R}(t = 0) = 0 \,, \quad t = 0$$

For not too high pressure amplitudes, p_m, the oscillations $R(t)$ became periodic with period T_R being an integral multiple of the period of the sound field, $T_S = 1/f$.

$$T_S = s \cdot T_R \quad (s = 1,2,3,\ldots)$$

The period T_R itself also can be an integral multiple of a "free oscillation" of the bubble with period T_B.

$$T_R = r \cdot T_B \quad (r = 1,2,3,\ldots)$$

The factor r/s is the order of the resonance. The cases with $s = 1$ and $r = 2,3,\ldots$ are the harmonic oscillations, the cases with $s = 2,3,\ldots$ and $r = 1$ are the subharmonic oscillations and the cases with $s = 2,3,\ldots$ and $r = 2,3,\ldots$ are the ultraharmonic (also called ultrasubharmonic) oscillations.

The pressure distribution in the surrounding liquid is calculated using (12) and (17) for many characteristics. The initial conditions for the function G are taken from the periodic solution of the radius-time curve $R(t)$. In this way one gets a three dimensional network of points, (p,r,t), which describes the pressure distribution around a spherical and periodically oscillating bubble. A typical result for steady-state solutions of (1) is shown in Fig. 1. The upper curve on the left shows the driving sound field, the curve below is the steady-state solution of the bubble wall motion (radius-time curve). The lines below show the first few spectral lines of the amplitude spectrum of the periodic oscillation. The curve on the right shows the corresponding pressure distribution with two amplitude spectra of the pressure-time curve for different distances from the bubble. The same situation is shown in Fig. 2 for an additional parameter set.

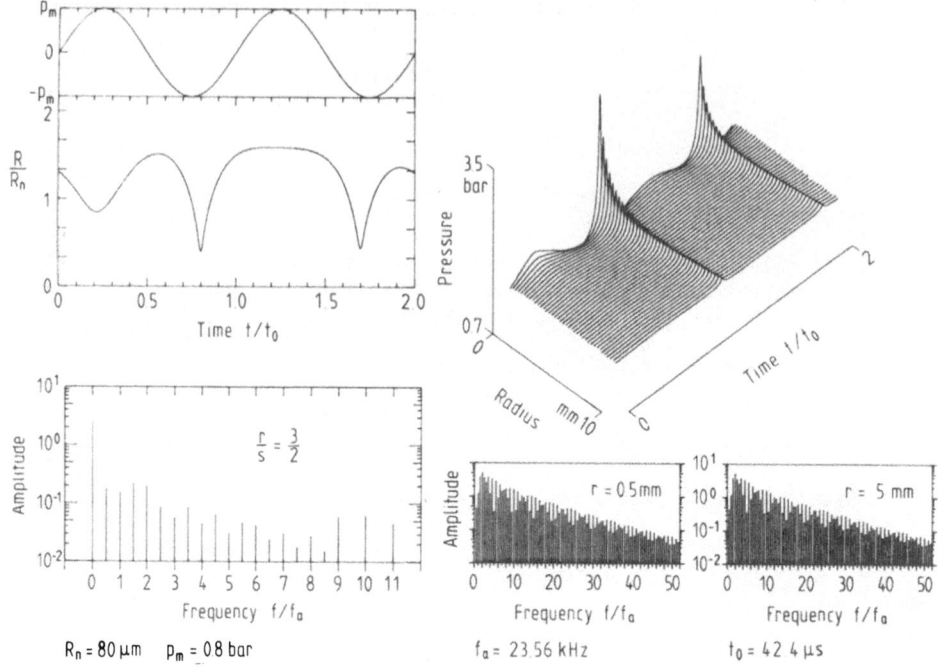

$R_n = 80\,\mu m$ $p_m = 0.8\,bar$

$f_a = 23.56\,kHz$ $t_0 = 42.4\,\mu s$

<u>Fig.1</u> Caption see opposite page

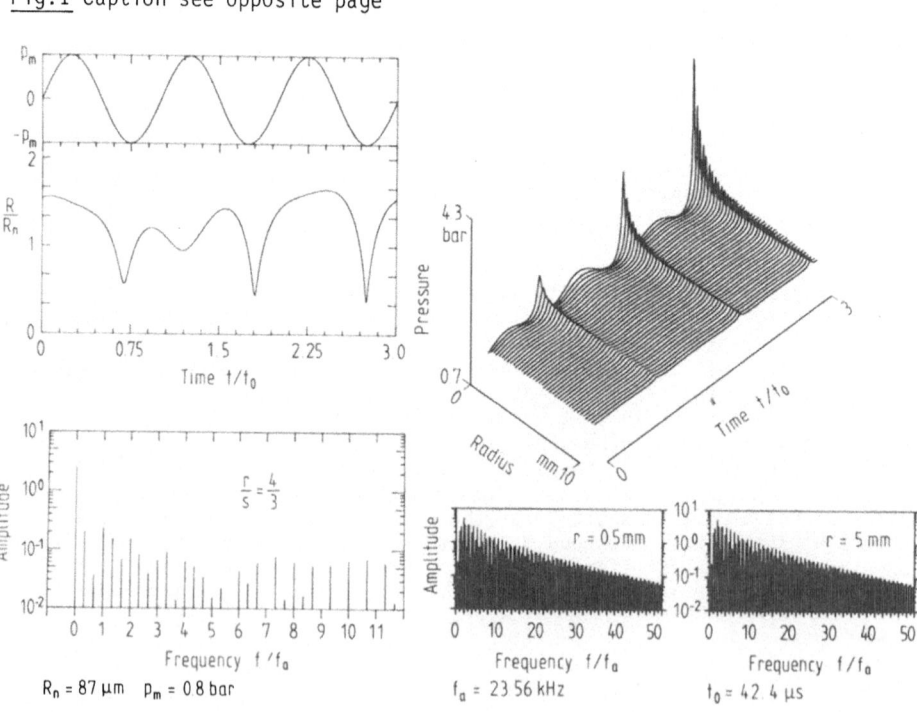

$R_n = 87\,\mu m$ $p_m = 0.8\,bar$

$f_a = 23.56\,kHz$ $t_0 = 42.4\,\mu s$

<u>Fig.2</u> Caption see opposite page

5. Resonance Curves as a Function of the Excitation Frequency with Fixed Bubble Radius

Frequency response curves for bubbles with two different radii of $R_n = 100\mu m$ (Fig. 3a) and $R_n = 10\ \mu m$ (Fig. 3b) were calculated. The maximum radius of the steady-state solution is plotted normalized to the radius at rest of the bubble, R_n, versus the frequency of the driving sound field normalized to the linear resonance frequency given by (18):

$$f_o = \frac{1}{2\pi R_n \sqrt{\rho}} \left[3\gamma\left(p_o + \frac{2\sigma}{R_n} - p_d\right) - \frac{2\sigma}{R_n} - \frac{4\mu^2}{\rho R_n^2} \right]^{1/2} \tag{18}$$

The numbers occurring above the peaks of the curves are the orders, r/s, of the resonances. All resonances lean over to lower frequencies. That is why during the calculation of the linear frequency (18) for strong oscillations it is not the radius at rest, R_n, which is important but the mean radius during oscillation. The reason for the jump phenomenon is the non-linearity of (1).

Of special interest are the ultrahormonic resonances of order 3/2, 5/2, 7/2 ... as they give rise to a subharmonic component at one half the driving frequency in the spectrum of the bubble oscillation (see Fig. 1). These subharmonic components may be responsible for the first subharmonic in the measured spectra of the acoustic cavitation noise.

For higher sound pressure amplitudes, p_m, the form of the resonance curves dissolves, and it is impossible to plot a connected curve. This happens for the 10 μm bubble for amplitudes higher than about 0.8 bar and for the 100 μm bubble for amplitudes higher than about 0.7 bar.

◄ Fig.1 Example of a steady-state solution at resonance in order 3/2. Left - Upper curves: driving sound field with frequency of 23.56 kHz and sound pressure amplitude of 0.8 bar, resulting radius-time curve of a bubble with radius of R_n = 80 μm. Lower curve: amplitude spectrum of the (periodic) solution of the radius-time curve. Right - Pressure distribution around the bubble. The two amplitude spectra of the pressure are taken at distances of 0.5 mm and 5 mm.

Fig.2 Example of a steady state solution at resonance in order 4/3. Left - Upper curves: driving sound field with frequency of 23.56 kHz and sound pressure amplitude of 0.8 bar, resulting radius-time curve of a bubble with radius of R_n = 87 μm. Lower curve: amplitude spectrum of the (periodic) solution of the radius-time curve. Right - Pressure distribution around the bubble. The two amplituae spectra of the pressure are taken at distances of 0.5 mm and 5 mm.

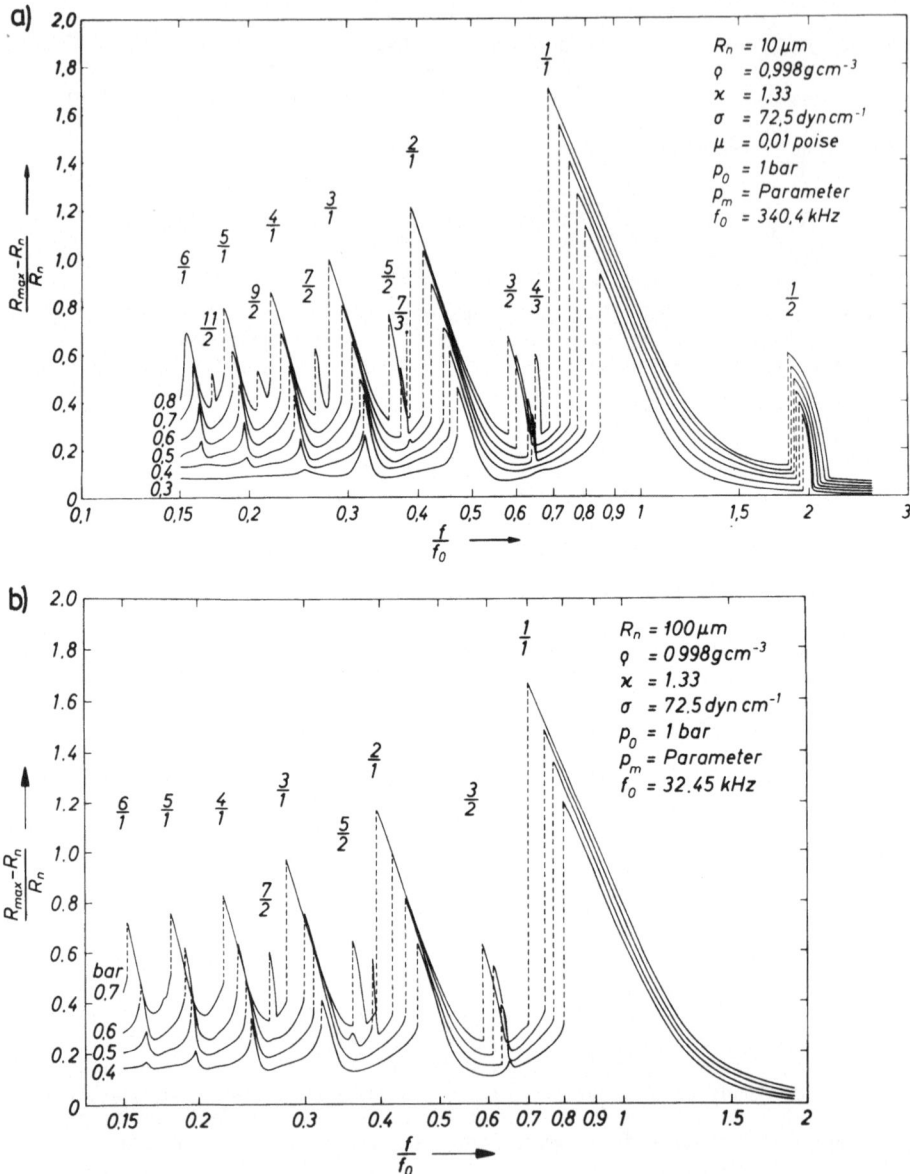

Fig.3a-b Response curve for a bubble with radius at rest of a) 10 μm
b) 100 μm in a driving sound field of various frequencies and sound pressure amplitudes p_m. The numbers occurring above the peaks of the curves are the order of the resonances. R_{max} is the maximum radius of the steady-state solution.

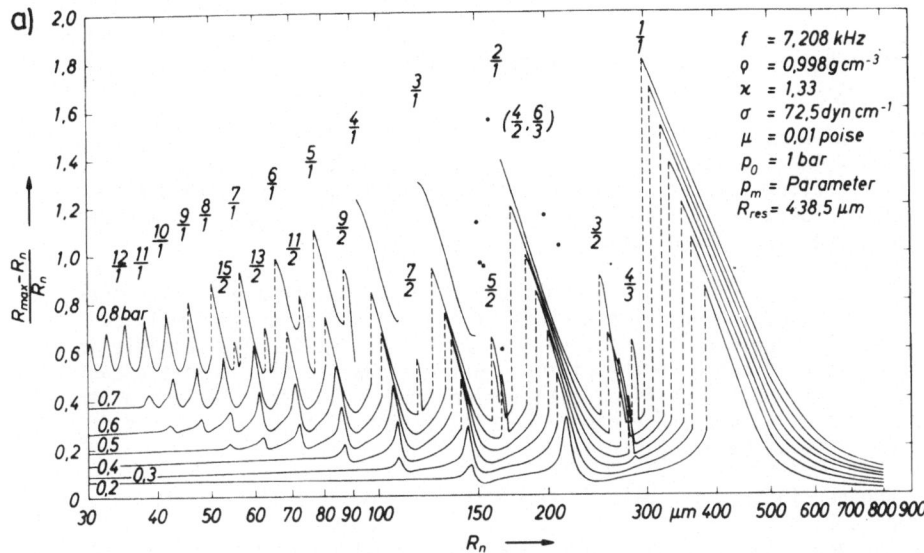

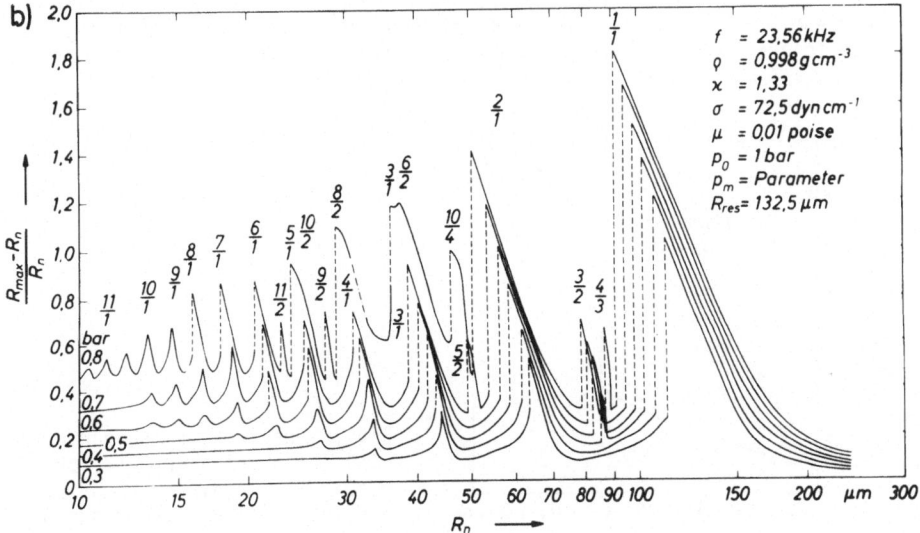

Fig.4a-b Response curves for bubbles with different radii in water at a driving sound field of a) 7.208 kHz and b) 23.56 kHz for different sound pressure amplitudes p_m. The numbers occurring above the peaks of the curves are the orders of the resonances. R_{max} is the maximum radius of the steady-state solution.

6. Response Curves as a Function of the Bubble Radius with Fixed Excitation Frequency (Resonance Curves (R_n))

A better picture of the experimental situation is given by the resonance curves (R_n). Here bubbles with various radii R_n are excited by a sound field of fixed frequency. In this case the frequencies f = 7.208 kHz (Fig. 4a) and f = 23.56 kHz (Fig. 4b) were taken, because these same frequencies were also used in experimentation [6]. The main difference in the resonance curves with fixed bubble radius is the fact that the damping of the oscillation declines faster with decreasing bubble radius than with declining frequency because all damping mechanisms except sound radiation increase with decreasing bubble radius.

The curves for a pressure amplitude of p_m = 0.8 bar are very interesting. In these cases the normal structure of the resonance curves simply dissolves. For the curve in Fig. 4a (f = 7.208 kHz) it is only possible in a few cases (indicated in the picture by individual points) to get steady-state solutions in the fixed calculation time of 150 oscillations of the sound field. In all other cases the oscillations jump between two different orders without becoming periodic.

Another interesting case is shown in Fig. 4b. Here the resonances of the order 5/2, 3/1, 4/1, and 5/1 change with increasing amplitude to resonances of the order 10/4, 6/2, 8/2 and 10/2. This is also shown for a special case in Fig. 5. A bubble with radius R_n = 39 μm excited by a sound field with frequency f = 23.56 kHz oscillates for low sound pressure amplitudes in the resonance of order 3/1. At a pressure of about 0.7 bar the oscillation changes to the order 6/2 and the amplitude increases violently.

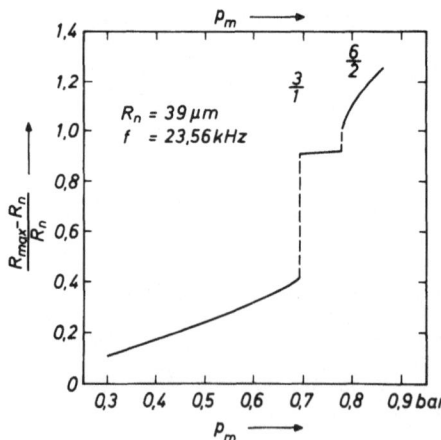

Fig.5 Onset of the subharmonic oscillation of order 6/2 for a bubble with radius R_n = 39 μm and sound field with frequency f = 23.56 kHz.

7. Comparison with Other Bubble Models

The bubble model used here takes into account damping through viscosity and in particular sound radiation. A comparison of corresponding response curves without sound radiation [1] and with Fig. 3a shows that only at high sound pressure amplitudes of p_m = 0.7 bar and p_m = 0.8 bar will marked differences occur. The response curve for a sound pressure of 0.8 bar cannot be drawn in the incompressible case [1]. However, a similar situation occurs for sound pressures higher than 0.9 bar in the compressible case. Another advantage in using the compressible model is the possibility of calculating larger bubbles where the only important damping mechanism is sound radiation.

8. Acknowledgement

Most of the calculations were done on the UNIVAC 1108, Gesellschaft für Wissenschaftliche Datenverarbeitung, Göttingen, Federal Republic of Germany. A few calculations were carried out on the CYBER 76 of the Regionales Rechenzentrum Hannover, Hannover, Federal Republic of Germany. This work was sponsored by the Fraunhofer Gesellschaft, München, Federal Republic of Germany.

I thank Prof. W. Lauterborn for many discussions and steady encouragement in pursuing this work.

9. References

1 Lauterborn, W., J. Acoust. Soc. Amer. 59 (2) (1976) 283.
2 Gilmore, F.R., Cal. Inst. Techn. Report No. 26-4 (1952).
3 Kirkwood, J.G., Bethe, H.A., OSRD Report No. 588 (1942).
4 Akulichev, V.A., in: Rozenberg, L.D. (ed.), High-intensity ultrasonic fields, Plenum Press, New York (1971).
5 Cole, R.H., Underwater explosions, Princetown Univ. Press (1948).
6 Cramer, E., PhD Thesis, Göttingen 1978.

Free and Forced Oscillations of Spherical Gas Bubbles and Their Translational Motion in a Compressible Fluid

H.J. Rath

Institut für Mechanik, Universität Hannover, Appelstraße 11
D-3000 Hannover, Fed. Rep. of Germany

1. Introduction

The fundamental nature of vibratory cavitation is the transient growth and collapse and the translational motion of individual gas bubbles in liquids. Recent investigations [1 - 9] have shown·that the influence of the fluid compressibility can be very strong for the free and forced oscillations of a gas bubble in a homogeneous sound field. The differential equation of the radial motion of a pulsating gas bubble in a liquid including the fluid compressibility given by HERRING is completed and solved numerically for free and forced oscillations. Furthermore an analytical solution of a nonlinearly freely oscillating gas bubble in a compressible liquid investigated by an asymptotic method will be presented [10]. Some experiments have shown that a gas bubble in an inhomogeneous sound field begins to move. The numerical solution of three differential equations [12, 13] shows that the bubble, besides oscillating radially, also moves differently strong in the translational direction in the liquid.

2. The Free and Forced Nonlinear Radial Oscillations

In the present investigation thermal problems and problems of gas diffusion are neglected. Included in the model are the effects of surface tension and the vapor pressure in the interior of the bubble. The gas inside the bubble is compressed according to a polytropic gas law. Neglecting the viscosity of the liquid the following normalized differential equation describes the free and forced radial pulsations of a gas bubble in a compressible liquid [1, 2]

$$RR''\left(1-2\frac{R'}{\alpha_1}\right)+\frac{3}{2}R'^2\left(1-\frac{4}{3}\frac{R'}{\alpha_1}\right)+\frac{\alpha_2}{R}\left[1-\frac{R'}{\alpha_1}\left(1-\frac{R'}{\alpha_1}\right)\right]-$$

$$-\alpha_3 R^{-3\varkappa}\left[1-3\varkappa\frac{R'}{\alpha_1}\left(1-\frac{R'}{\alpha_1}\right)\right]-\alpha_5\left[\sin\bar{\omega}\tau+\frac{R\cdot\bar{\omega}}{\alpha_1}\left(1-\frac{R'}{\alpha_1}\right)\cdot\right.$$

$$\left.\cdot\cos\bar{\omega}\tau\right]-\alpha_4+1=0 \tag{1}$$

with the following dimensionless quantities

$$R=\frac{r}{r_o}\quad,\quad\tau=\frac{t}{r_o}\sqrt{\frac{p_\infty}{\varrho}}\quad,\quad\bar{\omega}=\omega\frac{r_o}{\sqrt{p_\infty/\varrho}}\quad,\quad\alpha_1=c\sqrt{\frac{\varrho}{p_\infty}} \tag{2}$$

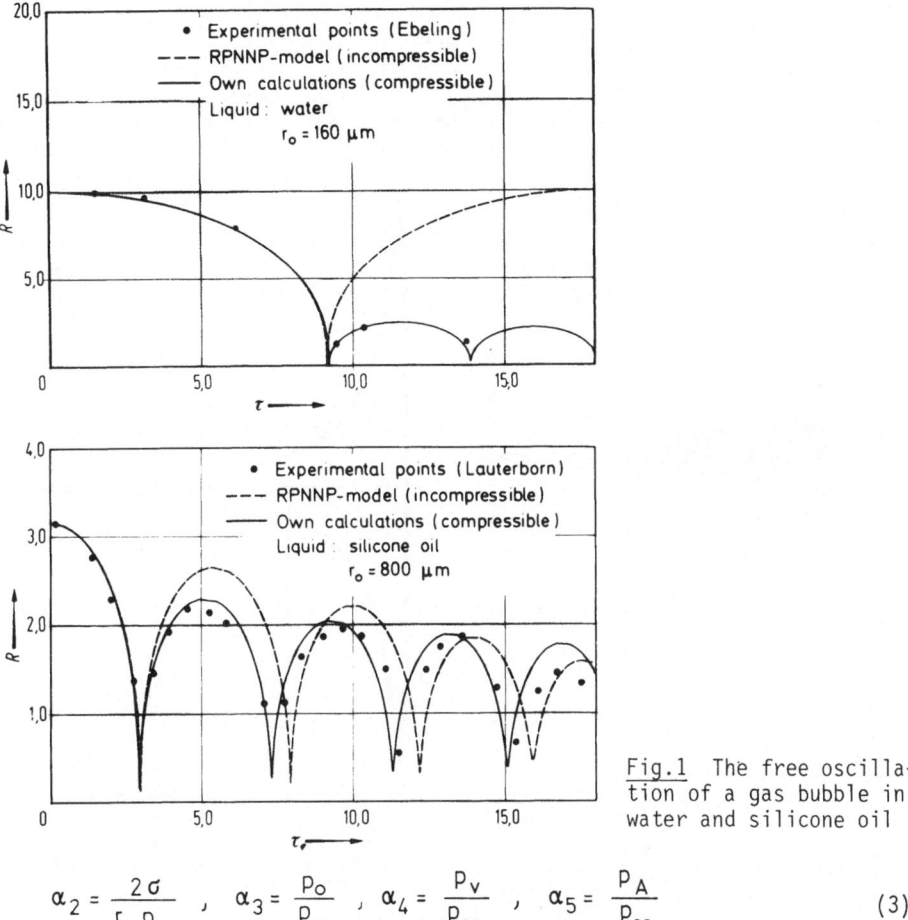

Fig.1 The free oscilla-
tion of a gas bubble in
water and silicone oil

$$\alpha_2 = \frac{2\sigma}{r_o\, p_\infty} \quad , \quad \alpha_3 = \frac{p_o}{p_\infty} \quad , \quad \alpha_4 = \frac{p_v}{p_\infty} \quad , \quad \alpha_5 = \frac{p_A}{p_\infty} \tag{3}$$

in which R, τ and $\bar\omega$ are the dimensionless radius of the bubble, the normalized
time coordinate and the dimensionless angular frequency of the sound field.
ϱ is the density of the liquid, t the time, r_o the radius of the bubble at
rest, r the momentary radius of the bubble, c the constant sound speed of the
liquid, σ the surface tension, $\varkappa$ the polytropic exponent of the gas inside the
bubble, p_∞ the pressure at infinity, p_v the vapor pressure, p_A the sound pressure amplitude, ω the angular frequency of the ultrasonic field and p_o is the
equilibrium pressure of the gas inside the bubble for $r = r_o$. A stroke denotes
a derivative with respect to time.

Equation (1) has been solved numerically. For the free oscillations ($p_A=0$),
($\omega =0$) we use the initial conditions($\tau =0$): $R = R_{max}$, $R' = 0$, and for the forced vibrations we begin with: $R = 1$, $R' = 0$. Figure 1 shows two examles for the
free oscillations of a gas bubble in water and silicone oil [1] . The marks
correspond to the experimental results by EBELING [3] and LAUTERBORN [5].
Accordingly we take: $p_\infty = 1$ bar, $\varkappa= 1,33$, for water: $p_v = 0,0233$ bar, $r_o =$
160 μm, $\sigma = 0,0725$ N/m, $c = 1460$ m/s, $\varrho = 998$ kg/m^3, $\mu = 10^{-4}$ kg/(sm), $r_{max}=1,6$mm,
for silicone oil: $p_v= 0$ bar, $r_o = 800$ μm, $\sigma = 0,0215$ N/m, $c = 1100$ m/s, $\varrho =970$

kg/m^3, $\mu = 0,485$ kg/(sm), $r_{max} = 2,5$ mm. Fig.1 shows that the compressible theory (Eq.(1)) is in a good agreement with the experiments. For the strong free oscillations of a bubble in water EBELING has calculated the same curves by the GILMORE theory [6]. The advantage of Eq.(1) compared with the GILMORE theory is that the computing time for the numerical solution of Eq.(1) is nearly the same as for the solution of the incompressible RPNNP- model [5] .

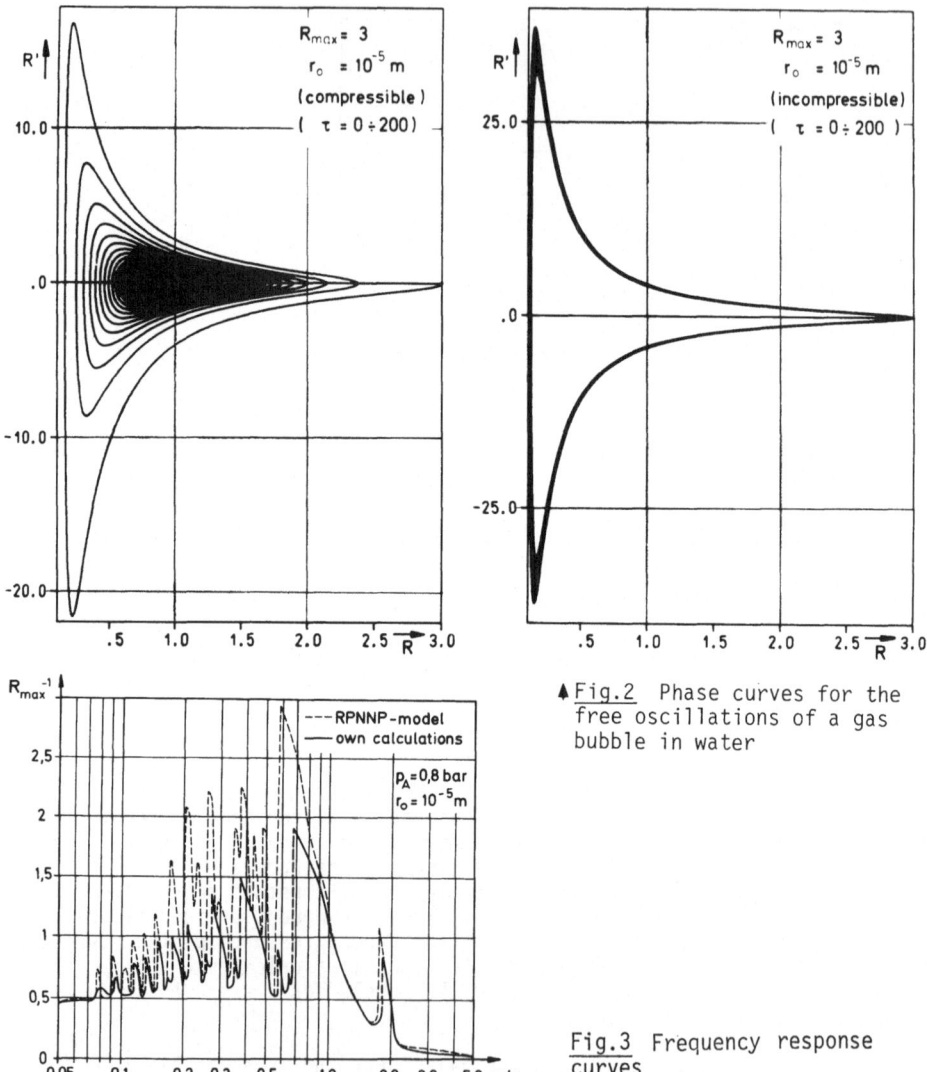

Fig.2 Phase curves for the free oscillations of a gas bubble in water

Fig.3 Frequency response curves

For the free vibrations of a gas bubble in water the phase portraits in Fig.2 are given for a compressible (on the left) and an incompressible liquid (on the right). Fig. 2 shows a very strong influence of the fluid compressibility, because the mean damping parameter is the sound radiation and not the viscosity effect. The resonance curves for the forced oscillations in Fig.3 show a lot of special features like harmonics, subharmonics and ultraharmonics

[2] . The dimensionless amplitude $R_{max} - 1$ is a function of the normalized frequency ω/ω_o where ω_o is the linear frequency given by MINNAERT [2] .

Nearly the same curves are calculated by CRAMER [4] by means of the GILMORE theory. In Ref. [7] the frequency dependence of cavitation thresholds are calculated and compared with the experimental boundary- lines given by ESCHE [8]. The results show that the fluid compressibility must be considered, particular for higher frequencies.

3. An Analytical Solution of a Nonlinearly Freely Oscillating Gas Bubble in a Compressible Liquid

The nonlinear free oscillations of a spherical gas bubble in a compressible fluid (water) are investigated by an asymptotic method [10]. We consider small deviations of the normalized radius of the bubble X from the equilibrium radius: $R = 1 + X$, with $X = \varepsilon \cdot x$ and $\alpha_1 = \varepsilon \alpha$ where $\varepsilon \ll 1$. If $R = 1 + X$ is substituted in Eq.(1) and a power series expansion in powers of x is carried out, we obtain the following equation neglecting higher terms in ε

$$\ddot{x} + \omega_o^2 x = \varepsilon \left[-\frac{3}{2} x'^2 + \beta_1 x^2 - \beta_2 \alpha x' \right] + \ldots \tag{4}$$

with

$$\beta_1 = \frac{9}{2} \varkappa (\varkappa + 1) - 2 \frac{2\sigma}{r_o P_o} \quad ; \quad \beta_2 = 3\varkappa - \frac{2\sigma}{r_o P_o} = \omega_o^2 \tag{5}$$

The approximate analytical solution of Eq.(4) by an asymptotic method is [10]

$$x(\tau) = a(\tau) \cos \psi + a^2(\tau) \left[\delta_1 - \delta_2 \cos 2\psi \right] \tag{6}$$

where

$$a(\tau) = a_o \cdot e^{-\frac{\beta_2 \alpha}{2} \tau} \tag{7}$$

$$\psi = (\omega_o - \delta_4) \tau + \frac{\delta_3}{\beta_2 \alpha} a_o^2 (1 - e^{-\beta \alpha \tau}) \tag{8}$$

with

$$\delta_1 = \frac{1}{2} \left(\frac{\beta_1}{\omega_o^2} - \frac{3}{2} \right) \quad ; \quad \delta_2 = \frac{1}{6} \left(\frac{\beta_1}{\omega_o^2} + \frac{3}{2} \right) \tag{9}$$

$$\delta_3 = \omega_o \left(-\delta_1 \frac{\beta_1}{\omega_o^2} + \frac{1}{2} \delta_2 \frac{\beta_1}{\omega_o^2} - \frac{3}{2} \delta_2 \right) \quad ; \quad \delta_4 = \frac{1}{2\omega_o} \left(\frac{\beta_2 \alpha}{2} \right)^2 \tag{10}$$

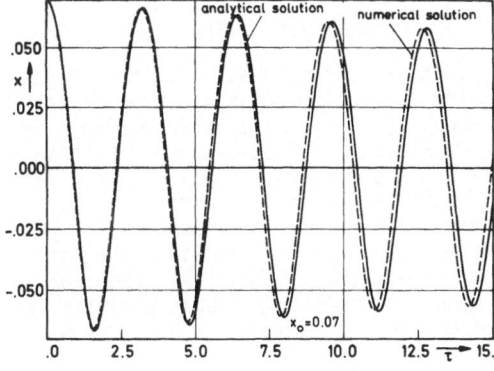

Fig.4 Comparison between the analytical and the numerical solution

67

Fig.4 shows a comparison between the analytical and the exact numerical result for the free vibrations. The quantities for water are already given in chapter 2. For an incompressible fluid, an analytical solution for the free and forced oscillations bas been calculated by PROSPERETTI [11].

4. The Nonlinear Translational Motion and Radial Oscillations of a Cavitation Gas Bubble in a Compressible Liquid

Some experiments have shown that a gas bubble in an inhomogeneous sound field begins to move. The external pressure outside of the bubble boundary is given by a standing sound wave of the form: $p = p_A \cos(2\pi x/\lambda) \sin \omega t$ where x is the local coordinate and λ the wavelength of the sound field. The following differential equation describes the translational motion of the bubble in a liquid of low viscosity [12, 13]

$$\left(\frac{2}{3} R^2 + 3\alpha_7 R\right) U_B' + \left[R'(2R + 6\alpha_7) + 6\alpha_6 \left(1 + \frac{1}{\alpha_7} R\right)\right] U_B -$$

$$- \left[R(2R + 6\alpha_7) + 6\alpha_6 \left(1 + \frac{1}{\alpha_7} R\right)\right] \cdot \frac{\alpha_8 \cdot \alpha_5}{\bar{\omega}} \sin(2\pi x/\lambda) \cdot \tag{11}$$

$$\cdot (1 - \cos(\bar{\omega}\tau)) - (2R^2 + 3\alpha_7 R) \alpha_8 \alpha_5 \sin(2\pi x/\lambda) \sin(\bar{\omega}\tau) = 0$$

The normalized equation for the radial bubble oscillation is [12]

$$RR'' \left[1 - 2\frac{R'}{\alpha_1}\right] + \frac{3}{2} R'^2 \left[1 - \frac{4}{3}\frac{R'}{\alpha_1}\right] + \frac{\alpha_2}{R} \left[1 - \frac{R'}{\alpha_1}\left(1 - \frac{R'}{\alpha_1}\right)\right] -$$

$$- \alpha_3 R^{-3\varkappa} \left[1 + 3\varkappa \frac{R'}{\alpha_1}\left(1 - \frac{R'}{\alpha_1}\right)\right] - \alpha_5 \left\{\cos(2\pi x/\lambda) \sin(\bar{\omega}\tau) + \right. \tag{12}$$

$$\left. + \frac{R}{\alpha_1}\left(1 - \frac{R'}{\alpha_1}\right) \cdot \left[\bar{\omega} \cos(2\pi x/\lambda) \cos(\bar{\omega}\tau) - \alpha_9 U_B \sin(2\pi x/\lambda) \cdot \sin(\bar{\omega}\tau)\right]\right\} - \alpha_4 + 1 = 0$$

For the x- direction the following relation is valid

$$d(x/\lambda) / d\tau = \alpha_{10} U_B \tag{13}$$

with the dimensionless quantities

$$\alpha_6 = \frac{\nu}{r_o} \sqrt{\frac{9}{P_\infty}} \quad , \quad \alpha_7 = \sqrt{\frac{2\alpha_6}{\bar{\omega}}} \quad , \quad \alpha_8 = \frac{2\pi r_o}{\lambda \cdot c} \sqrt{\frac{P_\infty}{9}} \tag{14}$$

$$\alpha_9 = \frac{2\pi r_o c}{\lambda \sqrt{P_\infty/9}} \quad , \quad \alpha_{10} = \frac{c \cdot r_o}{\lambda} \sqrt{\frac{P_\infty}{9}} \quad , \quad U_L = \frac{u_L}{c} \quad , \quad U_B = \frac{u_B}{c} \tag{15}$$

in which ν is the kinematic viscosity, u_B the translational velocity of the cavity and u_L the velocity of the liquid. The following initial conditions are valid for $\tau = 0$: $R = 1$, $R' = 0$, $U_L = 0$, $U_B = 0$, $x/\lambda = 1/8$. Accordingly we take: $p_v = 0,0233$ bar, $P_\infty = 1$ bar, $\sigma = 0,0725$ N/m, $c = 1460$ m/s, $9 = 998$ kg/m^3, $\mu = 10^{-4}$ kg/(sm), $\varkappa = 1,33$. As a result Fig.5 shows the normalized coordinate x/λ, the dimensionless translational bubble velocity U_B and the relative bubble radius R as a function of the normalized time The Parameters are $r_o = 10^{-5}$ m, $\omega/\omega_o = 2,0$ and for a time interval from 0 to 100. In Fig.6 the same parameters are present, only for a longer time interval. The time runs from 0 to 1000. Fig.5 shows that more rapid compressions of the bubble produces large accelerations and velocities of the translational motion in the x- direction.

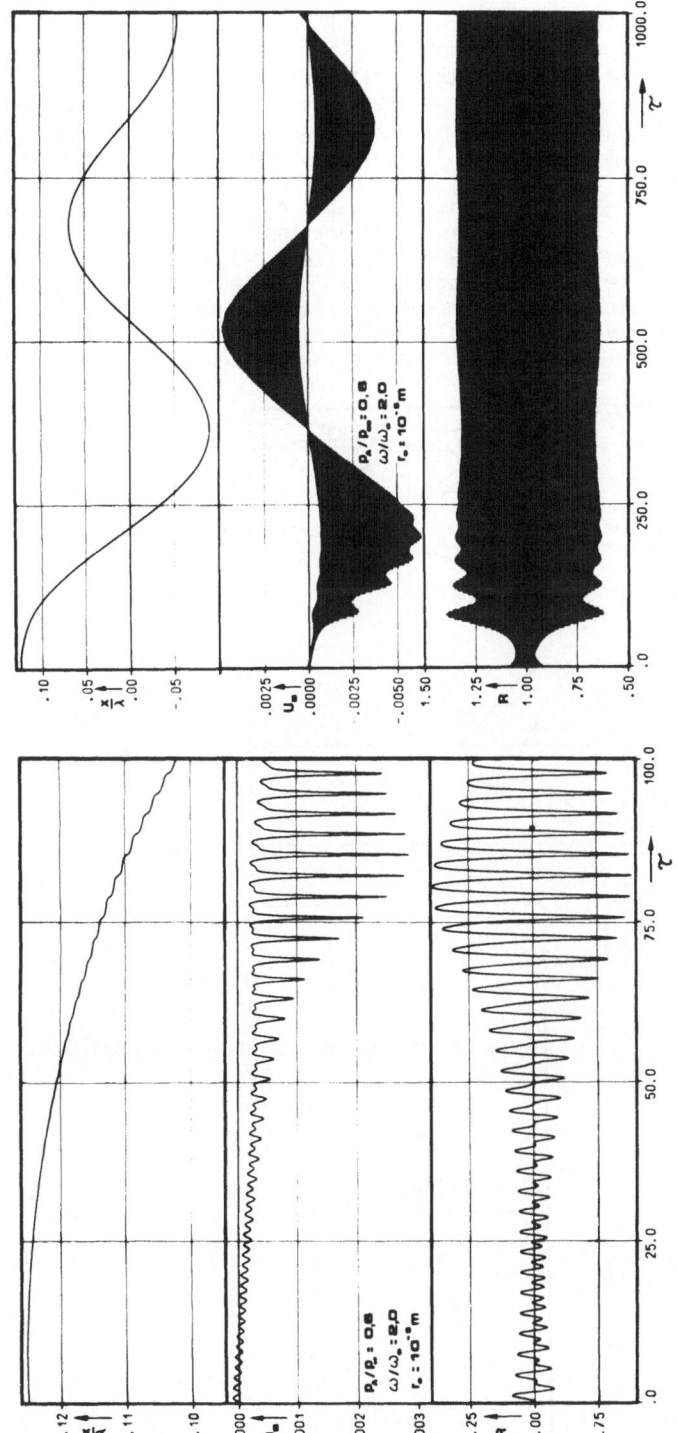

Fig.5 The local coordinate, the translational velocity and the bubble radius as a function of time

Fig.6 The local coordinate, the translational velocity and the bubble radius as a function of time

Fig.6 shows that x/λ and the envelopes of U_B and R for cavities in the vicinity of the first subharmonic resonance during the time tend to a quasi stationary state. The normalized local coordinate x/λ as a function of τ is given in Fig. 7. Parameter is the relative frequency ω/ω_o. For $\omega/\omega_o > 0{,}82$ the deviation of x/λ is negative and for $\omega/\omega_o < 0{,}82$ is positive. In both cases the average deviation (absolute value) from the initial condition is about $|x/\lambda| = 1/8$. In Fig.8 the dependance of U_B, $R_{max}-1$ and $\bar{\omega}_s$ from the relative frequency ω/ω_o are given, where $\bar{\omega}_s$ is the angular frequency of the envelopes of R for the steady state solution. The two upper curves are the maximal and minimal values of U_B during a time interval of 2000. The lines below are typical frequency response curves. The resonance regions lean over towards lower frequencies and show the typical jumping phenomena of a nonlinear system. The dependance of the frequency of the oscillations in the x-direction on the forced frequency are calculated in [13]. Further informations are given in [12-15].

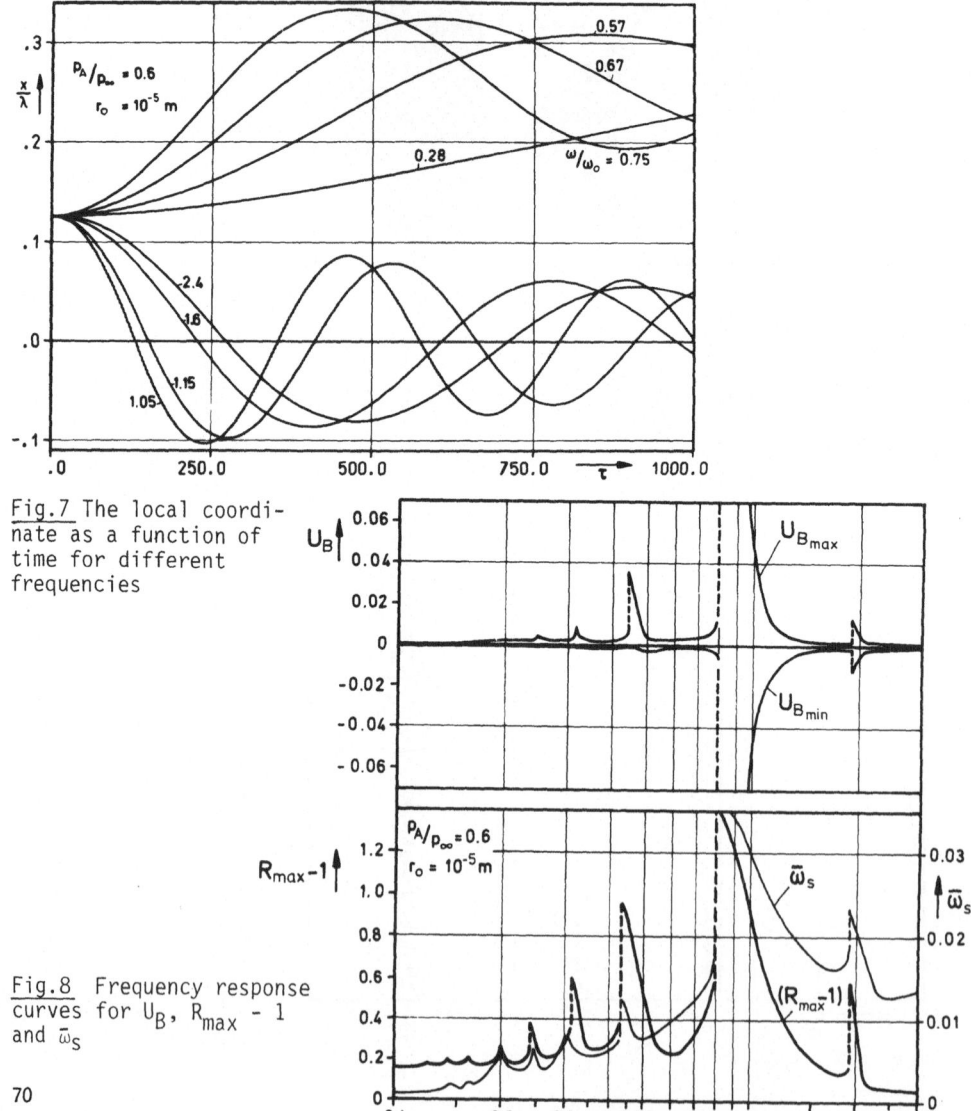

Fig.7 The local coordinate as a function of time for different frequencies

Fig.8 Frequency response curves for U_B, $R_{max} - 1$ and $\bar{\omega}_s$

70

5. References

1. Rath, H. J.: Zum Einfluß der Kompressibilität des Fluides bei sphärisch schwingenden Gasblasen in Flüssigkeiten. Ingenieur- Archiv 47, 383 - 390 (1978)

2. Rath, H. J.: Über nichtlineare Schwingungen sphärisch schwingender Gasblasen in Flüssigkeiten unter Berücksichtigung der Kompressibilität des Fluides. Zeitschrift f. angew. Mathem. u. Physik 30, 4 (1979)

3. Ebeling, K. J.: Hochfrequenzholografie lasererzeugter und akustisch erzeugter Kavitationsblasen. Dissertation Göttingen (1976)

4. Cramer, E.: Numerische und experimentelle Untersuchungen zum Kavitationsgeräusch. Dissertation Göttingen (1978)

5. Lauterborn, W.: General and basic aspects of cavitation. In: Finite- Amplitude Wave Effects in Fluids. Ed. by L. Bjørnø, Copenhagen, 240- 244 (1974)

6. Knapp, R. T., Daily, J. W., Hammit, F. G.: Cavitation. McGraw- Hill (1970)

7. Rath, H. J.: Zur Kinematik sphärisch kavitierender Gasblasen und Problematik der Kavitationsschwellen in einer kompressiblen Flüssigkeit. Acustica 43, 1 (1979)

8. Esche,R.: Untersuchung der Schwingungskavitation in Flüssigkeiten. Acustica 2, AB 208 (1952)

9. Rath, H. J.: Zum Einfluß hoher Schallwechseldruckamplituden bei sphärisch kavitierenden Gasblasen in einer kompressiblen Flüssigkeit. Forschung im Ingenieurwesen 45, 3 , 83 - 88 (1979)

10. Rath, H. J.: Eine analytische Lösung zur freien, nichtlinear schwingenden Gasblase in einer kompressiblen Flüssigkeit. Acta Mechanica (19807 (to be published)

11. Prosperetti, A.: Nonlinear oscillations of gas bubbles in liquids: transient solutions and the connection between subharmonic signal and cavitation. J. Acoust. Soc. Am. 57, 4, 810- 821 (1975)

12. Rath, H. J.: Zur translatorischen Bewegung einer nichtlinear schwingenden Gasblase in einer kompressiblen Flüssigkeit in Anwesenheit eines inhomogenen Ultraschallfeldes. Acustica 43, 4 (1979)

13. Rath, H. J.: The influence of an inhomogeneous sound field a the translational motion and nonlinear radial oscillation of a cavitation gas bubble in a compressible liquid. Mechanics Research Communications (1980) (to be published)

14. Agrest, E. M., Kuznetsov, G. N.: Instantaneous parameters of the motion of a cavitation bubble in an inhomogeneous sound field. Sov. Phys. Acoust. 19, 3, 212- 215 (1973)

15. Agrest, E. M., Korets, V. L.: Large- scale spatial oscillations of a cavity in a sound field. Sov. Phys. Acoust. 24, 1, 1- 5 (1978)

Acoustic Cavitation and Bubble Dynamics Due to a Tension Wave

R.A. Wentzell and G.J. Lastman

Department of Applied Mathematics, University of Waterloo
Waterloo, Ontario, Canada N2L 3G1

1. Introduction

The reflection of a shock pulse from a water surface, due to an underwater explosion, produces a tension wave which may produce cavitation. When cavitation occurs, the distinguishing features, as observed experimentally, are the severe elongation and attenuation of the tension wave followed by rapid damping, [1]. In order to explain these effects CUSHING [2] and WENTZELL [1] proposed a bulk model of cavitation based on spalled layers. However, the features of this acoustic cavitation depend on the dynamics of individual bubbles which are herein discussed.

The response of an adiabatic gas bubble to a tension wave is studied in (a) a viscous incompressible liquid, and (b) in an inviscid compressible liquid. Comparisons are made between corresponding results. The major observed effect is that viscosity is not sufficient to account for the rapid damping of bubble oscillations when cavitation occurs in water. However, rapid damping is observed in the inviscid compressible liquid, after cavitation occurs. In addition, a mathematical criterion for the onset of cavitation is given for a single bubble. The results are applied to the two bubble models studied.

2. Bubble Response in a Viscous Incompressible Liquid

The modified RNNP bubble model (RAYLEIGH [3], NOLTINGK-NEPPIRAS [4], PORITSKY [5]) consists of an adiabatic gas surrounded by a viscous, incompressible liquid with surface tension at the liquid-gas interface. The bubble has been hit by a tension wave at $t = 0$. Further details can be obtained in WENTZELL [6]. (Note that (1) is derived from Laplace's equation.)

$$x \frac{d^2x}{dt^2} + \frac{3}{2}\left(\frac{dx}{dt}\right)^2 + \frac{a}{x}\left(\frac{dx}{dt}\right) + 1 - (1+D)x^{-3\gamma} + Dx^{-1} = P_{sc}f(t) \tag{1}$$

where $x = R(t)/R_0$, $t = \left(\frac{\tau}{R_0}\right)\left(\frac{P_c}{\rho}\right)^{\frac{1}{2}}$, $a = \frac{4\mu}{R_0\sqrt{\rho P_c}}$, $P_{sc} = \frac{P_s}{P_c}$,

$$|f(t)| \leq 1 \quad , \quad D = \frac{2\sigma}{R_0 P_c} \quad ;$$

at $t = 0$, $x = 1$, $\frac{dx}{dt} = 0$; $R(t)$ is the bubble radius for $t \geq 0$, with $R(0) = R_0$. The variable τ is time, in seconds. For the model studied, $\rho = 1$ gm/cm^3 , $P_c = 10^6$ dyne/cm^2 (1 atm.), $\sigma = 72.8$ dyne/cm, $\gamma = 1.4$, $P_s = 5$ atm., $\mu = 10^{-2}$ dyne sec/cm^2 .

72

To study the motion of the bubble for $t \geq 0$, (1) is converted to a system of first-order ordinary differential equations by defining $y_1 = x$ and $y_2 = dx/dt$.

With $f(t) = e^{-t/t_0}$, where $t_0 = \dfrac{2.9 \times 10^{-5}}{R_0} \left(\dfrac{P_c}{\rho}\right)^{\frac{1}{2}} = \dfrac{0.029}{R_0}$, the

tension wave represents the reflection of a typical underwater explosion shock wave. In general, the forcing term $f(t)$ can be a periodic function, a step-function, or any function that decreases to zero as t tends to infinity. These forms for $f(t)$ represent physically realizable tension sources.

The pressure at the bubble's surface is $P = P_c P$, where P is the gauge pressure

$$\overset{\scriptscriptstyle\circ}{P} = (1+D)x^{-3\gamma} - (a \frac{dx}{dt} + D)x^{-1} - 1 . \tag{2}$$

Eq. (1) was numerically integrated for a typical cavitating bubble, $R_0 = 10^{-3}$ cm. The bubble response is given in Fig.1. These results clearly show the "bouncing effect", in that successive bubble-radius maxima are nearly equal. Consequently, viscosity has a negligible effect on the damping of bubble oscillations, [7], after cavitation occurs.

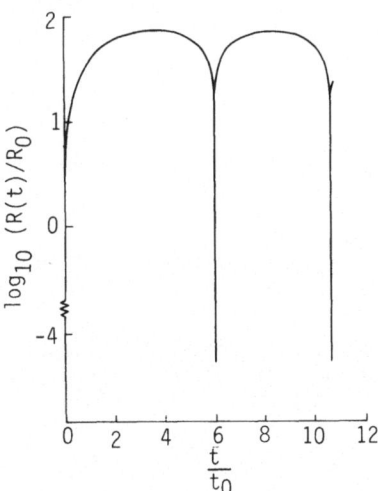

Fig.1 Bubble response in a viscous incompressible liquid

3. Bubble Response in an Inviscid Compressible Liquid

Because viscosity is insufficient to account for rapid damping of bubble oscillations in an incompressible liquid we will exclude viscosity from our bubble model. We are replacing a model of an incompressible viscous liquid with a model of a compressible inviscid liquid.

The bubble model to be considered here consists of an adiabatic gas surrounded by an inviscid compressible liquid. There is surface tension at the liquid-gas interface. The bubble has been hit by a tension wave at time zero. A mathematical model of a bubble with these properties can be obtained by modifying the model of KELLER [8] to introduce surface tension and a forcing-function tension wave.

The potential ϕ satisfies the wave equation

$$\frac{1}{r^2} \frac{\partial}{\partial r} \left(r^2 \frac{\partial \phi}{\partial r}\right) - \frac{1}{c^2} \frac{\partial^2 \phi}{\partial \tau^2} = 0 . \tag{3}$$

Eq. (3) is to be considered as giving a first order approximation for compressibility. Subsequently, it is regarded as the exact model equation giving the nonlinear bubble behaviour. By assuming a solution of the form

$$\phi = r^{-1} F[\tau - c^{-1}(r-R_0)] \quad , \quad R_0 = R(0) \tag{4}$$

we follow the development of [8] to obtain the differential equation

$$\left(\frac{dx}{dt} - v\right)\left[x \frac{d^2x}{dt^2} + \frac{3}{2}\left(\frac{dx}{dt}\right)^2 + Dx^{-1} - (1+D)x^{-3\gamma}\right]$$

$$= \left(\frac{dx}{dt}\right)^3 + \left(\frac{dx}{dt} + v\right)[1 - P_{sc}f(t)] - xb\, P_{sc} \frac{df(t)}{dt}$$

$$+ \frac{dx}{dt}\left[(3\gamma-2)(1+D)x^{-3\gamma} + Dx^{-1}\right] \tag{5a}$$

with initial conditions

$$x(0) = 1 \quad , \quad \frac{dx(0)}{dt} = 0 . \tag{5b}$$

The dimensionless quantities x, t, D and P_{sc} are the same as in (1), while $v = (P_c/\rho)^{-\frac{1}{2}} c$ and $b = R_0(P_c/\rho)^{-\frac{1}{2}}$.
Letting $v \to \infty$ in (5a) gives an equation which agrees with (1) when $a = 0$ (inviscid liquid). Setting $D = 0$ and $P_{sc} = 0$ yields the bubble equation of KELLER [8].

The forcing function f in (5a) can represent many physically realizable cases as discussed in Section 2. Here, we need only require that f be differentiable for $t \geq 0$ and that $|f(t)| \leq 1$. A particularly useful form for f, representing a shock wave, is

$$f(t) = e^{-t/t_0} \quad \text{where} \quad t_0 = (2.9 \times 10^{-5}/R_0)(P_c/\rho)^{\frac{1}{2}} . \tag{6}$$

The remaining quantity of interest is the gauge pressure at the bubble surface

$$P = (1+D)x^{-3\gamma} - Dx^{-1} - 1 , \tag{7}$$

while the actual pressure is $P_c P$.

74

For purposes of numerical integration (5a) is converted into a system of two first-order ordinary differential equations by defining $x_1 = x$ and $x_2 = dx/dt$.

The various constants in (5a), (6) were fixed as follows: $\gamma = 1.4$, $D = 1.456 \times 10^{-4}/R_0$, $t_0 = 0.029/R_0$ and $v = 148.5$ (corresponding to a speed of sound of 1485 metres per second). For a typical bubble, $R_0 = 10^{-3}$cm, with $P_{sc} = 5$, we obtained the results shown in Fig.2.

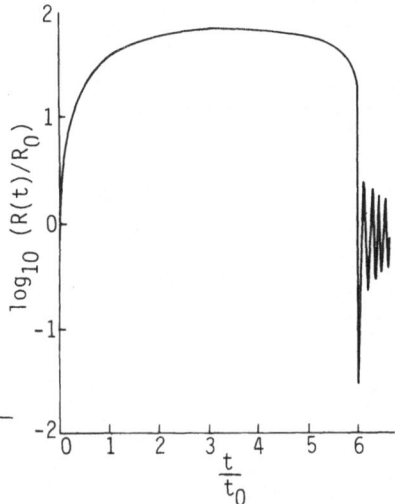

Fig.2 Bubble response in an inviscid compressible liquid

Fig.2 shows that the bubble oscillations are rapidly damped in the inviscid compressible liquid. Compare this with a bubble of the same initial radius in a viscous incompressible liquid where nearly the same bubble-radius maxima and minima are attained on successive oscillations; see Fig.1. This confirms the conjecture in [7], that the compressibility of the liquid accounts for the rapid damping of bubble oscillations. This type of rapid damping was also observed by KELLER [8] for a bubble model without surface tension and without a forcing function simulating a tension pulse. Following the first bubble-radius minimum the bubble oscillations are similar to those observed in [8]. This is because the forcing term is nearly zero after the first bubble-radius minimum, following the cavitation phase of the response.

Computer runs with other values of R_0 with $P_{sc} = 5$ were made. For those initial radii where cavitation occurred rapid damping of the oscillations was observed, reconfirming the importance of liquid compressibility in the damping process.

4. Threshold Conditions for the Onset of Cavitation

Cavitation occurs only for bubbles above a certain minimum size when the local pressure drops to a threshold level. The concept of cavitation threshold is physically a simple one, but its definition is mathematically vague. In this section the concept of cavitation is given a precise meaning in terms of the initial critical bubble radius, R_0^* , below which cavitation

75

does not occur. These results are obtained from an analysis of the phase plane trajectories of the relevant nonlinear differential equations.

In previous work on the onset of cavitation, NEPPIRAS [9] and FLYNN [10] find a definite lower threshold in R_0 below which cavitation does not occur for a periodic pressure wave or a pressure pulse. They found that cavitation does not occur because surface tension prevents the bubble from growing large enough. This can be summarized in the statement that cavitation will not occur for $R_0 < \frac{4\sigma}{3|P_s - P_c|}$. This prediction is from simple equilibrium theory and is in good agreement with the prediction of the dynamical theory.

We assume that the bubble dynamics can be described by a second-order ordinary differential equation in terms of the dimensionless variables y and t:

$$\frac{d^2y}{dt^2} = F(y, \frac{dy}{dt}, t ; R_0) \qquad (8)$$

where $y(0) = 1$, $\frac{dy(0)}{dt} = 0$. Note that the bubble radius is $R_0 y(t)$. Eq. (8) includes (1) and (5) as special cases. Define $y_1 = y$ and $y_2 = dy/dt$.

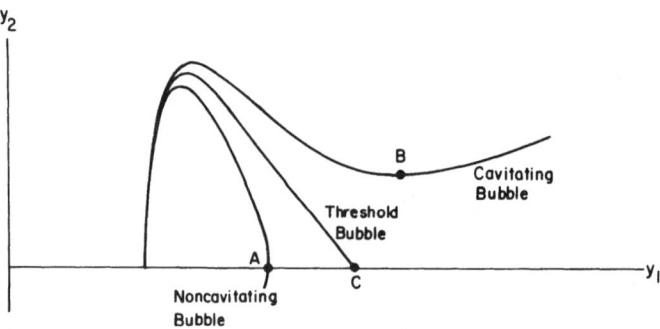

Fig.3 Trajectories for noncavitating, cavitating and threshold bubbles

Figure 3 shows typical trajectories for noncavitating and cavitating bubbles, with initial radii near R_0^* . A bubble of initial radius R_0^* is called a threshold bubble and R_0^* is called the threshold radius. A noncavitating bubble has the property that both y_1 and y_2 decrease as the trajectory evolves from point A in Fig.3. Consequently, for each noncavitating bubble we have $y_2(t_A) = 0$ and $F(y_1(t_A) , y_2(t_A) , t_A ; R_0) < 0$ when $R_0 < R_0^*$. On the other hand, a cavitating bubble will have $y_2(t_B) > 0$ and $F(y_1(t_B) , y_2(t_B) , t_B ; R_0) = 0$ for $R_0 > R_0^*$. As R_0 tends to R_0^* the trajectories for noncavitating and cavitating bubbles tend to the trajectory of the threshold bubble, also shown in Fig.3. For the threshold bubble we have

$$y_2(t_c) = 0 \tag{9a}$$

$$F(y_1(t_c) , y_2(t_c) , t_c ; R_0^*) = 0 . \tag{9b}$$

The motion of a threshold bubble may continue beyond point C (cavitation or noncavitation) or it may cease at point C. The latter condition could occur if $\frac{\partial F}{\partial t} \equiv 0$ at C.

Eqs. (9a), (9b) are two of the conditions that define the threshold radius R_0^*. There is an additional condition, which we will now determine. Define $y_1^{(max)}$ to be the first value of y_1 such that $y_2 = 0$ for $t > 0$. Values of $y_1^{(max)}$ are shown in Table 1 for a typical bubble model, the modified RNNP model discussed in Section 2; (1) with the forcing function given by (6), and $P_{sc} = 5$. From Table 1 we see that as R_0 increases to R_0^* the values of $y_1^{(max)}$ become very large; hence $\partial y_1^{(max)}/\partial R_0$ becomes large as R_0 increases to R_0^*. We approximate this condition by writing

$$\partial y_1^{(max)}/\partial R_0 \Big|_{R_0 = R_0^*} = \infty . \tag{9c}$$

Eqs. (9a), (9b), (9c) define the threshold bubble radius R_0^* .

Table 1

R_0 [cm]	$y_1^{(max)}$	R_0 [cm]	$y_1^{(max)}$
1.5×10^{-5}	1.43	2.0×10^{-5}	3.59×10^3
1.6×10^{-5}	1.528	3.0×10^{-5}	2.37×10^3
1.7×10^{-5}	1.75	10^{-3}	73.4
1.8×10^{-5}	3.995×10^3	10^{-2}	8.2

It is interesting to note that the existence of an equilibrium solution to (8) requires that $\frac{dy_1}{dt} \equiv 0$ and $\frac{dy_2}{dt} \equiv 0$; hence $y_2 = 0$ and $\lim_{t \to \infty} F(y_1, y_2, t; R_0) = 0$. These conditions are very similar to (9a), (9b) and can be used to obtain an estimate for R_0^*. As y_1 , y_2 are functions of both t and R_0 , t_c and R_0^* will satisfy the two-point boundary-value problem given by (8), (9a), (9b). The solution of this boundary-value problem will be complicated by (9c) and will be a difficult numerical computation. However, we can obtain solutions by a direct search procedure.

Setting $f(t) = 1$ in (1) yields $R_0^* \approx 1.696 \times 10^{-5}$ cm ; with $f(t)$ given by (6) $R_0^* \approx 1.701 \times 10^{-5}$ cm. The corresponding value predicted by NEPPIRAS [9] is 2.4×10^{-5} cm. A lower bound estimate for R_0^* can be obtained by setting $f(t) = 1$ and $a = 0$ in (1); this yields the estimate $R_0^* \approx 1.5365 \times 10^{-5}$ cm. (All of the above are for $P_{sc} = 5$.)

For the bubble model in an inviscid compressible liquid, (5a), an estimate of R_0^* can be obtained by assuming that $v^{-1} \cdot \max(|dx/dt|, |dx/dt|^3) \ll 1$, $x(vt_0)^{-1} P_{sc} \ll 1$, and setting $f(t) \equiv 1$ in (5a). The first two assumptions are justified for noncavitating bubbles, and thus will hold for a bubble on the threshold of cavitation. If cavitation occurs at R_0 when $|f(t)| < 1$ it will certainly occur at the same initial radius when $f(t) = 1$: consequently the estimate of R_0^* obtained with $f(t) = 1$ will be a lower bound. This leads to an equation which agrees with (1) when $a = 0$. Consequently, the lower bound estimates on R_0^* will agree with those obtained for the incompressible liquid model: $R_0^* \approx 1.5365 \times 10^{-5}$ cm at $P_{sc} = 5$. With $f(t)$ given by (6), a direct search procedure, using (5), yields $R_0^* \approx 1.543 \times 10^{-5}$ cm.

5. References

1. R.A. Wentzell, H.D. Scott and R.P. Chapman, J. Acoust. Soc. Am., 46, 789 (1969).

2. V. Cushing, Report, Office of Naval Research, AD615625, Washington, D.C. (1961).

3. J.W. Rayleigh, Philos. Mag., Ser. 6, 34, 94 (1917).

4. B.E. Noltingk and E.A. Neppiras, Proc. Phys. Soc., London, B63, 674 (1950).

5. H. Poritsky, Proc. First U.S. Natl. Congr. Appl. Mech., New York, 813 (1952).

6. R.A. Wentzell and G.J. Lastman, Can. J. Phys., 56, No. 5, 485 (1978).

7. G.J. Lastman, R.A. Wentzell and A.C. Hindmarsh, J. Comput. Phys., 28, 56 (1978).

8. J.B. Keller and I.I. Kolodner, J. Appl. Phys., 27, 1152 (1956).

9. E.A. Neppiras and B.E. Noltingk, Proc. Phys. Soc., London, 64 Sec B, 1032 (1951).

10. H.G. Flynn, Physical Acoustics, 1, Part B, Edited by W.P. Mason, Academic Press, 57 (1964).

Some New Results on Cavitation Threshold Prediction and Bubble Dynamics

R.E. Apfel

Department of Engineering and Applied Science, Yale University
New Haven, CT 06520, USA

1. Cavitation Prediction Charts

1.1 Introduction

In the course of preparing a chapter on "Acoustic Cavitation" [1] for a volume on "Ultrasonics" (Peter Edmonds ed.) for the series *Methods of Experimental Physics*, (Academic Press), I noticed that one could calculate analytically an approximate expression for the "transient" cavitation threshold. When this is plotted along with thresholds for nucleation (by BLAKE) and thresholds for rectified diffusion (by SAFAR), one gets useful graphs for the prediction of the type and violence of cavitation as it depends on initial bubble size and acoustic pressure.

1.2 The Various Cavitation Thresholds for Air Bubbles in Water

1.2.1 The "BLAKE" Threshold

Table I below presents the three cavitation thresholds. The first is the "BLAKE" threshold [2], which predicts the acoustic pressure P_B needed to produce the growth of a gas bubble of initial radius R_B in a liquid of surface tension σ at an ambient pressure P_0.

1.2.2 Rectified Diffusion

The second is a result by SAFAR [3] which predicts the minimum acoustic pressure P_D that will cause a gas bubble of relative saturation C and radius R_D to grow by rectified diffusion of gas. The acoustic frequency is given by f, and f_D is the frequency at which a bubble of radius R_D is in resonance [4].

1.2.3 Transient Threshold: An Analytic Approximation

One often used criteria for "transient" cavitation is that the collapse velocity of the bubble approaches the speed of sound. From the appendix of the classic paper by NOLTINGK and NEPPIRAS [5], one can estimate that the ratio of the maximum radius, R_M, to the initial radius, R_0, for an air bubble in water will be about 2.3 if the collapse velocity of the bubble just reaches supersonic value. One can estimate [6] the maximum radius R_M achieved by a bubble of initial radius R_0 in a liquid of density ρ when exposed to a sound field of acoustic pressure amplitude P_T and frequency f. The transient threshold P_T is then defined from this relation when R_M = 2.3 R_0, as given in the table below.

Notice that there is a constraint on the validity of this expression: namely that the derivation assumes that inertial and viscous effects are negligible. Bubbles of radius greater than R_I given in the Table are inertially controlled. For air bubbles in water the transition to inertially controlled behavior occurs at $P_T/P_0 \approx 11$ bars.

Table I Cavitation Thresholds

Threshold	Expression
"BLAKE" nucleation threshold, P_B, for bubble of radius R_B; $[X_B = 2\sigma/P_o R_B]$	$$\frac{P_B}{P_o} = 1 + \frac{4}{9} X_B \sqrt{\frac{3X_B}{4(1+X_B)}}$$
Threshold for rectified diffusion, P_D, for bubble of radius, R_D; $[X_D = 2\sigma/P_o R_D]$	$$\frac{P_D}{P_o} = \frac{[3\kappa(1+X_D)-X_D][1-(f/f_D)^2]}{[6(1+X_D)]^{\frac{1}{2}}[1-C+X_D]^{-\frac{1}{2}}}$$
Transient threshold, P_T, for bubble of radius R_T; appropriate only if R_T is less than R_I for a given P_T	$$R_T = \frac{0.13}{f}\sqrt{\frac{P_o}{\rho}}\left\{\frac{p-1}{\sqrt{p}}[1+\frac{2}{3}(p-1)]^{\frac{1}{3}}\right\}$$ $$R_I = \frac{0.13}{f}\sqrt{\frac{P_o}{\rho}}\sqrt{\frac{2}{3}(p-1)}$$ $$p \equiv P_T/P_o$$

1.3 Synthesis of Thresholds into Cavitation Prediction Charts

The various thresholds and constraint are plotted in Figure 1 for the cases of 20 kHz and 1 MHz acoustic frequency, respectively. The ordinate is the appropriate threshold radius normalized by the resonance radius [4]. The abscissa is the acoustic pressure threshold normalized by the ambient pressure. These results are for air-saturated water, C = 1. The hatch marked regions delineate regimes of different cavitation behavior. In region A, bubbles are inertially controlled and bubble growth is only by rectified diffusion. In continuous wave systems such bubbles will grow to resonance size, at which time they will undergo large oscillations. However such bubbles are pushed by radiation pressure forces away from high pressure regions in the sound field. Bubbles in region C will be transient; note the constraint implied by curve R_I. In region B growth both by dynamic means and by rectified diffusion is likely, although transient cavitation is unlikely unless growth by rectified diffusion to resonance size occurs.

1.4 The Violence of Transient Events

The ability for a transient cavitation event to do damage is strongly related to the maximum size reached by the cavity, which depends on pressure amplitude and frequency, but which depends only slightly on initial size if inertial and viscous effects are not important. For example, a 10 µm bubble that grows to about 23 µm will just be transient, as will a 100 µm bubble that grows to 230 µm. The latter bubble however stores far more energy in the fluid; this energy is given back in the kinetic energy of collapse and thus represents the potential for damage.

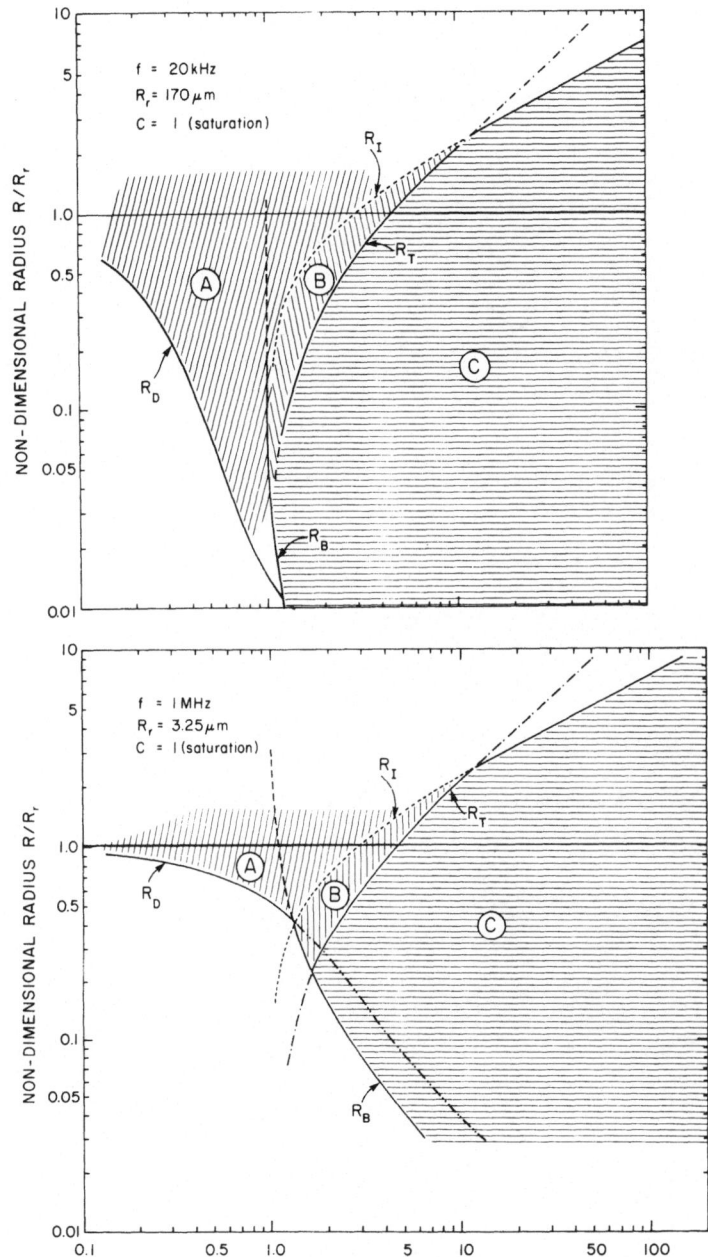

Fig. 1 Acoustic cavitation predictor graphs

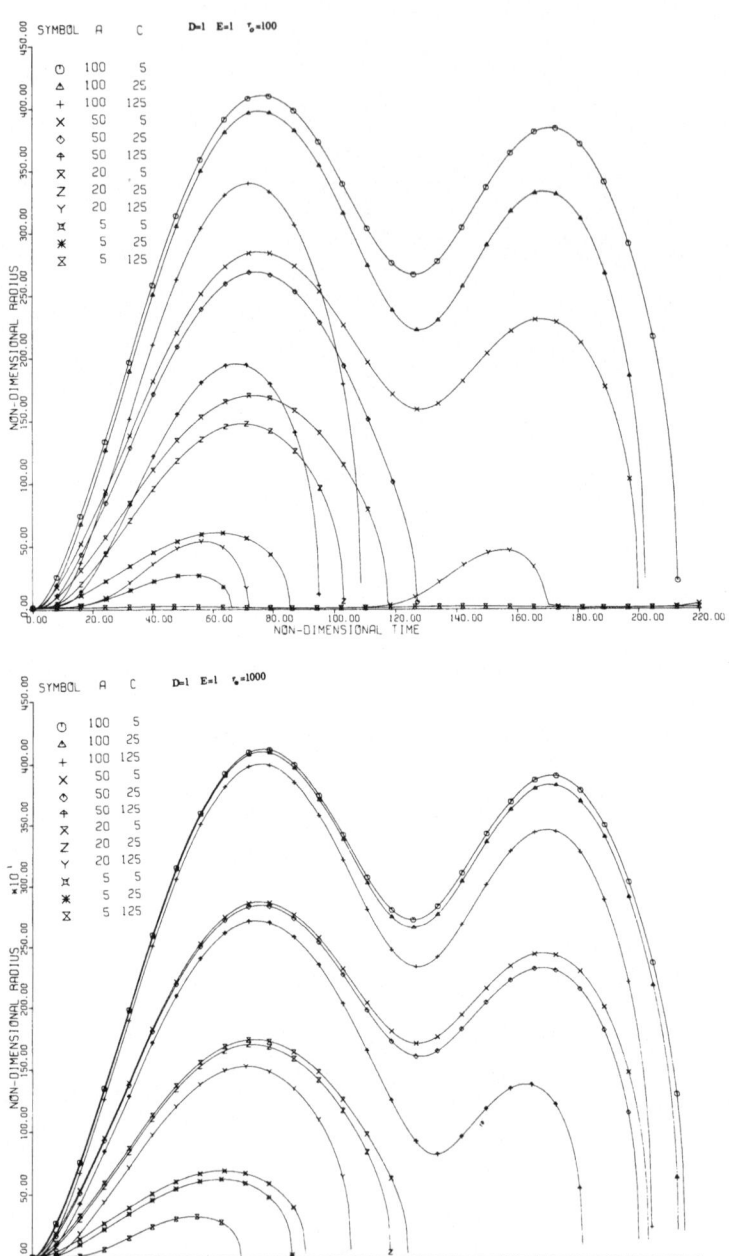

Fig. 2 Non-dimensional bubble dynamics graphs

2. Non-dimensional Bubble Dynamics Graphs

Figure 2 gives non-dimensional radius versus non-dimensional time graphs for bubble dynamics in a viscous liquid with surface tension included. The equation solved is the familiar one assuming an incompressible liquid [7,8],

$$\beta\,\ddot{\beta} + \frac{3}{2}\,\dot{\beta}^2 + C\,\dot{\beta}/\beta + D/\beta + E(\beta) + 1 = -A(\tau); \quad \dot{\beta} = d\beta/d\tau$$

where

$$\beta = R(t)/R_0; \quad \tau = t\,\sqrt{(P_0 - P_V)/\rho}\,/R_0; \quad C = 4\mu/R_0\,\sqrt{(P_0 - P_V)/\rho}\,;$$

$$D = 2\sigma/R_0(P_0 - P_V); \quad E(\beta) = -P_G(\beta)/(P_0 - P_V); \quad A(\tau) = P(\tau)/(P_0 - P_V);$$

and $P(t) = -P_A \sin t$, with P_A the acoustic pressure amplitude, P_V the vapor pressure, ρ the liquid density, μ the liquid viscosity, and P_G the pressure of the gas given by the appropriate adiabatic gas law.

The two cases shown are for non-dimensional periods $\tau_0 = 100$ and 1000. The only unique features of these graphs are their non-dimensional character and thus their applicability to liquids other than water.

Acknowledgments

I want to thank Dr. Ernest Neppiras for his help during the development of some of the cavitation prediction work, and the U.S. Office of Naval Research, Physics Program, for its support.

References

1. R.E. Apfel: In *Methods of Experimental Physics*; Vol. on *Ultrasonics*; *Acoustic Cavitation* (Academic Press, New York to be published)
2. F.G. Blake, Jr.: Technical Memo. 12, Acoust. Res. Lab., Harvard Univ., 1949
3. M.H. Safar: J. Acoust. Soc. Am. 43, 1188 (1968)
4. See, for example, C. Devin, J. Acoust. Soc. Am. 31, 1654 (1949) or ref. 1, Eq. 3.5
5. B.E. Noltingk and E.A. Neppiras: Proc. Phys. Soc. (London), B63, 674 (1950)
6. Ref. 1 Section 3.4
7. H.G. Flynn in *Physical Acoustics*, Vol. 1 B; *Physics of Acoustic Cavitation in Liquids*, ed. by W.P. Mason (Academic Press, New York 1964) Chap. 9
8. W. Lauterborn: J. Acoust. Soc. Am. 59, 283 (1976)

Acoustic Cavitation Thresholds in Water

L.A. Crum

Department of Physics, University of Mississippi
University, MS 38677, USA

Introduction

The acoustic cavitation threshold of water has been measured by several
observers and has been determined to be substantially below the theoretical threshold for inception based upon homogeneous nucleation of a
cavity within the liquid. Recent progress has been made toward a theory
of heterogeneous nucleation and results will be given in this paper which
demonstrate that a Harvey-type [1] model can lead to accurate predictions
of the cavitation threshold of water for a wide range of variables.

Model for heterogeneous nucleation

It is assumed that there exists within the bulk of the liquid numerous
nonpolar solid impurities that contain cracks and crevices in which gas
pockets may be stabilized against dissolution. Although various models
have been suggested for microbubble stabilization schemes, the defini-
tive experiments of GREENSPAN [2] have shown that these nuclei likely re-
side on macroscopic particles that can be removed by filtration. Accordingly, a model for cavitation inception in water is used that was first
introduced by HARVEY [1], and modified by such observers as STRASBERG [3]
and APFEL [4].

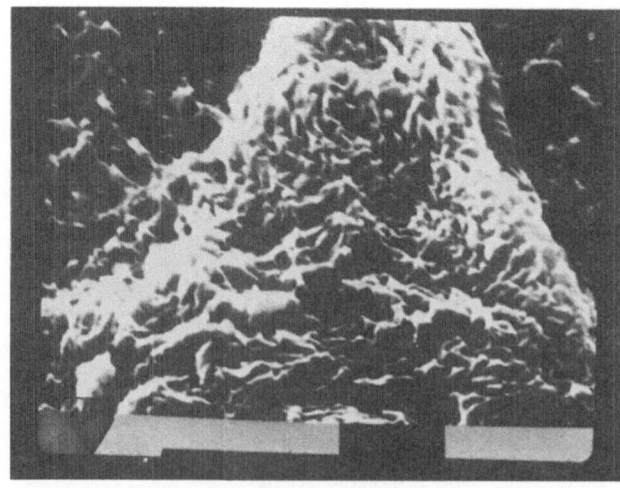

Fig. 1 Scanning
electron micrograph of
a solid impurity. The
particle was removed by
filtration of distilled
water with a 0.45μm
pore size filter, shown
as the mottled area in
the background. Note
the ragged appearance
of the surface showing
the numerous possible
nucleation sites. The
dark scale bar at the
bottom of the figure is
4μm.

Figure 1 shows a scanning electron micrograph of a particle removed by filtration from distilled water. Note the numerous possible nucleation sites. It is assumed that these small cracks and crevices can harbor pockets of gas that can serve as preferential sites for cavitation inception. I shall now modify a previously developed heterogenous nucleation theory [4] in order to obtain an equation for predicting acoustic cavitation thresholds in water.

Figure 2 shows a mathematical model for nuclei stabilization. A quantity of gas is trapped in a crevice within a nonpolar particle that is present in the liquid. If the liquid is degassed, the gas-liquid interface will advance until the stabilization condition is reached, namely, when the following equation is obeyed:

$$P_h = P_v + P_g + \frac{2\sigma}{R} , \qquad (1)$$

where P_h is the hydrostatic pressure, P_v is the vapor pressure, P_g is the equilibrium gas pressure of the gas dissolved in the liquid,

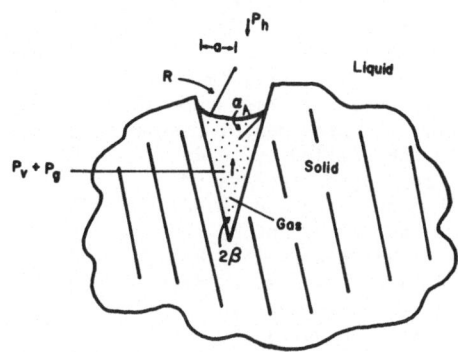

Fig. 2 Mathematical model for stabilization of gas pocket on a solid impurity.

σ is the liquid-vapor surface tension and R is the radius of curvature of the gas-liquid interface as shown in Fig. 2. The half-width, a, of the crevice can then be expressed as

$$a = \frac{2\sigma}{P_h - P_v - P_g} \left| \cos(\alpha_A - \beta) \right| . \qquad (2)$$

The angles α_A and β are defined in Fig. 2. It is possible to qualitatively see the effect of changing various liquid and environmental parameters. If the liquid is further degassed, the interface will advance until the equilibrium condition expressed in (2) is reestablished. If the liquid is pressurized the interface will similarly advance. If a surface-active agent is added to the liquid, however, the position of the interface, as determined by the half-width a, will not change, although the radius of curvature of the interface typically will be different. This fact allows us to express the quantity $\left| \cos(\alpha_A - \beta) \right|$ as a constant, δ, dependent only on the initial value of the surface tension, when considering data for the variation of the cavitation threshold with surface tension.

Consider next the nucleation condition. If the pressure in the liquid is reduced sufficiently, the configuration of the model will take the form as shown in Fig. 3. If the pressure in the liquid is sufficiently negative, the interface will recede out of the crevice with a receding contact angle α_R, and a free gas bubble will nucleated. One can thus define the conditions necessary for interface recession as the conditions necessary for cavitation inception.

According to APFEL [4] then, the condition for cavitation inception is expressed as

$$P_A = (P_h - P_v - \gamma P_g) + (P_h - P_v - P_g) \left| \frac{\cos(\alpha_R - \beta)}{\cos(\alpha_A - \beta)} \right| . \tag{3}$$

The acoustic pressure amplitude P_A is, of course, the negative peak amplitude required to induce cavitation; the condition $\alpha_R > \beta$ must be met, and γ is defined in [4].

It has been previously impossible to apply this equation directly to experimental results due to the difficulty in determining appropriate values for the advancing and receding contact angles α_A and α_R. CRUM [5] has discovered that this equation can be modified so that recently published values for the variation of the contact angles with surface tension can be utilized. Specifically, it has been shown [6] that the equilibrium contact angle α_E for water upon nonpolar solids can be described by a simple relationship for a wide variety of materials and surface tensions by the equation

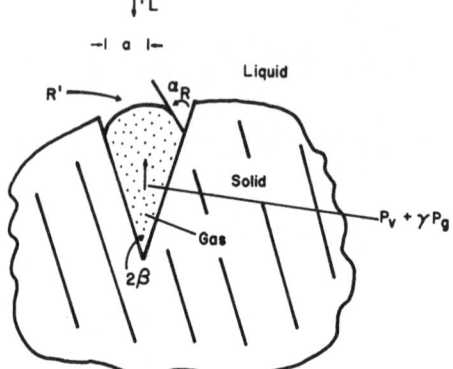

Fig. 3 Mathematical model for nucleation of a free gas bubble from a gas pocket on a solid impurity.

$$\cos \alpha_E = -1 + C/\sigma , \tag{4}$$

where C is a constant that depends upon the surface properties of the solid. For nonpolar solids such as parafin and beeswax, $C \approx 50$. The advancing and receding contact angles can be expressed in terms of the equilibrium contact angle α_E and a hysteresis angle α_H [5]. In terms of measureable parameters, the acoustic cavitation threshold can be expressed as

$$P_A = (P_h - P_v - \gamma P_g) + \frac{(P_h - P_v - P_g)}{\delta} \{\cos \phi \ (C/\sigma - 1) + \sin \phi \ [1 - (C/\sigma - 1)^2]^{1/2}\} \tag{5}$$

where $\phi = \alpha_H + \beta$.

Experimental measurements and application of theory to experiment

The incipient threshold was measured for acoustic cavitation inception in distilled water. A hollow piezoelectric cylindrical transducer was used that was open at one end and closed at the other with a thin glass window (thickness 0.01 cm). The thin glass window approximated a pressure-release boundary and the cylinder was driven in a $(r, \theta, z) = (3, 0, 3)$ resonant mode that generated cavitation only along the axis of the cylinder and only at the middle antinode. The height of the cylinder was

7.5 cm, the inside diameter was 7.0 cm and the thickness 0.5 cm. The criterion used for cavitation inception was an audible pop and an associated mechanical detuning of the resonator by the cavitation shock wave so that the input current to the system was appreciably changed. Often, visual confirmation of transient events were made in the form of those previously photographed [7]. The variation of the cavitation threshold with gas content, surface tension and temperature is shown in Figs. 4-6 respectively. In order to apply the theoretical expression given in (5) to the experimental results, it is necessary to specify the

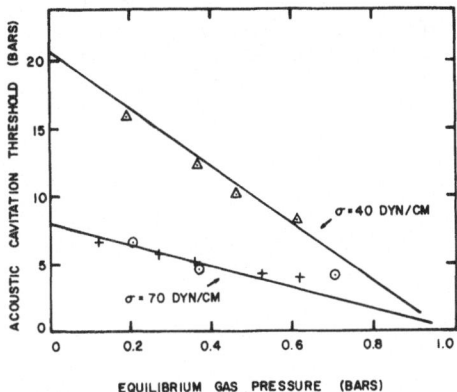

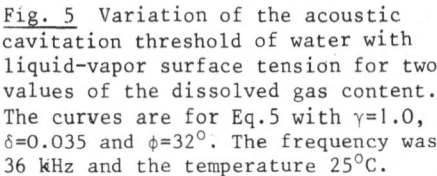

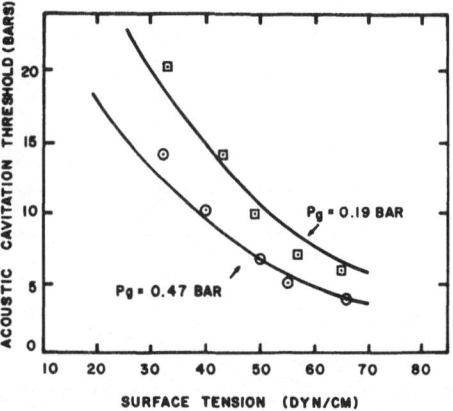

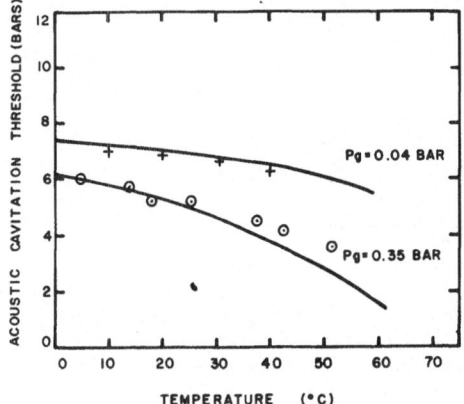

Fig. 4 Variation of the acoustic cavitation threshold of water with dissolved gas content for two values of the liquid surface tension. The curves are for Eq. 5 with $\gamma=1.0$, $\delta=0.035$ and $\phi=32°$. The frequency was 36kHz and the temperature 25°C. The circles and triangles are experimental measurements from this study and the +'s are results from STRASBERG [3].

Fig. 5 Variation of the acoustic cavitation threshold of water with liquid-vapor surface tension for two values of the dissolved gas content. The curves are for Eq.5 with $\gamma=1.0$, $\delta=0.035$ and $\phi=32°$. The frequency was 36 kHz and the temperature 25°C.

Fig. 6 Variation of the acoustic cavitation threshold of water with temperature for two values of the dissolved gas content. The curves are for Eq. 5 with $\gamma=1.0$, $\delta=0.035$ and $\phi=32°$. The temperature dependence of the vapor pressure P_v and the equilibrium gas pressure P_g have been inserted into (5). The +'s are normalized values from GALLOWAY[13].

87

unknown parameters γ, δ, and φ. γ is 1.0 for situations in which the gas
pressure in the nucleating cavity is maintained at equilibrium with that
in the fluid. Since a precise measurement of this constant is impossible,
it is assumed that equilibrium will be maintained and thus γ = 1.0. δ is
a parameter with measures the initial value of the advancing contact angle,
and is given by δ = |cos (α_A - β)|. Values for the advancing contact
of water on nonpolar solids are on the order of 106°. BARGEMAN [8] lists
values of α_A for four nonpolar solids with an average of 106°, standard
deviation 6°; water on parafin wax is known to be approximately 106° [9].
α_A was thus selected to be 106°. The crevice angle β is assumed to be
small; crevices need small angles for stabilization; a value of β = 14°
was selected to give δ = 0.035. Values of the hysteresis angle of water
plus surfactant on nonpolar solids were measured by FURMIDGE [10]. He
obtained values ranging from 14° - 44° for water plus surfactants on
white beeswax. In order to obtain a good fit to the data, α_H was chosen
to be 18°.

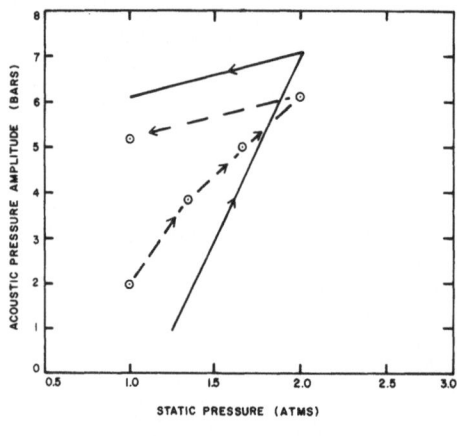

Fig. 7 Variation of the acoustic
cavitation threshold of water with
static pressure. The arrows indicate the direction of the change.
The circles (and dashed line) are
values obtained by STRASBERG [3];
the curve is for Eq.5 with γ=1.0,
δ=0.035 and φ=32°.

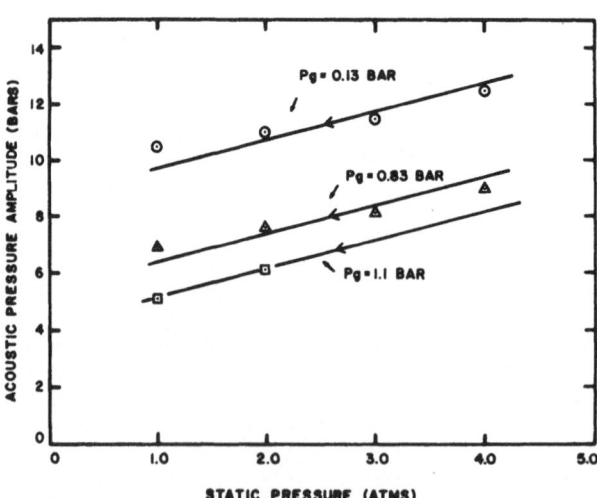

Fig. 8 Variation of the acoustic cavitation threshold with decreasing
static pressure for various values of the dissolved gas content. The
circles are values obtained by STRASBERG [3]; the curves are lines with
slopes of 1 bar/atm as predicted from Eq. 5. Compare with Fig. 7.

88

It is seen that good agreement is obtained between theory and experiment for a wide range of liquid parameters. Further, the theory also approximated reasonably well the observed dependence upon liquid prepressurization as observed by STRASBERG [3] (see Figs. 7 and 8). There is no frequency dependence to the theory, because it has been shown that the dependence of the threshold on frequency involves cavity dynamics [4,11], rather than nucleation conditions. Accordingly, the experimental measurements were made in the low kilohertz region where the cavitation threshold is frequency independent [12].

Conclusion

It has been demonstrated that the observed dependence of the acoustic cavitation threshold of water on such parameters as dissolved gas content, liquid-vapor surface tension, temperature, and hydrostatic pressure can be predicted by a simple theory based on heterogeneous nucleation from solid particulate matter.

Acknowledgements

The author wishes to acknowledge the support of the U.S. Office of Naval Research, and to W. T. Coakley and the University of Wales for assistance in obtaining the scanning electron micrographs.

References

1. Harvey, E. N. et.al. J. Cell. Comp. Physiol. $\underline{24}$, 1 (1944).
2. Greenspan, M. & Tschiegg, C. J. Res. NBS $\underline{71C}$, 299-312 (1967).
3. Strasberg, M. J. Acoust. Soc. Amer. $\underline{31}$, 163-176 (1959).
4. Apfel, R. E. J. Acoust. Soc. Amer. $\underline{48}$, 1179-1186 (1970).
5. Crum, L. A. Nature $\underline{278}$, 148-149 (1979).
6. Bargeman, D. & Van Voorst Vader, F. J. Coll. Sci. $\underline{42}$, 467-472 (1973).
7. Crum, L. A. & Nordling, D. A. J. Acoust. Soc. Amer. $\underline{52}$, 294-301 (1972).
8. Bargeman, D. J. Coll. Sci. $\underline{17}$, 309-324 (1972).
9. Adamson, A. W. Physical Chemistry of Surfaces, 113-139 (Interscience, New York 1967).
10. Furmidge, C. G. J. Coll. Sci. $\underline{17}$, 309-324 (1962).
11. Lauterborn, W. Acustica $\underline{22}$, 48-53 (1969).
12. Esche, R. Acustica, $\underline{2}$, AB208-AB218 (1952).
13. Galloway, W. I. J. Acoust. Soc. Amer. $\underline{26}$ 849 (1954).

The Influence of Modest Overpressures on the Persistence of Air Bubbles in Water

A. Evans

Decompression Sickness Research Laboratory, Royal Victoria Infirmary
Newcastle upon Tyne, United Kingdom

In recent years there has been much activity beneath the North Sea to discover and extract oil which has led to a vast expansion in deep sea diving. The diver faces many problems in the hostile environment beneath the waves, one of which is particularly concerned with his return to the surface. While he is submerged, he breathes gas at the local hydrostatic pressure, so that additional gas dissolves in the tissues of his body. During his ascent he is inevitably supersaturated, and bubbles of free gas may then arise in his body to give the signs and symptoms which we call decompression sickness. In our laboratory we are particularly concerned to discover where and why these macroscopic bubbles occur, and therefore how we may try to prevent them.

During the nineteen-forties HARVEY [1] investigated this problem in the United States, and showed clearly that some sort of free-gas nucleus was essential before large bubbles could form. This concept is now also widely accepted in other areas of liquid stress which are relieved by the appearance of bubbles, such as ultrasonic or hydrodynamic cavitation. Water which is completely free from gas micronuclei will remain in a meta-stable state without the appearance of bubbles even if it is grossly super-saturated with gas, superheated to well above its customary boiling point, or subjected to single or repetitive transient tensions, as in ultrasonic cavitation. Furthermore, it will also withstand a sustained mechanical tension, or negative pressure.

Clearly, before we can study the formation of bubbles induced by any of these means, it is first necessary to eliminate the gas micronuclei, which may be in the form of small bubbles in suspension, or attached to suspended particles or the container walls, which are always present with unprepared water. The method recommended by HARVEY [1] was to apply a hydrostatic pressure of the order of 1000 Atm for 30 min and similarly large hydrostatic pressures have been adopted by several subsequent investigators. For instance, in an investigation of the influence of the magnitude and duration of excess pressure on limiting static tensions, REES and TREVENA [2] used pressures from about 90 to over 400 Atm, for periods of 15 to 120 min. If it is essential to use such high pressures to dissolve all free gas, the investigator is restricted either to very small useful volumes of test liquid, or to bulky and expensive pressure chambers in which it is difficult to inspect and monitor the contents. I am going to suggest that it is possible to reduce Harvey's recommended excess pressure by at least one and possibly two orders of magnitude, applied for similar times, and still drive into solution in water quite large bubbles of air so that the water becomes completely free from gas micronuclei.

During the course of our studies of possible mechanisms for the nucleation of the bubbles associated with decompression in the living body, it became clear that it would be an advantage if we could apply the stress as a sustained mechanical tension rather than the more obvious gaseous supersaturation which we tried originally [3]. In order to be able to see where the bubbles come from, a transparent apparatus was necessary, and of the established tensiometers only the original technique introduced by BERTHELOT [4] allowed us to do this. The conventional Berthelot tensiometer is a stout glass tube, sealed at both ends, containing water with a small bubble of air. The tube is heated in a water bath to dissolve the air, and then cooled down to bring the water into tension.

In order to overcome some problems associated with calibration, we have formed the working tube into a coil, an idea previously used by MEYER [5], to make a self-indicating Berthelot tensiometer. Internal pressure makes the coil unwind slightly, like the element of a Bourdon pressure-gauge, and internal tension has the opposite effect. By making both ends of the coil point the same way, and extending them for some distance we obtain sufficient deflection to monitor with a distance meter, and so give a record of the variation of pressure within the apparatus during the course of a run [6]. The initial air bubble can be watched as it dissolves with a low-power microscope. Thus on each run, not only do we discover something about the limiting tension which the water will withstand, but also the behaviour of the bubble which has to be eliminated before this can happen.

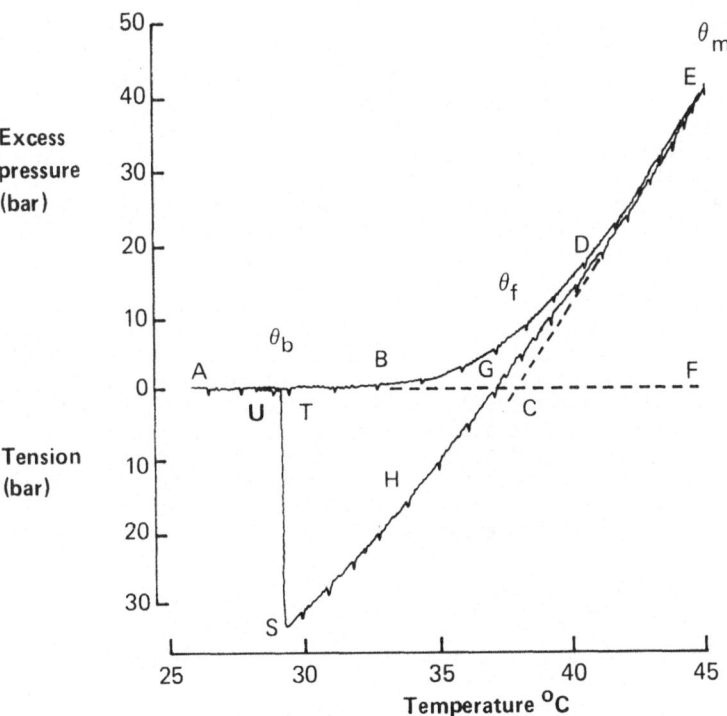

Fig.1 Variation of internal pressure with temperature within a Berthelot tensiometer during the course of a run

The course of a typical run with our tensiometer is depicted in Fig.1. Here the ordinate indicates the internal pressure: upwards for positive pressure, and downwards for tension. The two domains are separated by the horizontal 'zero pressure' line AF. The abscissa indicates the system temperature. The run commences at 'A', with water and a bubble of air in the tube. Heating causes a greater expansion of the water than of the cavity in the glass, so that the bubble is compressed somewhat, though up to point 'B' this gives only a slight rise of internal pressure. Further heating beyond 'B' leaves even less space for the gas bubble, so that the internal pressure starts to rise increasingly steeply round the curved section BD, by which point on this run the excess pressure tending to force the bubble into solution was about 20 bar.

By point D on the figure the volume of the (highly compressed) free gas is so small that its influence on the variation of pressure with temperature of the system is negligible. Further heating DE therefore follows the line for the fully filled tube. Heating is stopped at some point E which gives the required excess pressure, measured up from the zero pressure line as FE, and the terminal stages of the dissolving bubble can then be watched with the microscope.

Although in theory there should be no upper limit to the excess pressure attainable in this way, the strength of the glass tube limits us in practice to about 100 bar. It is perhaps obvious that the higher the excess pressure, the quicker the bubble should dissolve. Although this is true, the effect is far from linear. Certainly to a first approximation the differential gas tension between bubble and liquid rises linearly with the pressure; unfortunately the Boyles Law contraction of the gas gives a corresponding reduction in bubble surface area available for gas exchange which is approximately proportional to the $2/3$ power of the pressure. The net result is that the initial rate of dissolution of a fixed mass of gas varies as the cube root of the excess pressure, so that eight times the pressure should be required to double the rate at which gas exchange occurs. The dissolving bubble may therefore be much less sensitive to the magnitude of the excess pressure than might at first be supposed.

When we consider the later stages of the life of the bubble, in our apparatus we have to consider the air which has already dissolved. The bore of our tube is only 2 mm, and so it is clear that if all the air as a single bubble is forced into solution without moving it to different parts of the tube, the tension of dissolved gas will build up rapidly. As it is only the difference between the hydrostatic air pressure and the dissolved tension which is available for further solution, it is clear that for quick elimination of the bubble there must be some minimum excess pressure for any given bubble to dissolve. It is, therefore, clear that our technique will tend to over-estimate the required excess pressure and application time, rather than the reverse.

We have two ways to assess the completeness of the elimination of the bubble. We watch it grow smaller, and below about 0.1 mm dia it appears to shrink increasingly rapidly to extinction. However, it is not possible, even with dark ground illumination, to be sure that there is nothing smaller than about 50 μm with our microscope, and even with especially pure water there is often a granular or perhaps gelatinous deposit left behind where the bubble disappeared. Deposits of this sort have been described several times before, for example by MANLEY [7].

We therefore assess the integrity of the liquid column in a quite different way - namely by cooling the apparatus to see if it will withstand tension. It there is still some free gas, the cooling trace will resemble DB on Fig.1, though a completely filled tube will pass continuously down to and through the zero pressure line. Further cooling gives an increasing tension, which at some limiting value leads to a failure of the liquid with release of tension by the appearance of one or more vapour bubbles. As a liquid with micronuclei will stand no tension at all, we take the subsequent development of tension as a clear indication that the initial gas bubble has been properly eliminated. Immediately after the liquid column has snapped, however, the apparatus is not in a suitable state for the problem under investigation, as the bubble is composed largely of vapour with little air. The results which I shall give are only from long-standing bubbles, which may be expected to consist largely of air.

The figures given in the table were all obtained with the same sealing of one tube, in which there was about 0.6% by volume of air to dissolve. The various columns show the run number, the excess pressure employed, the application time, and the tension subsequently obtained. The early results in the series are not shown, as no attempt was made at that stage to investigate the effect of low excess pressures.

Table

Run No.	EP [bar]	Time [min]	Tension [bar]
14 ξ 26	44	30	42
14 ξ 32	27	40	40
14 ξ 43	18.5	24	41
14 ξ 53	21-13	25	30
14 ξ 66	12.5	55	35
14 ξ 77	36	10	40
14 ξ101	20	14	43

It will be seen that quite low excess pressures are sufficient to eliminate the last traces of free air from the water. Certainly no more than 20 bar is necessary if you are prepared to wait for half an hour, and 35 bar is seen to have been effective in only 10 min. It is clear that it is not necessary to think in terms of a thousand, or even a hundred bars, for the elimination of gas micronuclei, which means simpler and cheaper pumps and pressure chambers which do not have to be made of metal, even for specimens of substantial volume.

As expected, freshly-formed bubbles which have a much higher vapour content disappear even more easily than those which are mostly air; excess pressures of about 5 bar applied for about 15 min have given substantial tensions, and after only 3 bar applied for 18 min a limiting tension of 46 bar was recorded. With shorter times, 7 bar for 7 min and 8.5 bar for only 3.5 min were also effective.

There is an interesting conclusion from these results for our divers, many of whom develop decompression sickness even though they have been exposed to pressures in excess of 20 bar for times well in excess of half an hour. Unless there is a very effective mechanism to stabilise bubbles in the body, gas nuclei must be produced in vivo during the course of a dive. One possible source is from the spontaneous nuclear fission of the natural uranium in the body [8] , though we do not believe that this can be the only mechanism, and the search for other sources of suitable nuclei with the aid of the tensiometer is still in progress.

(This work is supported by the British Medical Research Council).

References

1. Harvey, E.N. et al. J. Cell Comp. Physiol. 24, 1 (1944).
2. Rees, E.P. and Travena, D.H. Nature 203, 396 (1964).
3. Evans, A. and Walder, D.N. J. Phys. E. 7, 879 (1974).
4. Berthelot, M. Ann. Chim. Phys. (3) 30, 232 (1850).
5. Meyer, J. Abh. D. BunsenGes. Band III No. 1 (1911).
6. Evans, A. J. Phys. E. 12, 276 (1979).
7. Manley, D.M.J.P. Br. J. Appl. Phys. 11, 38 (1960).
8. Walder, D.N. and Evans, A. Nature 252, 696 (1974).

On the Collapse of Cavity Clusters in Flow Cavitation

K.A. Mørch

Laboratory of Applied Physics I, Technical University of Denmark
DK-2800 Lyngby, Denmark

1. Introduction

The formation of fixed cyclic cavities in flow cavitation was studied by
KNAPP [1], and he ascribed the erosion of solid surfaces to the travelling
cavities collapsing in the stagnation zone at the downstream termination of
the fixed cavities [2]. Meanwhile, it is difficult to support that cavitation
erosion should be connected to the stagnation zone [3]. It is more likely
that it is produced in the final collapse stage of the cyclic cavities. KNAPP
found that cyclic cavities detach and are convected with the flow after being
filled, or partly filled with liquid containing travelling cavities. They are
then cavity clusters which collapse at a downstream position. VAN WIJNGAARDEN
[4] considered theoretically the collapse of a layer of small cavities at a
solid surface and found that the momentum of the liquid towards the surface
results in a considerable increase of pressure at the surface. Calculations
based on the shock waves from collapse of cavities in a cluster likewise indicate that large peak-pressures can be expected at the cluster centre by
concerted collapse of the cavities [5] and this is supported by erosion investigations [6,7]. In the present paper the collapse of cavity clusters
generated as fixed cyclic cavities is considered experimentally, and a theoretical approach to the problem of cavity cluster collapse is presented.

2. The cavitation equipment and its operation mode

The experiments are performed in a closed cavitation tunnel in which demine-
ralized water at 20 ^{0}C is circulated. Cavitation is generated at the downstream edges of a d = 14 mm diameter hole through a plate of 5 mm thickness,
the plate being aligned with the flow. The plate is mounted centrally and
horizontally, and with the hole at the throat of a two-dimensional venturi
nozzle of width 22 mm (Fig.1). For photographic purposes a transparent plate
can be mounted vertically instead. The hole generates a local flow instability of frequency f proportional to the flow velocity v in the throat, and
with an amplitude of the pressure fluctuations proportional to v^2. The
largest pressure oscillations occur at the downstream edges of the hole. At
given static pressure p at the throat, increase of the velocity v beyond a
value determined by the cavitation number at inception K_0 leads to cavity
formation at these positions ($K = (p-p_v)/\frac{1}{2}\rho v^2$ where p_v is the vapour pressure and ρ is the density of the liquid), and fixed cavities composed from
large numbers of small cavities are build up alternatingly on each side of
the plate during the low-pressure period of the oscillation. These cavity
clusters detach at both edges at the frequency f, and move downstream to the
collapse zone, while new fixed cavities develop at the edges, (Fig.1). The
cavitation number at inception is $K_0 \simeq 0.8$. The Strouhal number $S = f \cdot d/v$

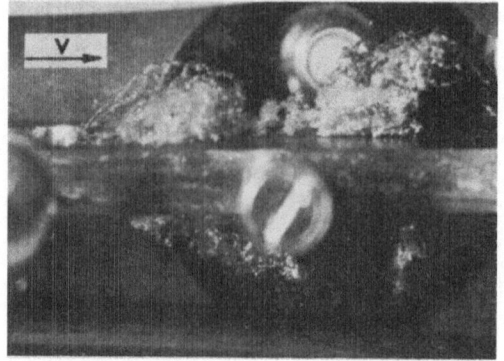

≈ 0.57 for K > 0.5 and it decreases to S = 0.50 at K = 0.2. A quartz transducer (Vibrometer 6 QP 500a with a resonance frequency of 180 kHz) for measurement of pressure fluctuations is mounted in the upper channel wall off the collapse position of the cavity clusters. Instead of the transducer a window can be inserted to allow transmission of light for photographing of the cavity cluster collapses.

Fig.1 Cavity clusters formed at
$K = 0.25$, $p = 0.31 \cdot 10^5$ Pa.

3. The collapse of cavity clusters

The collapse of the cavity clusters was photographed with a Beckman & Whitley Dynafax model 350 high-speed camera at $(16-19) \times 10^3$ frames/s. In Fig.2a,b two examples of a cavity cluster collapse are shown, one photographed parallel to the plate, the other perpendicular to the plate. After detachment the clusters move to the collapse zone where they become nearly stationary during collapse. It is characteristic that the clusters are convected close to the solid surface, separated from it by a thin liquid layer only (Fig.2a) and that a cluster breaks up into subclusters during the collapse (Fig.2b). The collapse is initiated at the outer (liquid) boundary of the cluster and proceeds towards the solid surface, the dimensions of the cluster decreasing at increasing rate during the collapse.

Pressure fluctuations as shown in Figs.3a,b,c were recorded with the pressure transducer in the tunnel wall, c. 10 mm from the collapse centre of the clusters. The transducer is not able to measure the shock wave emitted from the individual cavities in the cluster, but it registers the pressure effects due to the cluster collapse. It is apparent that the intensity of the peak pressure in each cavity cluster collapse varies significantly from one collapse to another, Fig.3 , the maximum value being of the order of 3×10^5 Pa at the position of the transducer while the mean static pressure driving the collapse is only 0.3×10^5 Pa. It is found that each pressure pulse is in general the resultant of a number of smaller pressure sub-pulses of very short duration. When these sub-pulses are not simultaneous the peak of the pressure pulse is small, Fig.3b, but when they develop simultaneously, peaked pressure pulses of intensity significantly above the static pressure are produced, with a total duration of up to a few hundred μs but with a much shorter duration of the peak pressure, Fig.3c. The interpretation is that the sub-pulses are produced by the sub-clusters during their collapse (possibly also by large individual cavities), but due to the irregular character of flow-produced cavity clusters, the break down is strongly statistical. This is supported by the photographs of the collapse process. The collapse intensity of the individual cavities is governed by the ambient pressure during their collapse. Therefore, those cavities collapsing latest and under the influence of the pressure field developed by the collapse of other cavities, will perform a strongly intensified collapse.

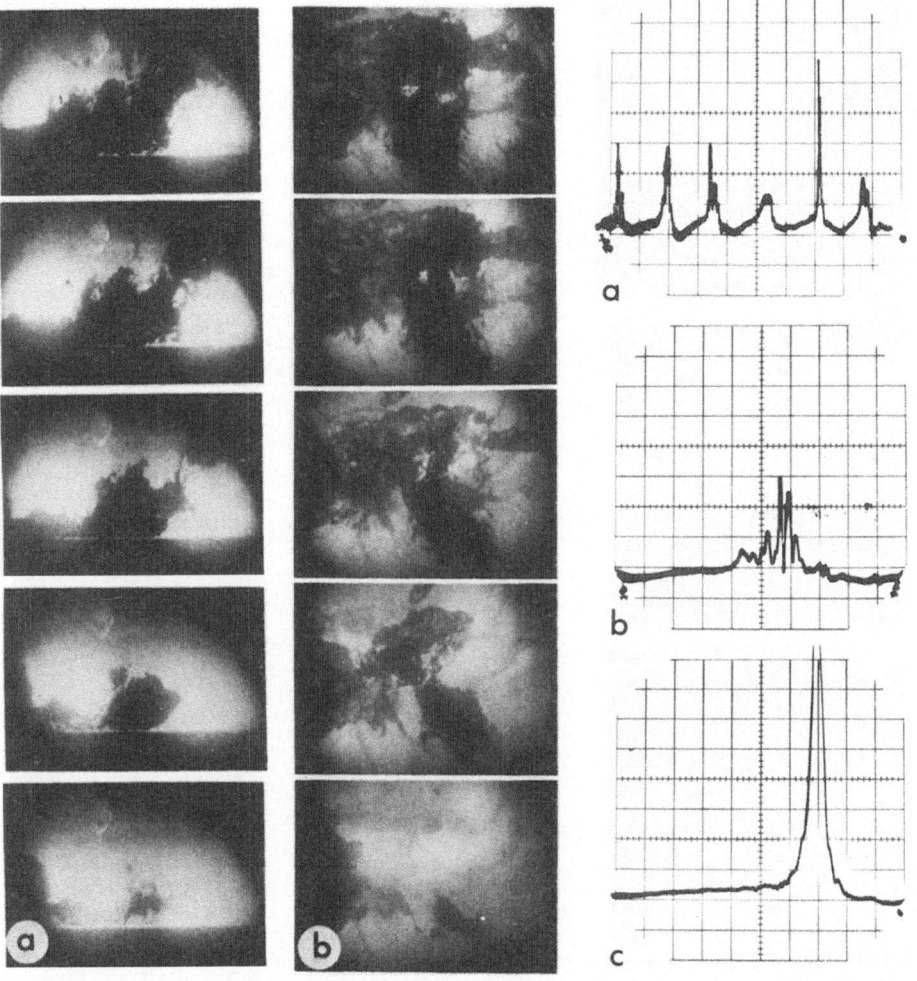

Fig.2 The collapse of cavity clusters seen
a) parallel to the plate, p = 0.29·10⁵ Pa and
b) perpendicular to the plate, p = 0.45·10⁵ Pa.
Time interval between the pictures 110 μs.
f = 550 Hz. Flow velocity from left to right.

Fig.3 Pressure fluctuations
at the channel wall off the
collapse position. p = 0.34·10⁵
Pa. a) x:1 ms/div, y:0.3·10⁵
Pa/div. b,c) x:0.1 ms/div,
y:0.3·10⁵ Pa/div.

4. Cluster collapse theory

Let us consider a spherical cluster of cavities, initially of radius R_0 and
in equilibrium at the pressure p_1 in an unbounded liquid space. The cluster
is a two-phase fluid and if the volume fraction of the cavities β_1 is not
very close to one or to zero the sound velocity inside the cluster c_m [8] is

$$c_m^{\,2} = \frac{p_1}{\rho \beta_1 (1-\beta_1)}. \tag{1}$$

If the cluster is subjected to an ambient pressure p_∞ a pressure wave is propagated as a shock wave into the cluster at the Mach number

$$M_1{}^2 = \frac{P_2}{P_1} \tag{2}$$

where subscripts 1 and 2 refer to conditions in front of the shock and behind the shock, respectively, while capital letters are used for conditions at the cluster boundary. The thickness of such a shock is of order $a_0/[\beta(1-M_1{}^2)]^{\frac{1}{2}}$, where a_0 is the initial radius of the individual cavities. For values of β of the order of 0.1 and $P_2/P_1 \geq 3$ the thickness becomes of order as the cavity diameter, and in general it is much smaller than the cluster dimensions. It is assumed that the cavities are virtually annihilated after the shock ($\beta_2 = 0$). Then the velocity of the cluster boundary equals the shock velocity

$$V_{sh} = c_m M_1 = \dot{R} = \sqrt{\frac{P_2}{\rho \beta_1 (1-\beta_1)}} \tag{3}$$

and the particle velocity outside the cluster at radius r becomes radial and is given by

$$v_r = V_r \frac{R^2}{r^2} = \beta_1 \dot{R} \frac{R^2}{r^2} \tag{4}$$

The liquid motion is governed by

$$\frac{\partial v_r}{\partial t} + v_r \frac{\partial v_r}{\partial r} = - \frac{1}{\rho} \frac{\partial p}{\partial r} \tag{5}$$

which, by use of (4) and integration gives

$$p = p_\infty + \rho \left[\beta_1 \frac{2R\dot{R}^2 + R^2\ddot{R}}{r} - \beta_1{}^2 \frac{R^4\dot{R}^2}{2r^4} \right]. \tag{6}$$

For the boundary of the cavity cluster (3) and (6) give

$$R\ddot{R} + (1 + \frac{\beta_1}{2})\dot{R}^2 = - \frac{p_\infty}{\rho\beta_1} \tag{7}$$

which is the cavity cluster collapse equation. If $\beta \to 1$, (7) is seen to become the ordinary equation of motion for the wall of a single bubble. However, for most cavity clusters $\beta_1 < 0.1$-0.2 can be expected, and then the cluster collapse can be adequately described by

$$R\ddot{R} + \dot{R}^2 = - \frac{p_\infty}{\rho\beta_1}. \tag{8}$$

Assuming a pressure at infinity

$$p_\infty = p_{\infty,max}(1-e^{-kt}) \text{ for } t \geq 0 \tag{9}$$

and using dimensionless quantities

$$R^* = \frac{R}{R_0}, \qquad t^* = \frac{t}{R_0} \sqrt{\frac{p_{\infty,max}}{\rho\beta_1}}, \qquad p^* = \frac{p}{p_{\infty,max}} \qquad (10a,b,c)$$

the solution becomes

$$R^{*2} = 1 - t^{*2} + \frac{2}{k^*} t^* - (1 - e^{-k^* t^*}) \frac{2}{k^{*2}} \qquad (11)$$

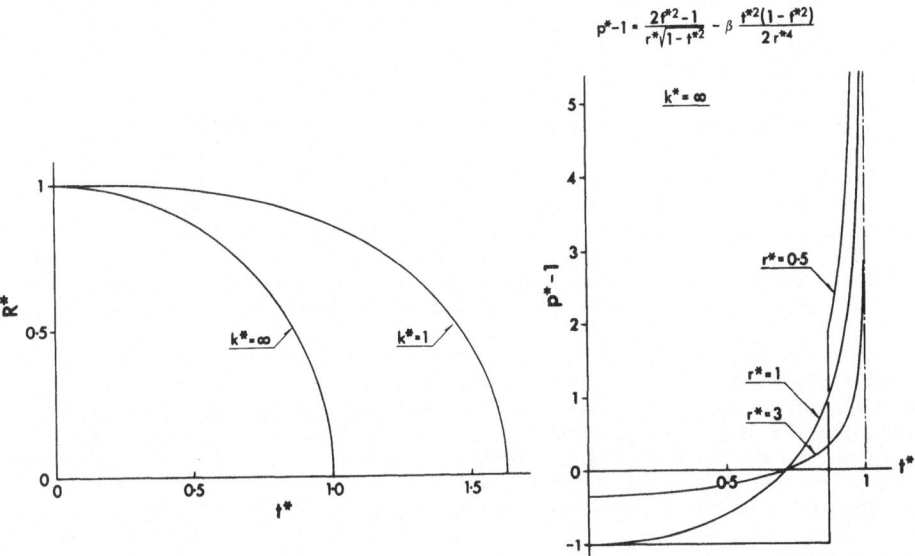

Fig.4 The radius R^* of a spherical cavity cluster during collapse at ambient pressure $p_\infty^* = 1 - e^{-k^* t^*}$

Fig.5 Pressure-time curves at fixed positions during the collapse of a spherical cavity cluster

The cavity cluster collapse is shown in Fig.4 for $k^* = \infty$ and $k^* = 1$. Further the pressure variation in the liquid during the collapse for $k^* = \infty$ is calculated from (6) and shown in Fig.5 at different radii $r^* = r/R_0$. It is seen that close to the centre of the cluster the cavities collapse at a strongly increased pressure. At $r^* = 3$, corresponding to the position of the pressure transducer in the experiments (Fig.1), and if $k^* = \infty$, pressure increase above p_∞ is to be expected for $0.7 < t^* < 1$. With $R_0 \simeq 4$ mm, $p_\infty = 0.03$ MPa, $\beta \simeq 0.2$ this is an interval of 100 µs, while the total collapse time for the cluster $\tau \simeq 300$ µs. According to Fig.3c the calculated duration of the pressure growth period is nearly correct but the collapse time (Fig.2a) is underestimated by a factor of 2. Meanwhile, $k^* = \infty$ corresponding to an instantaneous rise of the pressure at infinity is not realistic and a much better agreement for τ would be obtained e.g. for $k^{x} = 1$. Further it must be noted that the present theory describes the collapse of a spherical cavity cluster in an unbounded liquid space, while the experimentally obtained results are from the collapse of irregularly shaped cavity clusters close to a solid surface in a flow channel of dimensions only a few times larger than the cluster diameter.

The potential energy connected to the spherical cavity cluster is initially

$$E = p_\infty \, \beta_1 \, \frac{4\pi}{3} \, R_o{}^3, \tag{12}$$

but if we calculate from (4) and (11) the kinetic energy of the liquid at the moment of total collapse it turns out to be zero. This is a result of the potential energy of the individual cavities being converted into kinetic energy during their collapse and subsequently being partly radiated from the cluster with the shock waves resulting from each cavity collapse. Thus, as the cluster radius vanishes, most of the energy is radiated already, but a high pressure exists at the cluster boundary due to the rapid change of momentum in the liquid.

The present theoretical approach to the cavity cluster collapse, which is in principle equivalent to the theory presented in [4], is based on the momentum in radial direction relative to the centre of the cluster, but it does not treat the shock waves from the individual cavity collapses. These shock waves, which are pressure pulses of very small thickness, transport energy into the cluster. In a two-phase medium the wave propagation is dispersive, and it is to be expected that these high-frequency waves are propagated faster than the cluster collapse wave. Their absorption as well as their propagation velocity depend on the resonance frequency of the cavities, which depends on the temperature. It is conjectured that the influence of temperature on cavitation erosion is related to the difference in propagation velocity of the cavity cluster boundary and the shock waves from individual cavity collapses. If these are propagated ahead of the cluster boundary they will initiate the collapse of the cavities inside the cluster, their velocity and absorption thus determining the thickness of the cluster boundary shock wave, and the collapse process of the cluster will be significantly affected.

References

1 R.T. Knapp, Trans. ASME 77, 1045 (1955).

2 R.T. Knapp, J.W. Daily and F.G. Hammitt, Cavitation (McGraw-Hill, New York, 1970).

3 K.A. Mørch, in Treatise on Materials Science and Technology, Vol.16: Erosion, edited by C.M. Preece (H. Herman). In the press.

4 L. van Wijngaarden, Proc. 11th Int. Congr. Appl. Mech., edited by H. Görtler (Munich, Germany, 1964), p.854.

5 K.A. Mørch, Proc. Acoustic Cavitation Meeting, Institute of Acoustics (Poole, Dorset, U.K., 1977), p.62.

6 I. Hansson, K.A. Mørch and C.M. Preece, Proc. Ultrasonics International 77, (Brighton, U.K., 1977), p.267.

7 I. Hansson and K.A. Mørch, Proc. Ultrasonics International 79, (Graz, Austria, 1979), in the press.

8 L. van Wijngaarden, in Ann. Rev. Fluid Mech., Vol.4 (1972), p.369.

Effect of Polarization on Electric Pulses Produced by Cavitation Bubbles

G. Gimenez and F. Goby

Laboratoire de Génie Electrique de Paris, 33, avenue du Général Leclerc
92260 Fontenay-aux-Roses, France

1. Introduction

The motivation for the work described below was observation of electric
pulses caused by bubble collapse [1]. We felt it necessary to test what
seemed to be the most plausible hypothesis in explaining the physical origin of these electric pulses.

We describe briefly the characteristics of the experimental apparatus
and the main phenomena which it allows one to observe. Ultrasonic cavitation
is produced in distilled water by a transducer (frequency $\simeq$ 20 kHz) whose
active area is 1 cm in diameter. The cylindrical stainless steel measuring
cell is 8 cm wide and its effective length is 28 cm. The water temperature
is 20^0C and is in equilibrium with its surrounding air (α/α_s = 1). When an
electric probe is placed in a cavitation field one can observe pulses which
occur at the same time as the shock waves produced by the collapse of cavitation bubbles. At this stage we have to stress the fact that to avoid a
spurious signal which is present the probe must comprise three electrodes:
two of them (called "active") are connected to a differential amplifier,
whereas the third one acts as a screen. The active electrodes and their
associated electronics must be as symmetrical as possible; if not, spurious
signals appear and electric pulses are no longer observable.

With the aim of explaining the physical origin of these electric pulses
we have suggested that the shock waves produced by the collapse of cavitation bubbles disturb the electric double-layers of the electrodes. In effect, an electrode immersed in the distilled water of the measuring cell
makes up a solid-liquid interface at the ends of which an electrochemical
potential difference appears. This is caused by the presence of electric
charges on both sides of the interface. Fig.1a (from [2]) gives an example
of a distribution of such electrical charges, whereas Fig.1b shows the corresponding variation of the potential. After the collapse of a cavitation
bubble, the subsequent shock wave could disturb the layer of ions and then,
briefly, the potential (i.e. an electric pulse should appear). In short,
the electric probe would act as a condenser microphone.

There is a way to test the above hypothesis. If the potential difference
across the interface is changed, the charge of the interface is also changed.
Generally, such a modification could be achieved with an external potential
applied by means of a reference electrode. Fig.1c (from [3]) shows the
variation of surface tension versus potential difference measured with respect to an hydrogen electrode for mercury in contact with 1.0 N HCl. The

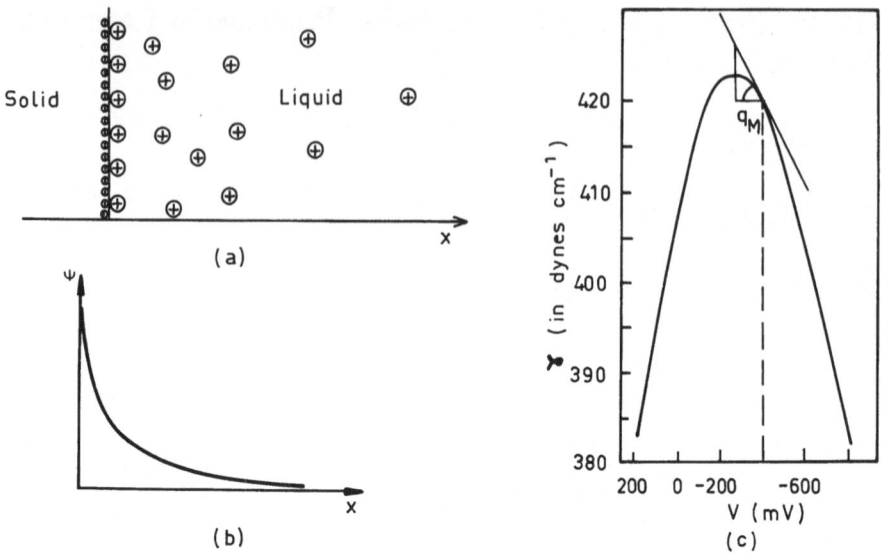

Fig.1 (a) Ions at a negatively charged metal interface. (b) Typical potential variation with distance from surface with no specific interaction. (c) Surface tension versus the potential difference

charge density q_M on the electrode at a particular value of the potential difference is given by the slope of the curve at that potential. One may notice that the charge on an electrode starts off with one sign and then changes its sign after passing through a zero-charge value. The potential difference across the system at which the charge on the electrode is zero is known as the potential of zero charge (pzc). If electric pulses are due to the disturbance of the charge on the liquid side of interface the height of these pulses must change with the value of the applied voltage. Thus for a particular value of the voltage (pzc) the pulses must vanish. This is why we felt it interesting to study the effect of a polarizing voltage on electric pulses.

2. Experimental Set-up

2.1 The Electric Probe

Fig.2 shows the electrical probe employed. The electrodes are of stainless steel, the diameter of the two inner ones is 1 mm, and 3 mm/5 mm for the third one (the screen).

Measurement with an impedance bridge indicates that the probe behaves, in air, as 1.5×10^{10} Ω in parallel with 6 pF. In the measuring cell, in working position (with the probe 1 mm from the transducer), the resistance is about 10^7 Ω and the capacitance 9 pF. We should note that the low value of this capacitance is due to the presence of a conductor, in fact the outer shields of the coaxial cables, with a fixed potential (the ground one) between the two active conductors. This low value would be still

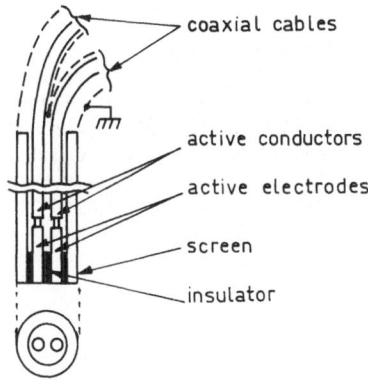

Fig.2 Cross section of the three-electrode electric probe

lower if the coaxial cables extended into the screen, up to the active electrodes.

2.2 The Polarizing Generator

Fig.3 shows why a 'conventional' voltage generator is not useful for observing the effect of a polarizing voltage on the electric pulses. Within the dotted lines in Fig.3 the probe is represented by its equivalent circuit. The resistance R_p and the capacitance C_p are those which were men-

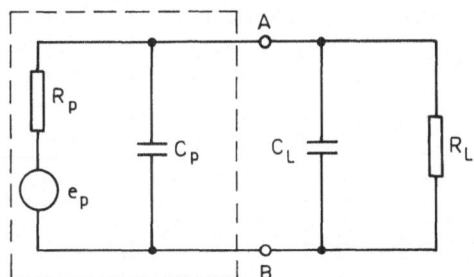

Fig.3 Equivalent electric circuit for the probe connected to a measuring instrument

tioned in 2.1, and e_p is the generator which delivers the electric pulses. When the probe is connected to a measuring instrument (e.g. an oscilloscope), the pulse height is proportional to its input resistance R_i. If a voltage generator is used in parallel with the measuring instrument for applying a voltage V_{AB} (actually V_A-V_B) between the two active electrodes the resistance in parallel with the probe is very low. Hence the pulses are no longer detectable.

The diagram of the electronic circuit which maintains the voltage between the electrodes and permits observing the pulses delivered by generator e_p is shown in Fig.4. One can demonstrate that the potential difference, V_{AB}, applied to the active electrodes of the probe depends only on the potentials of the positive inputs of the operational amplifiers; consequently,

V_{AB} does not depend on the electrical processes which affect the electrodes. Furthermore, one can see that the occurrence of a positive step voltage output at e_p produces a negative step voltage at s (s: potential difference between the outputs of the operational amplifiers), the ratio between s and e_p being Rm/Rp (cf. Fig.3 and Fig.4). Thus, the presence of the resistance Rm allows observation of electrical pulses. Let us remark that with this circuitry the pulses are not detectable with an instrument connected in parallel to the probe because, of course, the potential here is absolutely constant.

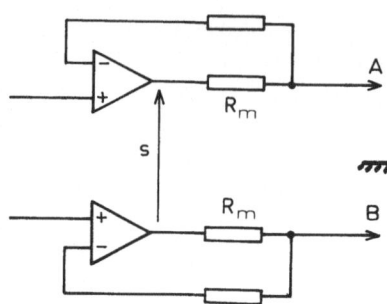

Fig.4 Simplified diagram of the polarizing generator

3. Experimental Results

When the electric probe is directly connected to the differential input of an oscilloscope, the signal in Fig.5a is obtained. We see some interference between successive pulses: the response time is much more important than the frequency of the phenomenon.

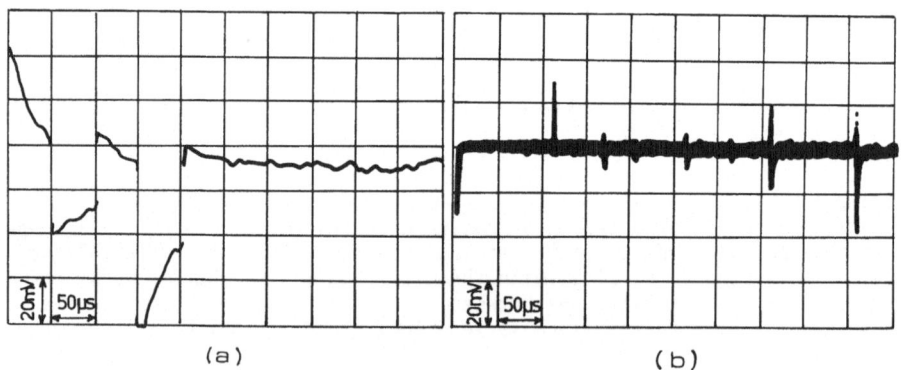

(a)   (b)

Fig.5 (a) Electric pulses when the active electrodes are directly connected to a differential amplifier. (b) Same but 10 kΩ in shunt

One can estimate the time constant as about 50 µs, this being supported by the values mentioned above (2.1). To reduce the time constant a 10 kΩ resistance can be put in shunt between the active electrodes. This produces the oscillogram of Fig.5b. We note the following: (i) the pulses are either positive or negative; (ii) the pulse heights are different; (iii) the elapsed time between the running pulses is not less that 50 µs, i.e. not less than an ultrasonic period.

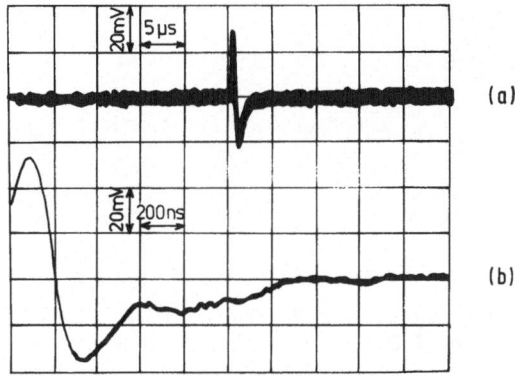

(a)

(b)

Fig.6 Electric pulses obtained with the polarizing generator

The oscillograms of Fig.6 are obtained when the active electrodes are connected to the voltage generator described in 2.2, with R_m = 10 kΩ and V_{AB} = 0 V. Two kinds of pulses are shown on Fig.6. Fig.6b allows estimation of the pulse (positive side) duration: about 100 ns. It should be notieced that this value is due to the electronic circuitry, and hence must not be attributed to the original pulse in the liquid. By increasing the speed of the circuitry (faster operational amplifiers, decrease of the resistance R_m), shorter pulses would be expected.

Let us underscore an experimental observation: though the two active electrodes are identical and the electronic device quite symmetrical, the amount of pulses of one polarity is different from the amount of pulses of the other polarity. More precisely, at the beginning of the service life of the electric probe, in the case of low power P fed into the transducer (P < 17 W) and for one given experiment, there is just one polarity for the pulses; this polarity could change from one experiment to one another (i.e. when the ultrasonic power is switched off, then on). In the case of higher power levels (P > 17 W), there are pulses of both polarities but in different amounts; the ratio between the two kinds of pulses changes during one experiment and, of course, from one experiment to another. As time passes, the disparity increases and finally there are only pulses of one polarity whatever the power may be. At this stage it is very interesting to observe the signals between each active electrode with regard to the ground instead of the differential signal. Then it appears that one electrode delivers the spurious signal (see 1.) on which are superposed negative pulses (Fig.7) and the other electrode produces only the spurious signal. The amplitude of this signal is 1.82 mV (peak to peak) and obviously it vanishes under differential measurement.

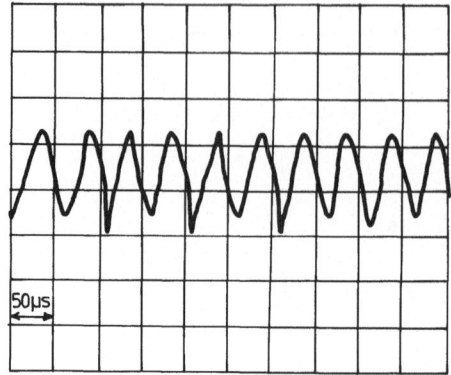

<u>Fig.7</u> Negative electric pulses superposed on the spurious signal

To study the effect of the polarizing voltage V_{AB} on electric pulses, we used a multichannel analyser having 1024 channels. After adequate shaping, pulses of a particular polarity are sorted out according to their height. The classes are about 7 µV in height, which agrees with the overall instrumentational accuracy. When the power is 20 W and the polarizing voltage V_{AB} is 0 V, the process produces the histogram of Fig.8.

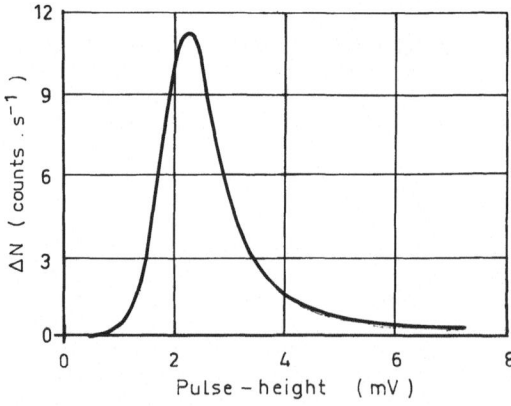

<u>Fig.8</u> Histogram of the pulse heights. $N = \Sigma \Delta N = 2425$ counts. s^{-1}

When V_{AB} is not 0 V, the spurious signals between each electrode and ground are identical, thus a spurious signal exists even under differential operation. To reject this 20 kHz signal and to allow pulses to be fed to the analyser a high pass filter is added. With this circuitry modification we varied V_{AB} from -4 V to +4 V, which is wide enough to cover the range encountered in electrochemistry (in particular for obtaining the pzc of Fig.1c). No significant effect on pulse height was observed. Thus, not only do the pulses not vanish for a particular value of V_{AB}, but neither does the polarizing voltage influence the pulse height.

4. Discussion

In our opinion the existence of pulses of both polarities could be explained by: (i) that the original pulses in liquid are negative; (ii) the collapse of cavitation bubbles sometimes have an effect on one electrode, sometimes on the other. The aging of the probe is probably due to the modification which occurs on its surface: cavitation develops from original cracks of the surface because they trap gas nuclei; then, erosion increases the number of cracks round the original one which strengthens cavitation at this place and so on. After sufficient time had elapsed cavitation occurs only at certain places. At this stage only one of the two active electrodes delivers pulses. As they are negative, we must conclude that collapse of cavitation bubbles produce negative electric pulses. Let us recall that the duration of these pulses is less than 100 ns.

Since a polarizing voltage, applied to the active electrodes, had no influence on the pulse height, the phenomenon does not depend on the solid-liquid interfaces of the electrodes. Consequently, the pulses may be due to electric charges carried by the bubbles. For example, WHYBREW et al. [4] study the electrification of small air bubbles in air water and estimate the apparent surface electric charge density to 0.8 esu.cm^{-2}. Theses values have been corroborated by IRIBARNE and KLEMES [5]. Thus, the electric charges located on the surface of the bubbles should be conveyed to one active electrode when the collapse occurs.

5. REFERENCES

1 G. Gimenez, J. Phys. D: Appl. Phys. 12, L 25 (1979).
2 J.T. Davies and E.K. Rideal, Interfacial Phenomena (Academic Press, New York, 1963), p. 88 and p. 103.
3 J. O'M. Brockris and A.K.K. Reddy, Modern Electrochemistry (Plenum Press, New York, 1970), Vol. 2, p. 702.
4 W.E. Whybrew, G.D. Kinzer, and R. Gunn, J. Geophys. Res. 57, 459-471 (1952).
5 J.V. Iribarne and M. Klemes, J. Chem. Soc. Faraday Trans. 70, 1219-1227 (1974).

Cavitation Effects at Megahertz Frequencies

P.W. Vaughan, S. Leeman, M. Hedges, E. Graham, and P. Sutton

Department of Medical Physics, Royal Postgraduate Medical School
and Hammersmith Hospital
London, W12 0HS, United Kingdom

Introduction

This experimental programme, using plane travelling-wave ultrasound with the sample liquid in the first axial maximum of the acoustic far-field, has the following objectives:-

1. To find threshold intensities for the onset and extinction of subharmonic oscillation as the acoustic intensity is raised and lowered in the sample.

2. To find the equivalent thresholds for sonoluminescence in the sample, and compare these with 1.

3. To study the effect of sample gas content on 1 and 2.

4. To investigate driving the transducer in an A.C. bridge, using the off-balance voltage of this bridge as an indication of changes in the motional impedance of the transducer, and compare these values with the R.F. voltage of the drive amplifier when coupled directly to the transducer.

5. To measure the sound velocity in the sample liquid by pulse-echo technique, looking for changes in bulk modulus at the subharmonic threshold.

6. To find the acoustic threshold for release of molecular iodine from $KI/CC\ell_4$ samples which have varying gas content, and compare these thresholds with 1 and 2.

7. Compare a recently measured biological end-point with the above thresholds.

Equipment and Instrumentation

The irradiation cell consists of 10 cm x 10 cm x 6 cm x 10 mm thick clear acrylic units bolted together and filled with coupling medium. This module provides a plane travelling-wave acoustic system, with a space centrally to accommodate the sample holder, at the first axial maximum of the sound field. Drive transducer is of plane PZT4 type, resonant at 1.5 MHz, and driven by a signal generator, power amplifier, either via an A.C. bridge or direct. The subharmonic detector is another transducer mounted orthogonally opposite the sample holder, and output from this goes to a filter/amplifier system from which the mixed and filtered signal (at half drive frequency) can be displayed on an oscilloscope. A rectified component of the filtered signal (the

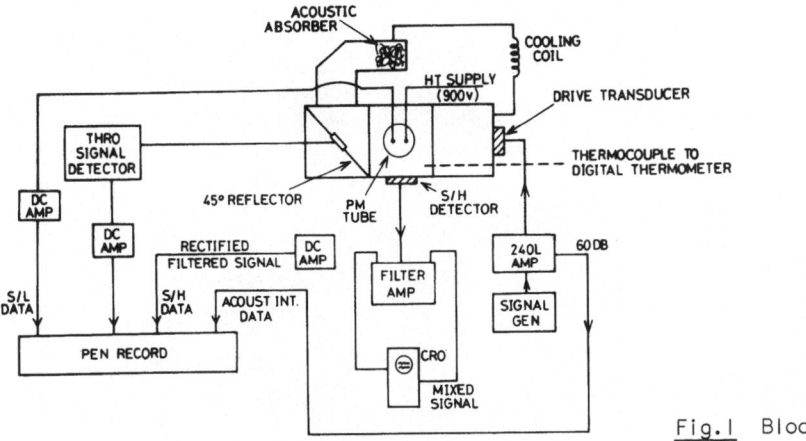

Fig.1 Block diagram

subharmonic) can be collected on the multipen chart recorder. Any sonoluminescence generated within the sample is detected by a photomultiplier tube
mounted above it (Fig.1), and an amplified signal collected on the chart,
together with the subharmonic, and also the voltage of the power amplifier
which is calibrated in terms of spatial-average acoustic power (Table I).
There is also a cooling system for the coupling medium, capable of holding
temperatures to within 1°C during experiments.

Experimental

Conduct of Experiments

Trials with various coupling media led us to select a light mineral oil,
viscosity 10 Cs, as being most suitable. It was found that no subharmonication or sonoluminescence was generated in this oil over the power range available.

 The experimental procedure was to select the sample liquid, and subject it
to increasing then decreasing acoustic stress levels on a regulated time scale.
The automatic control of the signal generator could be used to increase to
maximum over 1.5, 3, or 4 min periods, and the tests were all conducted under
these standardised conditions.

 The subharmonic (S/H) and sonoluminescent (S/L) data were collected simultaneously, and Figs.2-5 show typical data.

 As a result of a large number of experiments we are able to establish a
threshold for S/H and S/L in gassy water which is more consistent on decreasing acoustic stress than on increasing stress. The two effects have a very
closely associated threshold, and the effect of degassing by shaking under
vacuum is to remove both S/H and S/L. Reduction of both effects results
from repeated sonication, and also the effects are decreased as the gas content of the sample is reduced. There is a very consistent pattern of hysteresis in both the S/H and S/L effects. One "idiosyncrasy" of the equipment
was the way twisting the sample-holder during an experiment could sometimes
stimulate subharmonication from an otherwise "quiet" sample. We believe this
effect is due to the geometry of the sound-field being disturbed.

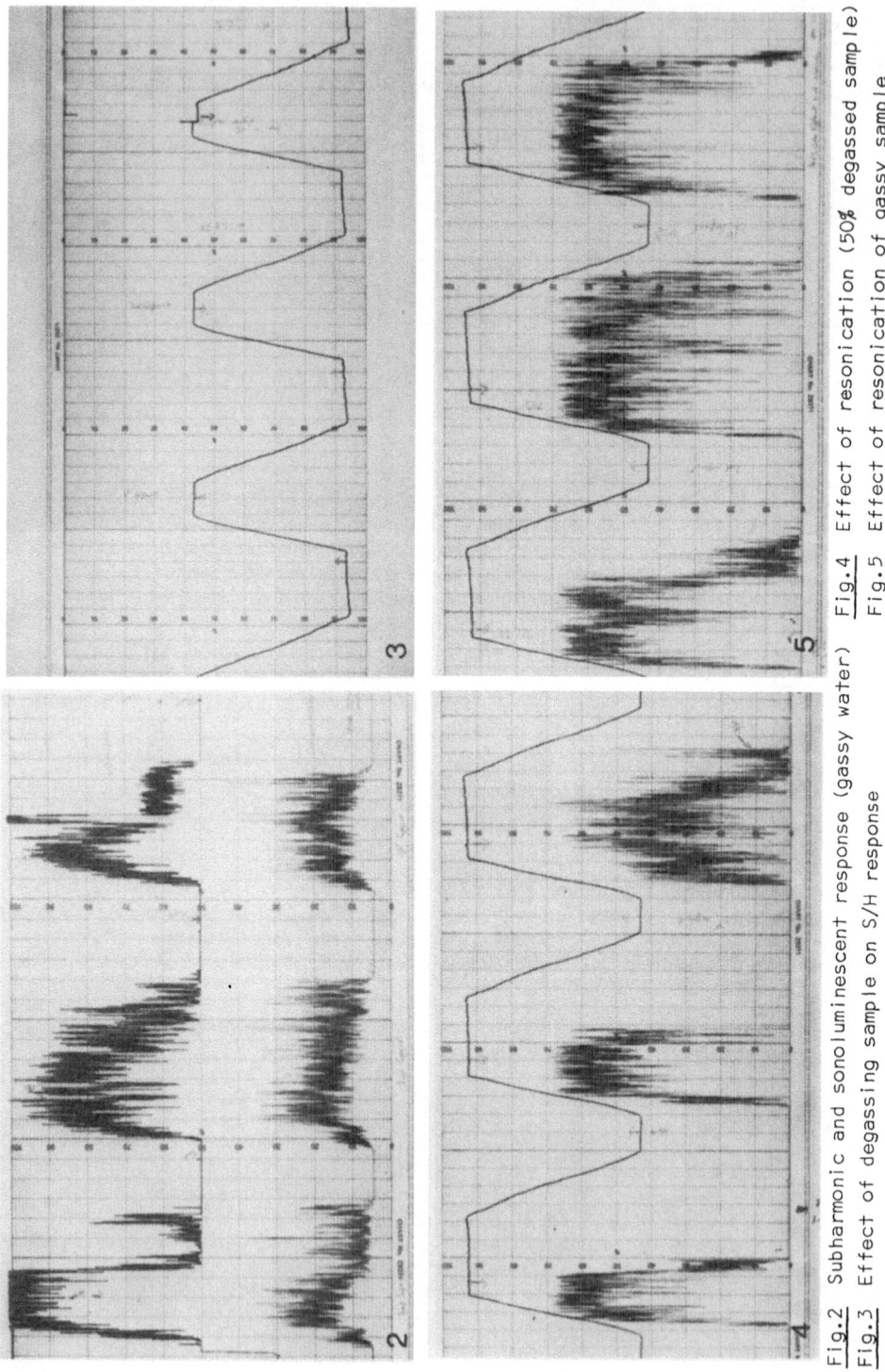

Fig.2 Subharmonic and sonoluminescent response (gassy water)
Fig.3 Effect of degassing sample on S/H response

Fig.4 Effect of resonication (50% degassed sample)
Fig.5 Effect of resonication of gassy sample

110

For the sonochemical experiments, 0.1 M KI was shaken with CCl_4, and then, gassed with an air bubbler for 5 mins, allowed to separate and the aqueous layer used as gassy KI stock. Degassed KI was obtained from this by shaking under vacuum for 45 mins. The sonochemical data is presented in Table 2, and we conclude that the quality of iodine released relates to acoustic intensity in a gassy sample, and no iodine is released from a degassed sample.

Table I   Output calibration

Volts	15	20	30	38	49	55	60
W/cm^2	.50	.75	1.67	1.92	2.49	3.01	3.51

Table II   Iodine release (relative values) against acoustic intensity

Volts		15	20	30	32	37	40	40	40	46	47
I_2-release	G A S S Y	.02		.05	.09	.02	.38	.49			.32
S/H		No		No	Yes	No	Yes	Yes			Yes
S/L		No		No	Yes	No	Yes	Yes			Yes
I_2-release	D E G A S S E D		.02						.005	.01	
S/H			No						No	No	
S/L			No						No	No	

Some measurements of "through signal" were made, and we found no distinct change in this value at the S/H and S/L threshold. Above these thresholds, the signal became very variable, but the "onset" threshold was not easily discernible and the technique was not developed any further at this stage.

The off-balance voltage when the transducer was driven in A.C. bridge mode was also examined when increasing acoustic stress through the S/H and S/L thresholds, and we find that there is no indication from this signal of any changes in the acoustic impedance of the liquid in the sample. As the bridge circuit imposed a limit on the electrical power to the transducer, we reverted to direct R.F. output voltage as a measure of acoustic intensity into the sample.

The sound velocity measurements also gave negative results. This means that we cannot detect any change in velocity within 2%, which would imply no change in K within 4%.

Discussion

We consider cavitation as being the non-linear motion of bubbles in the liquid, which produce certain typical effects [1]. The first of these effects is subharmonication. A strong signal at the half-frequency is found to be generated in acoustically strained liquids. It is apparently associated with the motion of gas bubbles, as degassing removes the signal. The quantity of subharmonic emission is proportional to the gas content of the liquid, and there is strong evidence for hysteresis. "Tweaking" the sample holder (not rotation) has the effect of stimulating subharmonic emission in an otherwise "quiet" sample, and this is considered to result from disturbing the sound field geometry. The extinction threshold proves to be a more reliable threshold for subharmonication than the onset threshold. We have strong evidence that repeated sonication extinguishes subharmonication, sonoluminescence and so presumably cavitation.

Experiments with supersaturated liquids show that the subharmonic signal is greatly reduced where bubble collapse is suppressed.

Comparison of thresholds ·for subharmonication and sonoluminescence and one particular sonochemical effect, and the observation that degassing suppresses these effects, lead us to conclude that these effects are consequences of gaseous cavitation activity in the sample.

If we assume that these effects are due to non-linearity in the bubble motion, we reach the conclusion that the conditions for subharmonication to occur also correspond to those for sonoluminescence. Under our experimental conditions, the spatial average intensity is around 1.2 Wcm^{-2}. We note that the subharmonic signal is steady and sustained rather than of short duration, and reach the tentative conclusion that the cavitation is of stable rather than transient type, in the terminology of FLYNN [2], and both types of emission are from vibrating gas bubbles in the sample whose motion has reached sufficient amplitude.

References

1. Hedges, M., Leeman, S. and Vaughan, P. Paper E7, 95th Meeting of The Acoustical Society of America, Providence, New Jersey, May, 1978.

2. Flynn, H. G. in "Physical Acoustics", Vol. 1B, pp 57-172, ed. W. P. Mason, Academic Press, N.Y., 1964.

Nonlinear Sound-Scattering by Small Bubbles

P.M. Tilmann

Fraunhofer-Institut für Hydroakustik
D-8012 Ottobrunn b. München, Fed. Rep. of Germany

Abstract
The flow of a nonviscous compressible medium is governed by a nonlinear PDE for the velocity potential and by a generally nonlinear dependence of the density on the pressure. Focussing interest on the Langevin-nonlinearity, we assume constant speed of sound. Then, for the case of spherical symmetry a near field second order Mach-number expansion of the velocity potential is derived. With its help, an equation of motion of the bubble wall is established also including the second order terms with respect to Mach-number. Comparison with the Gilmore-equation which is based on the Kirkwood-Bethe hypothesis shows a difference in the second order terms. No arguments could be found to prove that the difference in general be small.

1. Introduction

We want to investigate the domain of validity of the Gilmore-equation for the radial motion of a spherical bubble in an inviscid compressible fluid. We cannot compare the Gilmore-equation with an exact one because the latter is not yet available. But we can compare tow Mach-number expansions: one directly of the governing equations and one of the Gilmore-equation and look up to what order these two agree.

2. Governing equations

Spherical symmetry and isentropic change of state assumed, the equations of continuity and motion are

$$\frac{1}{c^2} (h_{/t} + v h_{/r}) + v_{/r} + 2 \frac{v}{r} = 0 \qquad (2.1)$$

and

$$v_{/t} + v v_{/r} + h_{/r} = 0, \qquad (2.2)$$

where c is the speed of sound, h is the enthalpy, v is the radial motion and the suffix-slash, followed by an independent variable, indicates partial differentiation with respect to this variable.
Introducing the velocity potential φ by

$$v \equiv \varphi_{/r} \qquad (2.3)$$

we integrate the Euler-equation (2.2) to get the Bernoulli-equation

$$\varphi_{/t} + \frac{1}{2} \varphi_{/r}^2 + h = f(t), \qquad (2.4)$$

where f(t) is an arbitrary function of time only. We complete the definition of φ by

$$f(t) \equiv 0 . \tag{2.5}$$

From (2.4) and (2.2) we get $h_{/t}$ and $h_{/r}$ expressed by the velocity potential φ . These expressions inserted into (2.1) give us

$$\Delta\varphi - \frac{1}{c^2} \varphi_{/tt} = \frac{1}{c^2} N\varphi , \tag{2.6}$$

where $N\varphi$ is the term of convective or Langevin-nonlinearity,

$$N\varphi = 2\varphi_{/r} \varphi_{/rt} + \varphi_{/r}^2 \varphi_{/rr} . \tag{2.7}$$

We will focuse our interest to this convective nonlinearity and assume

$$\dot{c^2} = \text{const.} \tag{2.8}$$

3. Acoustical Approximation

Neglecting the nonlinear terms $N\varphi$ we have the wave equation

$$\Delta\varphi - \frac{1}{c^2} \varphi_{/tt} = 0. \tag{3.1}$$

The solution of this will give us insight into the action of an externally driven sound-field on the bubble.
The general solution of (3.1) is

$$\varphi_{ac} = \frac{1}{r} \left[A(t - \frac{r}{c}) + B(t + \frac{r}{c}) \right] + K_1 t + K_0 , \tag{3.2}$$

where A and B are functions of one variable and K_1, K_0 are constants. Inserting this solution into the Bernoulli-equation (2.6) and looking to the limit $r \longrightarrow \infty$ we get

$$h_\infty = -K_1 = \text{const.} \tag{3.3}$$

This shows that a time dependent pressure at infinity is impossible for compressible media. The action of the external sound field must be described using the incoming wave $\frac{1}{r} B (t + \frac{r}{c})$.

Consider the parallel case for which the fluid and the boundary-conditions and the external sound-field are the same but the bubble is removed. Then the fluctuating field may be expanded into spherical harmonics with respect to the former bubble center. Only the zero order waves, the incoming $\frac{1}{r} B_0 (t + \frac{r}{c})$ and the outgoing $\frac{1}{r} A_0 (t - \frac{r}{c})$ together, determine the enthalpy at the origin. To avoid singularity

$$A_0(t) = - B_0(t) \tag{3.4}$$

114

is necessary. Inserting this into the Bernoulli-equation, but now taking the limit $r \rightarrow o$, we get

$$\frac{2}{c} B_0'' (t) = h_\infty - h_0 (0,t) \qquad (3.5)$$

where the superfixed slash indicates differentiation with respect to the argument.

As the bubble is now replaced again (and boundaries are assumed far enough) the incoming wave must not be changed. Therefore

$$B''(t) = B_0''(t) = \frac{c}{2} \left[h_\infty - h_0 (0,t) \right] , \qquad (3.6)$$

where $h_o(o,t)$ exactly means the enthalpy which would arise at the location of the center of the bubble in the case of the bubble removed. It is not the enthalpy in a certain distance large to the radius!

4. The nearfield second-order Mach-number expansion

The acoustical approximation is a good one for large distances r and becomes worse approaching the bubble wall. Only for small amplitude oscillations($|\dot{R}^2| << |R \ddot{R}|$) the neglected nonlinear term $N\varphi$ is small compared to the featured linear term $\varphi_{/tt}$ at the bubble wall. But in general the acoustical approximation is not better there than the purely incompressible one.
By using an iterative procedure to improve the acoustical approximation according to

$$\varphi \approx \varphi_s = \varphi_{ac} + \frac{1}{c^2} \varphi_N , \qquad \Delta\varphi_N = N\varphi_{ac} \qquad (4.1)$$

but restricted to the zero order level of (4.1) we get an approximation which is as good as the acoustical one for large distances and is an improvement to second order in the Mach-number at the bubble wall.
The zero order terms of $N\varphi$ are of the following type:

$$Q = \frac{f(t \pm r/c) \cdot g(t \pm r/c)}{r^n} , \qquad n \geq 4 . \qquad (4.2)$$

there is no alteration in the zeros order when we replace Q by

$$\overset{*}{Q} = Q \underset{(a)}{\pm} \frac{2}{c(n-2) r^{n-1}} \left[f'_{<\pm r>} \cdot g_{<\pm r} \underset{(b)}{>} f_{<\pm r>} \cdot g'_{<\pm r>} \right] \qquad (4.3)$$

$$+ \frac{1}{c^2(n-2)(n-3) r^{n-2}} \left[f''_{<\pm r>} \cdot g_{<\pm r} \underset{(c)}{>} f'_{<\pm r>} g'_{<\pm r>} + f_{<\pm r>} \cdot g''_{<\pm r>} \right]$$

where $< r >$ indicates shifted argument,

$$f <r> \equiv f(t + r/c), \qquad (4.4)$$

and where the sign (a) agrees with that of the shift of g, the sign (b) with that of the shift of f and the sign (c) is positive if the two shift-signs are equal. The function

$$\varphi_Q = \frac{f <\pm r> \cdot g <\pm r>}{(n-2)(n-3)\, r^{\,n-2}} \tag{4.5}$$

is a bounded solution of

$$\Delta \varphi_Q = \overset{*}{Q} \tag{4.6}$$

for $r > R > 0$.

As for φ_{ac} according to (3.2) is given by

$$N\varphi_{ac} = \frac{2}{r^4} \left[A_{<-r>} + B_{<r>} \right] \cdot \left[A'_{<-r>} + B'_{<+r>} \right] \tag{4.7}$$

$$+ \frac{2}{r^7} \left[A_{<-r>} + B_{<r>} \right]^3 + \frac{1}{c}(\cdots),$$

the respective sum of the φ_Q-s of type (4.5) give

$$\varphi_N = \frac{1}{r^2} \left[A_{<-r>} + B_{<+r>} \right] \cdot \left[A'_{<-r>} + B'_{<+r>} \right] \tag{4.8}$$

$$+ \frac{1}{10\, r^5} \left[A_{<-r>} + B_{<+r>} \right]^3.$$

φ_N remains the same in zeros order when we approximate

$$\varphi_N \approx -r\,\varphi_{/r} \left(\varphi_{/t} + h_\infty \right) - \frac{r}{10}\,\varphi_{/r}^3 , \tag{4.9}$$

so that

$$\varphi_s = \frac{1}{r} \left[A_{<-r>} + B_{<+r>} \right] - h_\infty t - \frac{r}{c^2} \left[\varphi_{/r} (\varphi_{/t} + h_\infty) + \frac{\varphi_{/r}^3}{10} \right] \tag{4.10}$$

is our nearfield-improved approximation from which the equation of motion of the bubble wall can be derived.

5. The second order differential equation for the bubble-radius

Introducing the dot as the symbol of the substantial time-derivative convective with the bubble wall we get the following representations of the partial derivatives of φ at the bubble wall (by use of the governing equations (2.1), (2.2) and (2.4)):

$$\varphi_{/r}(R) = \dot{R} \tag{5.1}$$

$$\varphi_{/rr}(R) = -\left(2\,\frac{\dot{R}}{R} + \frac{1}{c}\,\dot{h}_R \right)$$

$$\varphi_{/t}(R) = -\left(\frac{1}{2}\,\dot{R}^2 + h_R \right)$$

$$\varphi_{/rt}(R) = \ddot{R} + 2\,\frac{\dot{R}^2}{R} + \frac{1}{c^2}\,\dot{R}\,\dot{h}_R$$

$$\varphi_{/tt}(R) = -\left(\dot{h}_R + 2\dot{R}\ddot{R} + 2\,\dot{R}^3/R + 1/c^2\,\dot{R}^2\,\dot{h}_R \right)$$

where h_R is a shorthand for $h(R)$.

These expressions inserted into (4.10) give

$$R^2\ddot{R} = -(A_{<-R>} + B_{<+R>}) - \frac{R}{c}(A'_{<-R>} - B'_{<R>})$$

$$- \frac{R^2}{c^2}\left[2\dot{R}^3 + \dot{R}(h_R - h_\infty) + R\dot{R}\ddot{R}\right] + \frac{1}{c^3}(\cdots). \tag{5.2}$$

We note

$$(A_{<-R>})' = A'_{<-R>} \cdot (1 - \frac{\dot{R}}{c}) \tag{5.3}$$

and

$$(B_{<+R>})' = B'_{<+R>} \cdot (1 + \frac{\dot{R}}{c})$$

and differentiate (5.2):

$$2R\dot{R}^2 + R^2\ddot{R} = -(A_{<-R>} + B'_{<+R>}) \tag{5.4}$$

$$- \frac{R}{c}\left[(1 - \frac{\dot{R}}{c})A''_{<-R>} - (1 + \frac{\dot{R}}{c})B''_{<R>}\right]$$

$$- \frac{R}{c^2}\left[(2\dot{R}^2 + R\ddot{R})(h_R - h_\infty) + R\dot{R}\dot{h}_R\right.$$

$$\left. + 4\dot{R}^4 + 9R\dot{R}^2\ddot{R} + R^2\ddot{R}^2 + R^2\dot{R}\dddot{R}\right] + \frac{1}{c^3}(\cdots).$$

Bernoullis equation (2.4), taken at the bubble wall, results in

$$A'_{<-R>} + B'_{<+R>} + \frac{1}{2}R\dot{R}^2 + R(h_R - h_\infty) + \frac{1}{c^3}(\cdots) \tag{5.5}$$

$$+ \frac{R}{c^2}\left[(2\dot{R}^2 + R\ddot{R})(h_R - h_\infty) + R\dot{R}\dot{h}_R + \frac{11}{5}R\dot{R}^2\ddot{R} + \frac{12}{5}\dot{R}^4\right] = 0,$$

the substantial time-derivative of it gives

$$\frac{R}{c}(1 - \dot{R}/c)A''_{<-R>} + (1 + \dot{R}/c)B''_{<+R>} + R\dot{R}\ddot{R} + \dot{R}^3/2 \tag{5.6}$$

$$+ \dot{R}(h_R - h_\infty) + R\dot{h}_R] + 1/c^3(\cdots) = 0.$$

Inserting (5.5) and (5.6) into (5.4) we get

$$R\ddot{R} + \frac{3}{2}\dot{R}^2 = h_R - h_\infty + \frac{2}{c}(1 + \dot{R}/c)B''_{<R>} \tag{5.7}$$

$$+ \frac{1}{c}\left[R\dot{R}\ddot{R} + \frac{1}{2}\dot{R}^3 + \dot{R}(h_R - h_\infty) + R\dot{h}_R\right]$$

$$- \frac{1}{c^2}\left[\frac{8}{5}\dot{R}^4 + \frac{34}{5}R\dot{R}^2\ddot{R} + R^2\ddot{R}^2 + R^2\dot{R}\dddot{R}\right].$$

Again consider the "parallel case" of fluid and boundary conditions being the same but the bubble removed ($R, \dot{R} = 0$). Equation (5.7) then repeates the result (3.5):

$$\frac{2}{c}B''(t) = h_\infty - h_0(0,t)$$

It may be convenient to expand the function h_o $(0, t + \frac{R}{c})$ with respect to the anticipation $\frac{R}{c}$ and to replace the $\dot{R}$ and $\ddot{R}$ in the second order term by the Rayleigh-equation. The final result is

$$R\ddot{R} + \frac{3}{2}\dot{R}^2 = h_R - h_0(0) + \frac{1}{c}\left\{R\dot{R}\ddot{R} + \frac{1}{2}\dot{R}^3 + \left[R(h_R - h_0(0))\right]'\right\} \qquad (5.8)$$

$$+ \frac{1}{c^2}\left\{- R\dot{R}\dot{h}_R + \frac{1}{5}\dot{R}^2(h_R - h_0(0)) - (h_R - h_0(0))^2 + \frac{7}{20}\dot{R}^4\right\}$$

which differs in the second order term from Gilmores equation according to

$$\text{eq.}(5.8) \text{ minus "Gilmore"} = \frac{1}{c^2}\left\{\frac{1}{5}\dot{R}^2(h_R - h_0) - (h_R - h_0)^2 + \frac{7}{20}\dot{R}^4\right\}. \qquad (5.9)$$

6. Domain of validity

To give estimates for the domain of validity we note that we have assumed

$$\frac{1}{c}Rf' << f, \quad f = A \text{ or } B \qquad (6.1)$$

for replacing Q by Q^*,

$$\frac{R}{c} << 1 \qquad (6.2)$$

to cut off (5.4) after the second order terms and

$$|h_0(0) - h_\infty| << |h_R - h_\infty| \qquad (6.3)$$

because of (3.5).

Let a bubble be driven from the interior, anyhow, so that $A(t)$ is oscillating harmonically with angular frequency ω. Evaluation of (6.1) at the bubble wall gives

$$\frac{\omega}{c}R = kR << 1. \qquad (6.1a)$$

This says that the radius of the bubble must always be small compared to the wavelength of the emitted frequency. Let $B(t)$ oscillate harmonically (incoming wave) than the relation (6.1a) holds for the wavenumber of the driving frequency. That is the reason why our theory is announced by the title to be restricted to small bubbles (otherwise a nearfield improvement is not possible). For a harmonically oscillating bubble (6.2) says that the amplitude must be small compared to the wavelength of the emitted sound. Of course that is a consequence of (6.1a), too. Collapse calculations most probably are wasted labour if they are done further beyond reaching $\dot{R}/C = \frac{1}{2}$.

The consequences of the formal order of magnitude of $h_0(0)$ according to (3.5) are not yet analyzed very well.

Dynamics of a Cylindrical Cavity in a Boundless Compressible Liquid

V.K. Kedrinskii and V.T. Kuzavov

Institute of Hydrodynamics, Siberian Branch of the USSR Academy of Sciences
Novosibirsk 630090 USSR

The problem considered in this paper has a number of interesting peculiarities. The main is the impossibility to describe
the behaviour of the cylindrical cavity by an exact equation
even within the framework of incompressible liquid. Meanwhile,
there is a class of problems of underwater and underground explosions and hydroacoustics connected with the necessity to
obtain simple and reliable estimates for the pulsation parameters of a cylindrical cavity with detonation products.

The equation of a one-dimensional cylindrical cavity pulsation was obtained in [1] , and its peculiarities were researched numerically and experimentally in [2,3] . Let us consider a
common approach to the problem based on the KIRKWOOD and BETHE
idea [4] for an arbitrary one-dimensional symmetry. The one-dimensional isentropic flow of liquid is described by the known
system of equations

$$\rho_t + v\rho_r + \rho v_r + \nu \cdot \rho \cdot v/r = 0, \tag{1}$$

$$v_t + vv_r + p_r/\rho = 0, \quad p = p(\rho)$$

where $\nu = 0,1,2$ (this value determines the flow symmetry).
Now the kinetic enthalpy function $\Omega = \omega + v^2/2$ ($\omega = \int dp/\rho$)
will be introduced and the potential flow ($v = -\nabla\varphi$) will be
considered. After some transformations the system (1) has the
form

$$c^{-2}\varphi_{tt} - (1-v^2/c^2)\varphi_{rr} - \varphi_r(\nu/r + 2c^{-2}\varphi_{rt}) = 0, \quad \varphi_t = \Omega$$

where c is the local sound velocity in liquid, or in the acoustic approximation

$$c_0^{-2}\varphi_{tt} - \varphi_{rr} - \nu\varphi_r/r = 0, \quad \varphi_t = \Omega .$$

Then after introducing the function $\Phi = r^{\nu/2} \cdot \varphi$ the last system has the more convenient form for analysis:

$$c_0^{-2}\Phi_{tt} - \Phi_{rr} - \frac{\Phi}{2r^2}\nu(1-\frac{\nu}{2})=0$$

$$\Phi_t = r^{\nu/2}\Omega \tag{2}$$

The first equation of (2) can be written as

$$c_o^{-2} \Phi_{tt} - \Phi_{rr} = 0 \qquad (3)$$

if $\nu = 0,2$. The same result can be obtained for $\nu = 1$ as an approximate one within the framework of an asymptotic approach with an accuracy of the term $\Phi/4r^2$ in (2). The solution of (3) in the form $\Phi = \Phi(t \pm r/c_o)$ allows the following relations

$$\Phi_{tt} = G_t, \quad \Phi_r = -\Phi'/c_o = -\Phi_t/c_o = -G/c_o, \quad \Phi_{rr} = -G_r/c_o$$

to be obtained for the case of diverging waves using the second equation of (2) in the form $\Phi_t = r^{\nu/2} \cdot \Omega = G$. The substitution of them to (3) leads to

$$G_t + c_o \cdot G_r = 0.$$

It means that the function $G = r^{\nu/2} \varphi_t$ is the invariant along the characteristics diverging with the velocity c_o. Under the assumption of the conservation of function G along the characteristics diverging with velocity $c + v$ [4] it can be written as follows:

$$G_t + (c + v) \cdot G_r = 0. \qquad (4)$$

This physically evident assumption of the KIRKWOOD-BETHE model proved to be of principal and allowed a rather simple method to be suggested for calculating the parameters of shock waves at underwater explosions of spherical charges based, for the most part, on the determination of function $G(t)$ on the cavity wall with the detonation products. Naturally, for this aim it is necessary to obtain the cavity pulsation equation substituting $G = r^{\nu/2} \cdot (\omega + v^2/2)$ into (4) and changing the partial derivatives by full ones from (1) when $r = R$ and $v = R$. Now it is obtained that

$$R(1-\dot{R}/c)\ddot{R} + \frac{3}{4}\nu\dot{R}^2(1-\dot{R}/3c) = \frac{\nu}{2}(1+\dot{R}/c)\omega_o + R\dot{\omega}_o(1-\dot{R}/c)/c \qquad (5)$$

where $\omega_o(R)$ is the value of enthalpy on the cavity wall, R is its radius, the derivative of t is denoted by a point. The substitution of ν into (5) allows the pulsation equation of a plane cavity ($\nu = 0$), either spherical ($\nu = 2$) [5] or cylindrical ($\nu = 1$) to be obtained. In this paper the last case ($\nu = 1$) will be researched.

At first, let us consider three various approaches to deriving the pulsation equation in the ideal incompressible liquid. The equation (5), if $c \to \infty$ and $\nu = 1$, gives an approximate analog of the pulsation equation in a boundless incompressible liquid

$$(y\ddot{y} + \dot{y}^2) \cdot 2 - \dot{y}^2/2 = p_o \cdot y^{-2\gamma} - 1$$

In fact the free surface always is present in the experiments and therefore it is interesting to consider the pulsation equation in the half-space of liquid which is practically exact if the explosion depth $h_0 = h/R_0 \gg y$ [2]

$$(y\ddot{y} + \dot{y}^2)\ln 2h_0/y - \dot{y}^2/2 = p_0 \cdot y^{-2\gamma} - 1$$

And, after all, for the same aim it is convenient to use the model of a cylindrical layer with an uncertain external radius r . This model leads to the following equation if $r \gg R^2$

$$(y\ddot{y} + \dot{y}^2)\ln r_0/y - \dot{y}^2/2 = p_0 y^{-2\gamma} - 1.$$

Here p_0 is the pressure on the cavity wall taken relative to P_∞, $R = yR_0$, $r = r_0 \cdot R_0$, $t = \tau\sqrt{\rho_0 p_\infty^{-1}} \cdot R_0$ and the new variable τ instead of t was introduced. If R_0 is the maximum radius of the cavity these equations as well as in the case of spherical symmetry allow the expressions for the pulsation period T of the cavity containing detonation products to be obtained:

$$T_1 = \sqrt{2/3} \cdot \Gamma(7/6) \cdot \Gamma(1/2) \cdot \Gamma(5/3) = 1,485; \quad T_{2,3} = \sqrt{\ln b}$$

where $b = 2h_0$ or r_0 and the index corresponds to the above-mentioned equations. According to the theoretical dependence for T_2 the period of the pulsation is the essential function of h. However, the experiments with the detonation cords (diameters are 0.65, 1.65 and 3 mm) carried out for the explosion depth interval from 0.21 to 3.5 m didn't confirm this dependence. The expression for T_1 proved to be rather exact. Since the values T_1 and R_0 are known from the experiments, it is not difficult to find r for the model of a cylindrical layer if $T_3 = T_1$. According to the experimental and theoretical data its value approximately equals $10^3 \cdot R_1$ [2] where R is the charge radius and determines the limited area which is involved in the motion at the cylindrical explosion in the liquid surrounding the charge.

The analysis of the asymptotic approximation assumed for the case $\nu = 1$ shows that it admits some arbitrariness for the numerical coefficient α in the equation of pulsation

$$R(1-\dot{R}/c)\ddot{R} + \alpha\dot{R}^2(1-\dot{R}/3c) = \frac{\omega_0}{2}(1+\dot{R}/c) + \frac{R\dot{\omega}_0}{c}(1-\dot{R}/c) \qquad (6)$$

According to some estimates made within the framework of incompressible liquid this coefficient α equals 0.75, 1 or 1.25 for the following approaches used for deriving the equation: $c \to \infty$ in (5), the energy conservation law or the substitution of the approximated expression for the velocity potential $\varphi \simeq 2R^{3/2}\dot{R}/r^{1/2}$ into the Cauchy-Lagrange integral and transition to the cavity wall.

For determining the cavity parameters and checking the pulsation equation the numerical calculation of (6) was carried out for three values of α and for two types of isentropes of detonation products: $\gamma = 3 = const$ and variable values of γ

taken from [6] and approximated by the following way

$$0.625 \leq \rho^{-1} \leq 1.66 \qquad \gamma = 2.78; \quad 1.66 \leq \rho^{-1} \leq 2.51 \qquad \gamma = 2.14$$
$$2.51 \leq \rho^{-1} \leq 5.0 \qquad \gamma = 1.73; \quad 5.0 \leq \rho^{-1} \leq 20.0 \qquad \gamma = 1.36$$
$$\rho^{-1} > 20 \qquad \gamma = 1.26$$

In this case $P_R = 1.295 \cdot 10^5$ atm corresponds to $\rho^{-1} = 0.625$. The calculated results are presented in Fig. 1 ($h = t\, c_0 / R_1$). Curves 1 – 1 correspond to $\alpha = 0.75 - 1.25$ and $\gamma = 3$. Curves 2 – 2" correspond to the same values of α but γ is variable in this case. The experimental value of maximum is marked by a cross. The dependence 2' is denoted by points, especially in order to emphasize the curve nearest to the experimental data. Every dependence is bounded from the right by a vertical line determing the stop moment of the cavity expansion. The inclination of the experimental dependence is shown by a dotted line for comparison.

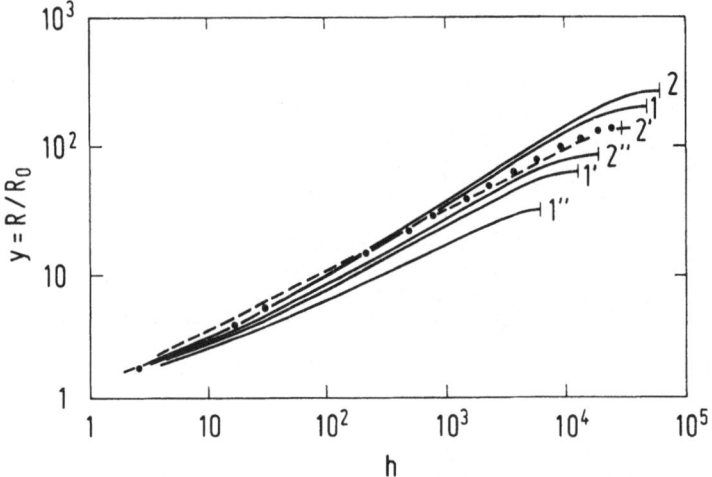

Fig.1 Influence of α and γ on character and parameters of cylindrical cavity pulsation

Presented in Fig. 2 is the same dependence 2 for three pulsations and the experimental data for $d = 1.65$ mm (x) and $d = 0.65$ mm (·). The calculated and experimental data within the region $y \geq 10$ are in satisfactory agreement. Below the main characteristics for the cylindrical cavity pulsation are presented (the experimental data are marked by *):

$$y_* \simeq 1.5 \cdot h_*^{0,45} \quad \text{for} \quad 30 \leq h_* \leq 10^4,$$
$$R_* \simeq 321 \cdot R_1^{0,55} t_*^{0,43} \text{cm}, \quad 2 \cdot 10^{-4} R_1 \leq t_* \leq 6.67 \cdot 10^{-2} R_1 c,$$
$$y_{*,1}^0 = 135, \quad y_1^0 = 141; h = 3.10 \quad / t_{*,1}^0 = 0.2\, R_0 c /, \quad h_1^0 = 3.10;$$

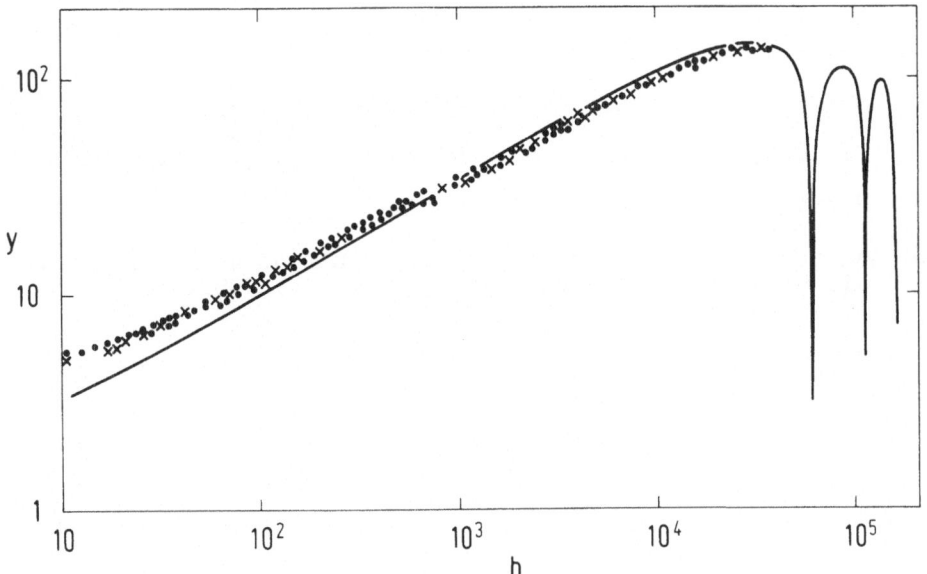

Fig.2 Comparison between experimental and theoretical data on cylindrical cavity dynamics in a boundless compressible liquid

$$E_{*_1} = 0.22 \ E \ , \quad E_1 = 0.218 \ E \ , \quad E_2 = 0.14 \ E_o, \quad E_3 = 0.11 \ E_o;$$

$$T_{*_1} = T_1 = 0.4 \ R_1 c, \quad T_2 = 0.33 \ R_1 c, \quad T_3 = 0.3 \ R_1 c.$$

The indexes 1 - 3 denote the number of the pulsation period, zero above is the maximum value of the parameter, R_1 is measured in centimeters, E_o is the heat of explosion per unit length of the charge, T is the pulsation period, E is the energy remaining in the detonation products after expansion. On the basis of the obtained data the expression for the pulsation period can be written in the form

$$T_i \simeq 1,635 \cdot (\rho_o \cdot \beta_i \cdot E_o)^{1/2} \cdot p_\infty^{-1}$$

where $\beta_i = 0.218, \ 0.14, \ 0.11$ is the portion of the explosive energy consumed for a radial motion during the first, second or the third pulsation.

REFERENCES

1. Kedrinskii, V.K., O pulsatsii tsilindricheskoi gasovoi polosti v bezgranichnoi zhidkosti. In: Dinamika sploshnoi sredi, Vip. 8, Izd. Instituta gidrodinamiki SO AN SSSR, 1971.

2. Kedrinskii, V.K., O nekotorikh priblizhennikh modeliakh odnomernoi pulsatsii tsilindricheskoi polosti v neszhimaemoi zhidkosti. FGV, 1976, No.5.

3. Kedrinskii, V.K., Kuzavov, V.T., Dinamika tsilindricheskoi polosti v szhimaemoi zhidkosti. PMTF, 1977, No.4.

4. Koul, R., Podvodnie vzrivi. Moskva, IL., 1950.

5. Gilmore, F.R., The growth and collapse of a spherical bubble in a viscous compressible liquid. Calif. Inst. of Tech.
   Hydrodyn. Lab. Rept. 26 - 4, 1952.

6. Kuznetsov, N.M., Shvedov, K.K., Izentropicheskoe rasshire-
   nie produktov detonatsii geksogena. FGV, 1967, No.2.

Part II

Sound Waves and Bubbles

Sound and Shock Waves in Bubbly Liquids

L. van Wijngaarden

Technische Hogeschool Twente
Enschede, The Netherlands

Introduction

When small bubbles containing air or other gases are dispersed
in water, the resulting acoustical properties differ a great
deal from those of water even if the gas concentration by volume
is only a few percent. The speed of sound is significantly lower
than in pure water and effects of dispersion and attenuation
brought about by the presence of the bubbles alter the propagation of sound waves. In what follows the most important features of propagation of sound in bubbly liquids are dealt with.
In many practical applications such as propagation of pressure
waves in steam-water mixtures or under water explosions, the
amplitude of the waves is not small. Attention therefore has to
be given to waves of finite amplitude as well. These finite
amplitude waves may develop into shock waves the structure of
which is discussed.

1. Acoustic Waves

When small air bubbles are dispersed in water the compressibility of the resulting liquid is much larger than that of the
pure liquid. The compressibility in its turn determines the
velocity of propagation of sound waves of low frequency, which
can be calculated as follows. Denoting densities of gas and
liquid with ρ_g and ρ_l respectively, concentration of air by
volume with α, we have under the assumption of no relative velocity between bubbles and liquid

$$\frac{\alpha \rho_g}{\rho_l(1-\alpha) + \rho_g \alpha} = \text{const.} \tag{1.1}$$

which expresses that in a unit mass of the mixture the mass of
gas is constant. In the denominator in (1.1) appears the density of
the mixture, which we denote with ρ,

$$\rho = \rho_l(1-\alpha) + \rho_g \alpha. \tag{1.2}$$

When the thermodynamic changes in the gas phase are isentropic
and the frequency is low enough for the local pressure in the
liquid to be equal to the local pressure in the gas we have

from (1.1)

$$\frac{\alpha p^{1/\gamma}}{\rho_l(1-\alpha)+\alpha\rho_g} = \text{const.}, \tag{1.3}$$

where γ is the ratio between specific heats in the gas. When taking $\rho_g \ll \rho_l$ and neglecting the liquid compressibility it follows from (1.1) that

$$\alpha \approx \frac{\rho_l-\rho}{\rho_l} \tag{1.4}$$

and that, from (1.3) and (1.4),

$$\frac{dp}{d\rho} = c_o^2 = \frac{\gamma p}{\rho_l\alpha(1-\alpha)}. \tag{1.5}$$

Of course this relation for the sound speed needs correction when α is very close either to zero or to unity but otherwise it provides an accurate prediction for the sound speed in liquid with dispersed gas. For water (1.5) gives at 1 bar for α = 4%, 8% and 10% the values c_o = 60 m/s, 44 m/s and 40 m/s. These velocities are substantially lower than the sound velocities in either air or water. See Fig.1.

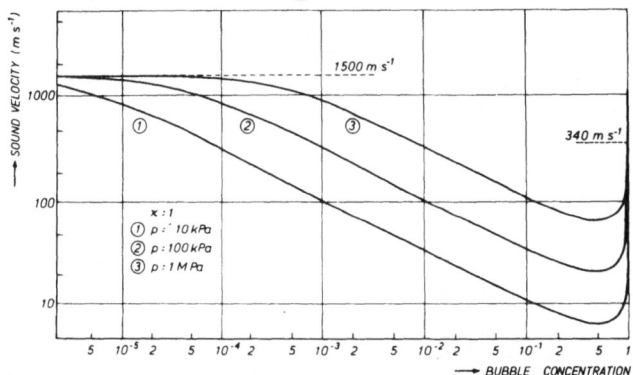

Fig.1 Sound velocity in water as function of free air concentration with pressure as parameter ($\kappa = \gamma$)

Equation (1.5) holds not only for bubbly flow but also for other topologies in which the liquid forms the continuous phase. There have been numerous experimental verifications of (1.5), for a survey see [1], and (1.5) may be considered as the fundamental relation for the propagation of sound waves in liquids containing gas. Corrections and modifications on (1.5) are necessary when physical processes need to be considered which are left out of account in the simple argument that leads to (1.5).

Relative Radial Motion

Brooks and other small free surface currents emit sound, first
described by MINNAERT in his famous paper "Musical Air Bubbles
and the Sound of Running water" [2]. The sound is due to volume
vibrations of small air bubbles. We consider, for convenience,
spherical bubbles in an incompressible viscous liquid. When R(t)
denotes the radius, motion in the liquid due to spherical vibrations $\dot{R}(t)$ of the bubble is described by the velocity potential

$$\phi = - \frac{R^2\dot{R}}{r} , \qquad\qquad\qquad (1.6)$$

where r is the distance of a point in the liquid to the centre
of the bubble. The pressure in the liquid can be calculated
with the aid of Bernoulli's theorem. The continuity of normal
stresses at the interface between air and liquid gives, p_o being
the pressure in the liquid far away from the bubble and p_g being
the pressure of the gas in the bubble,

$$p_g - p_o = \rho_l \{R\frac{d^2R}{dt^2} + \frac{3}{2}(\frac{dR}{dt})^2 + \frac{4\nu}{R}\frac{dR}{dt}\}. \qquad\qquad (1.7)$$

In this equation ν is the kinematic viscosity, 10^{-6} m^2/s in water.[1]
Further, surface tension has not yet been included[1]. From this
equation we can obtain the angular frequency of free volume oscillations of a bubble, by linearizing about an equilibrium state
characterized by p_o and R_o. One finds in this way for isentropic
behaviour of the gas content [2]

$$\omega_B = \frac{1}{R_o} (\frac{3\gamma p_o}{\rho_l})^{\frac{1}{2}}. \qquad\qquad\qquad (1.8)$$

The associated frequency $f_B = \omega_B/2\pi$ is for a bubble of 1 mm
radius 3.3×10^3 c/s. Bubbles with radii between o.1 mm and 10 mm
have natural frequencies within the audible range. The question
whether turbulence will excite resonant bubble oscillations must
be answered negatively [3], because at the resonance frequency
of the small bubbles that persist, the length scale of pressure
fluctuations is too small for the pressure oscillations to be coherent over the bubble surface.

[1]
The occurrence of a flow potential as in (1.6) to describe the
flow of a viscous liquid is no paradoxical situation: Such a flow
exactly satisfies the Navier-Stokes equations. It does not, in
general, satisfy the no slip condition at the surface of a body.
In the present case only the radial velocity is prescribed at
the interface, so that potential flow exactly satisfies the boundary conditions.

An important kind of small bubbles are cavitation bubbles, filled with vapour and air. When also surface tension is taken into account the quantity p_O in (1.8) should be replaced by $(p_O - p_V + 4/3\frac{T}{R_O})$, T being the coefficient of surface tension and p_V the vapour pressure. No real value of the frequency exists when

$$p_O < p_V - 4/3\frac{T}{R_O} , \qquad (1.9)$$

which is the wellknown threshold for vaporous cavitation. When the frequency of acoustic waves in bubbly liquids approaches the resonance frequency, as given in (1.8), the inertia of liquid radially accelerated and decelerated becomes important. As a result the velocity of propagation is no longer as in (1.5) but depends on the frequency. The radial motion is damped by viscosity, and taking (1.7) for the difference between the pressure in the liquid and the pressure in the bubbles leads for e.g. the velocity u in the mixture to

$$\frac{\partial^2 u}{\partial t^2} = c_O^2 \frac{\partial^2 u}{\partial x^2} + \frac{c_O^2}{\omega_B^2} \frac{\partial^4 u}{\partial x^2 \partial t^2} + \frac{4c_O^2 \nu}{\omega_B^2 R_O^2} \frac{\partial^3 u}{\partial x^2 \partial t} . \qquad (1.10)$$

Inserting in (1.10) solutions of the form exp i(kx-ωt) gives the dispersion equation

$$\frac{k^2}{\omega^2} = \frac{1}{c_l^2} + \frac{1 - \omega^2/\omega_B^2 + i\delta\omega/\omega_B}{c_O^2 \{(1-\omega^2/\omega_B^2)^2 + \delta^2\omega^2/\omega_B^2\}} . \qquad (1.11)$$

The first term on the right hand side of (1.11) $1/c_l^2$, where c_l is the velocity of sound in pure liquid, does not follow from (1.10) but must be included when the liquid is no longer regarded as incompressible. The damping mechanisms are accounted for by the logarithmic decrement δ, which is for the viscous term included in (1.10)

$$\delta_{visc} = \frac{4\nu}{\omega_B R_O^2} . \qquad (1.12)$$

In general damping of thermal nature with a corresponding δ_{th}, to be discussed below, is more important. Another cause for damping is the acoustic radiation of a bubble oscillating with frequency ω. The associated logarithmic decrement δ_{ac} is, see e.g. [4],

$$\delta_{ac} = \frac{\omega^2 R_O}{\{1 + (\frac{\omega R_O}{c_l})^2\}\omega_B c_l} . \qquad (1.13)$$

Of course, if a mixture contains not bubbles of the same size but a distribution of bubbles of different size the expression

corresponding with (1.11) is much more complicated and results
from integration of (1.11) weighed with the pertinent number
density over the bubble distribution. The real part of ω/k as
given by (1.11) constitutes the velocity of propagation of acoustic waves with angular frequency ω, the imaginary part gives the
damping. They are represented for some particular values of R_0
and α for air bubbles in water in Fig.2 and Fig.3, taken from
[5]

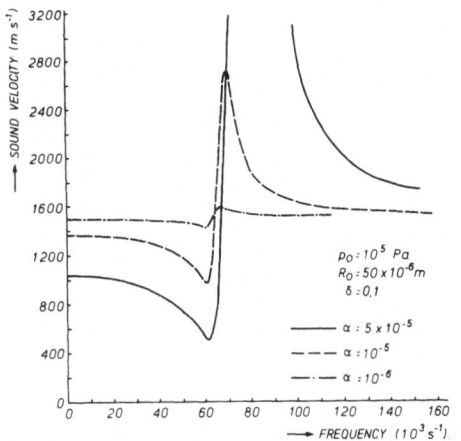 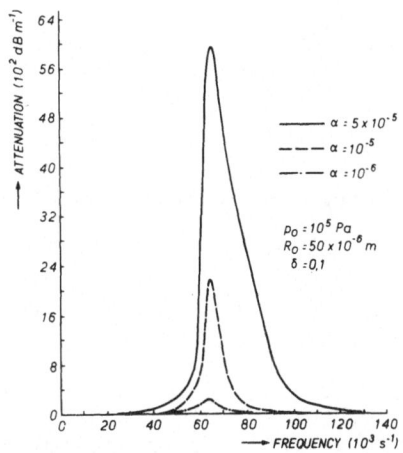

Fig.2 Sound speed in bubble   Fig.3 Attenuation of sound in
mixture as function of frequency   bubble mixture

These relations appear, see [1] to be fairly well supported by
experimental observations. Note that whereas the velocity of
sound tends to the value given in (1.5) for $\omega \to o$ it rises in the
vicinity of $\omega = \omega_B$ above c_l and tends to c_l for $\omega \to \infty$. The strong
dispersion of acoustic waves produced by the dynamic response
of bubbles is of importance for cavitation noise [5] and also
for the propagation of sound waves near the sea's surface where
air is entrained mostly in the form of bubbles.

Relative Translational Motion

Only very tiny bubbles will move with the liquid locally. In
general there will be as a result of the different densities
a relative motion. This relative motion has profound effects
on the mechanical behaviour of two phase flows in which only
very partial insight has been gained till thus far. The study
of these effects meets with severe difficulties even in the
study of simple configurations. We will discuss here some relatively simple aspects. In establishing conservation equations
for the suspension, the strategy is to deduce properties of the
suspension on a scale comprising many bubbles from the relative
flow around an individual constituent in the suspension. This
relative flow depends apart from the particle under consideration
also on all other particles. The difficulties referred to have
to do with this interaction. The interaction can be neglected
when the average distance between particles (or bubbles in our
case) is so large that each bubble is unaware of the presence

131

of others. For number density n the average distance between bubble centres is $n^{-1/3}$ and interaction may be neglected when this is very large with respect to a representative bubble radius R,

$$n^{-1/3} >> R, \text{ or}$$

$$\alpha = \frac{4}{3}\pi n R^3 << 1. \tag{1.14}$$

Under this restriction we have with bubble velocity v and liquid velocity u,

$$\frac{d}{dt} m (v-u) = -\rho_l V \frac{du}{dt}, \tag{1.15}$$

m being the virtual mass of a bubble and V its volume,

$$m = \tfrac{1}{2}\rho_l V. \tag{1.16}$$

Equation (1.15) expresses that with $\rho_g << \rho_l$ the force on a bubble partly due to the pressure gradient in the uncoming flow, partly due to the relative motion, is zero. The validity of (1.15) can be rigorously proven when u is uniform in space. For nonuniform flow, for example brought about by the presence of other bodies only approximate results [6] are available. Insertion of (1.16) in (1.15) gives v = 3u for rigid spheres in potential flow. If surface active agents are absent potential flow represents the flow around the bubble fairly accurately at high Reynolds numbers. The calculation of the velocity of sound leads with v = 3u to [7]

$$c_f^2 = \frac{\gamma p(1+2\alpha)}{\rho_l \alpha(1-\alpha)}. \tag{1.17}$$

In the derivation (1.1) cannot be used, but the conservation of mass for the gas has to be formulated as

$$\frac{\partial}{\partial t}(\rho_g \alpha) + \frac{\partial}{\partial x}(\rho_g \alpha v) = 0. \tag{1.18}$$

Relation (1.17) is valid for $\alpha << 1$. Its verification cannot be done directly by measurement of the velocity of propagation of sound waves because for α of the order of a few percent the difference between the right hand sides of (1.5) and (1.17) is within the experimental accuracy. There is indirect verification possible however. The motion described by (1.15) is impeded by a viscous force. For an air bubble in water this is for low Reynolds numbers

$$W = 4\pi\mu R(v-u), \tag{1.19}$$

as follows from the Hadamard-Rybczynski relation [8] and for high Reynolds number

$$W = 12\pi\mu R(v-u) \tag{1.20}$$

based on potential flow around the bubble, see [7] and [8]. In these expressions μ is the viscosity of the liquid. The relations (1.19) and (1.20) hold when the bubbles are spherical, and for (1.20) the liquid must be devoid of surface active agents. When We adopt in spite of these restrictions (1.20) as resistance law, bearing in mind that in many practical circumstances the Reynolds number for the relative motion is large (1.15) becomes

$$\frac{d}{dt} m(v-u) + 12\pi\mu R(v-u) = -\rho_{\underset{\sim}{l}} V \frac{du}{dt}. \tag{1.21}$$

The competition between the inertia term with the virtual mass and the viscous term gives a relaxation time τ for adjusting the bubble velocity v to the velocity u of the liquid, given by

$$\tau = \frac{R^2}{18\nu}. \tag{1.22}$$

When relative motion is started at t=o viscous resistance is uneffective at times

$$t \ll \tau. \tag{1.23}$$

During this time interval the relative motion obeys (1.15). Straight forward linearization of the governing equations including translatory motion as governed by (1.18), (1.21), the continuity and momentum equations for the mixture and the conservation of bubble numbers

$$\frac{\partial n}{\partial t} + \frac{\partial}{\partial x} (nv) = o, \tag{1.24}$$

but excluding relative radial motion, gives, [9], for the acoustic pressure $p_g - p_o = \tilde{p}$,

$$\frac{\partial}{\partial t} \{c_f^2 \frac{\partial^2 \tilde{p}}{\partial x^2} - \frac{\partial^2 \tilde{p}}{\partial t^2}\} + \tau^{-1} \{c_o^2 \frac{\partial^2 \tilde{p}}{\partial x^2} - \frac{\partial^2 \tilde{p}}{\partial t^2}\} = o. \tag{1.25}$$

In unsteady essentially one dimensional flow, phenomena with a time scale less than τ are governed according to this relation by a wave equation with wave speed c_f. At larger time scales waves propagate with speed c_o. The influence of the higher order terms in this case can be appreciated by writing (1.25) as

$$\frac{\partial}{\partial t} \{(c_f \frac{\partial}{\partial x} - \frac{\partial}{\partial t}) (c_f \frac{\partial \tilde{p}}{\partial x} + \frac{\partial \tilde{p}}{\partial t})\} + \tau^{-1}\{(c_o \frac{\partial}{\partial x} - \frac{\partial}{\partial t}) (c_o \frac{\partial \tilde{p}}{\partial x} + \frac{\partial \tilde{p}}{\partial t})\} = o.$$

Now we consider a wave which travels almost undisturbed to the right with velocity c_o. Inserting $c_o \, \partial/\partial x + \partial/\partial t = O(\varepsilon)$, $\varepsilon \to o$, in the above expression and requiring $p = p_o$ at $x \to \infty$, gives upon integration

$$\frac{\partial \tilde{p}}{\partial t} + c_o \frac{\partial \tilde{p}}{\partial x} - \tfrac{1}{2}\tau(c_f^2 - c_o^2) \frac{\partial^2 \tilde{p}}{\partial x^2} = o. \qquad (1.26)$$

This shows the interesting fact that the higher order terms have a diffusive effect on the wave. The diffusion coefficient is $\tfrac{1}{2}\tau(c_f^2 - c_o^2)$, which is, see (1.5) and (1.17) proportional to α for $\alpha \ll 1$, and therefore small. A linear wave will eventually diffuse completely as a result of this diffusion but in a non-linear wave this diffusion may be resisted by nonlinear effects as we shall see later.

Heat and Mass Transfer Between Bubbles and Liquid

In the derivation of a relation for the sound velocity (1.5) and for the natural frequency (1.8) it has been assumed that the thermodynamic changes within the bubbles are isentropic. Of course, this is a simplification because actually during compression and rarefaction of the gas inside the bubbles heat is exchanged with the surrounding liquid. Since the heat capacity of this liquid is huge in respect with the heat capacity of the bubbles, the liquid can be considered as being of constant temperature. Within the bubble appreciable temperature differences occur within a depth $(D_h/\omega)^{\frac{1}{2}}$ from the interface, where ω is again the angular frequency of the sound wave and D_h the thermal diffusivity, 2.10^{-5} m^2/s for air. When this is very small with respect to the radius R of the bubble,

$$(D_h/\omega R^2) \ll 1 \qquad (1.27)$$

the thermodynamic changes in the bubble are nearly adiabatic. In the opposite case where the depth of penetration of heat is large with respect to R no appreciable temperature differences occur inside a bubble which therefore behaves isothermally. In these limits the pressure p_g of the gas inside the bubbles is a function of ρ_g alone. In intermediate circumstances p_g is a function of ρ_g and the temperature T_g. In particular for an harmonic motion there is a phase difference between p_g and ρ_g. This makes in principle a representation as

$$p_g \sim \rho_g^n \qquad (1.28)$$

impossible. Nevertheless for practical purposes the thermodynamic behaviour can [4] be represented as in (1.28). As a result of heat exchange between bubbles and liquid acoustic waves are damped and hence there is a thermal contribution δ_{th} to the logarithmic decrement δ in (1.10). In fact this contribution to δ dominates over δ_{visc} in (1.12) and δ_{ac} in (1.13), as shown for $R_o = 10^{-5}$m and for $R_o = 10^{-3}$m in Fig. 4, taken from [4].

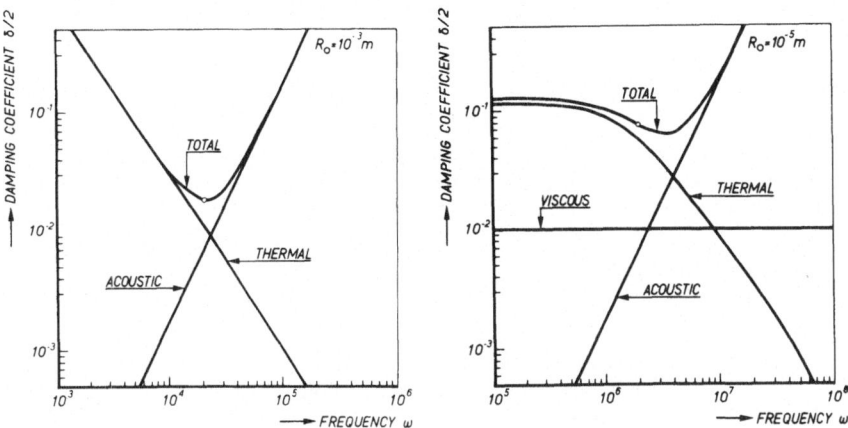

<u>Fig. 4</u> Damping coefficient δ for small amplitude forced oscillations of an air bubble in water (from [4])

In analogy with (1.12) the thermal contribution to δ is often represented by

$$\delta_{th} = \frac{4\nu_{th}}{\omega_B R_o^2} .$$

(1.29)

It is [10] however not possible to give ν_{th} as a simple analytical expression. An approximate expression is

$$\delta_{th} = \frac{3(\gamma-1)}{2(\omega/2 D_h)^{\frac{1}{2}} R_o} ,$$

(1.30)

obtained by PFRIEM, see [1]. Mass transfer is of importance in various ways though perhaps less for acoustic waves as for cavitation phenomena. In vaporous cavitation condensation and vaporization are, of course, of essential interest, but also mass transfer by diffusion affects cavitation phenomena significantly. Mass transfer by diffusion in sound waves is not of very great importance. Rectified diffusion leads [4] to a very slow growth of a bubble in a sound wave. In [4] it is shown that for a bubble of 1 mm to grow to twice this radius 10^6 sec are needed with a relative pressure amplitude $(P_{max} - P_{average})/P_{average} = 0.25$.

2. Waves of finite amplitude

While we have been discussing till thus far waves of infinitesimal amplitude, one should be aware that in many practical situ-

135

ations like in the case of cavitation or under water explosions
amplitudes are no longer small. For understanding the effects
of finite amplitude, it is convenient to leave out momentarily
the effects of relative motion and heat transfer. We then have
a homogeneous mixture with a sound velocity given by (1.5). To
such a mixture we may apply the theory of one phase gasdynamics
as described for instance in [11]. With the quantity

$$\sigma = \int \frac{c_o(\rho)d\rho}{\rho} \tag{2.1}$$

one of the outcomes of this is that the equations of motion can
be written in characteristic form as

$$\{\frac{\partial}{\partial t} + (u \pm c_o) \frac{\partial}{\partial x}\}\{u \pm \sigma\} = 0. \tag{2.2}$$

With the aid of (1.4) and (1.5) both ρ and c_o can be expressed
in terms of α and in this way σ can be found. With isothermal
behaviour and $\alpha \ll 1$ we find

$$\sigma = c_o(\alpha_o - \alpha). \tag{2.3}$$

Now we look in particular at a simple wave, that is a wave travelling towards an undisturbed medium. Along such a wave $u = \sigma$
because all 'left' characteristics, that is characteristics
coming from the undisturbed region, carry along them a zero
value of $u - \sigma$. Then along the right characteristics u and σ are
constant, whence these characteristics are straight. The simple
wave is described therefore, introducing

$$\frac{p-p_o}{p_o} = \tilde{p}, \text{ with} \tag{2.4}$$

$$\frac{\partial \tilde{p}}{\partial t} + c_o \frac{\partial \tilde{p}}{\partial x} + c_o \tilde{p} \frac{\partial \tilde{p}}{\partial x} = 0, \tag{2.5}$$

because $u + c_o = \sigma + c_o = c_o(1 + \frac{\alpha_o - \alpha}{\alpha}) = c_o(1 + \frac{p-p_o}{p_o})$, where we use that
with $\gamma = 1$ and $\alpha \ll 1$ (1.3) gives $p\alpha$ =constant. This indicates that
a wave of finite amplitude, while travelling forward, becomes
steeper because each disturbance $\tilde{p}$ travels with a speed $c_o(1+\tilde{p})$.
This is sometimes called amplitude dispersion. A wave of ex-
pansion continuously broadens by amplitude dispersion. Here,
as with acoustic waves, relative motion and dissipation due to
the presence of the bubble will affect the wave as it propagates
forward. A complete analysis of this is far beyond analytical
or numerical treatment but this becomes better if we restrict
to an approximation, apart from $\alpha \ll 1$, in which the wave ampli-

tude is of moderate magnitude so that we may neglect terms which
are of the third order in this amplitude. This has been done for
acoustic waves in gases and gravity waves on water of finite
depth [11],[12]. A similar analysis including amplitude dispersion
and frequency dispersion by relative radial motion leads [1] to

$$\frac{\partial \tilde{p}}{\partial t} + c_o \frac{\partial \tilde{p}}{\partial x} + c_o \tilde{p} \frac{\partial \tilde{p}}{\partial x} + \tfrac{1}{2} \frac{c_o^3}{\omega_B^2} \frac{\partial^3 \tilde{p}}{\partial x^3} = 0. \qquad (2.6)$$

In this equation c_o is the value following from (1.5), with $\gamma=1$,
in the undisturbed state. Equation (2.6) is known as the Korteweg-
de Vries equation, originally [13] derived for water waves but
later found to be of general validity for waves in which dispersion and nonlinearity compete [14]. Since all dissipative effects
are disregarded in (2.6) there is little chance that in practice
waves will obey this equation. Including dissipation as represented by δ in (1.11) for linear waves leads in an analogous
way to

$$\frac{\partial \tilde{p}}{\partial t} + c_o \frac{\partial \tilde{p}}{\partial x} + c_o \tilde{p} \frac{\partial \tilde{p}}{\partial x} + \tfrac{1}{2} \frac{c_o^3}{\omega_B^2} \frac{\partial^3 \tilde{p}}{\partial x^3} - \tfrac{1}{2} \frac{\delta c_o^2}{\omega_B} \frac{\partial^2 \tilde{p}}{\partial x^2} = 0. \qquad (2.7)$$

It is assumed here that the dissipative effect represented by
the last term on the left hand side of (2.4) is small in the
sense that it is comparable with the nonlinear term. Solutions
of (2.7), evolving from a given $\tilde{p}$ disturbance at t=o behave
roughly like those of (2.6) with the important difference that
(2.7) has *steady* solutions in the form of a shock wave in which
the tendency to steepen is balanced by dispersion. Such shock
waves, weak because the formulation of (2.6) and (2.7) rests
on the assumption of moderate amplitude, have the over all thickness of

$$d \sim \frac{R_o}{\alpha^{\frac{1}{2}}} (\frac{P_o}{\Delta p})^{\frac{1}{2}} , \qquad (2.8)$$

as follows from balancing the corresponding terms on the left
hand side of (2.7), and the general behaviour [9] of an undular
bore. The predictions of the extensive theory [11] , [15] for
(2.7), have been verified experimentally in the case of bubbly
flows both for initial conditions in the form of a step , see [1]
as for more general forms of initial pressure profile from which
for example,[16],a finite number of solitons evolves. We have
seen during the discussion of waves of small amplitude in section
2 that relative translational motion leads to the existence of
a second characteristic speed, c_f in (1.18), next to c_o defined
in (1.5). Moreover we found that at times t>τ the higher order
terms, associated with c_f, act as diffusion in the wave governed
by the lower order terms as made plausible by (1.26).

A linear wave is eventually diffused by this but in a nonlinear wave this diffusion may be balanced by nonlinear steepening. Such a balance can in bubbly liquids only occur at small pressure differences over the wave, as the following argument shows. A steady uniform wave travels in an essentially adiabatic flow with velocity U given by its Mach number as

$$U^2/c_o^2 = \frac{p_1/p_o - 1}{1-(p_1/p_o)^{1/\gamma}} ,$$

(2.9)

where p_1 and p_o are the pressures at both ends of the wave, $p_1/p_o > 1$. When this wave is not a shock wave but is smoothed out by diffusion the small wavelets in front, see Fig. 5,

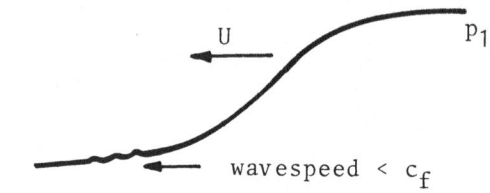

p_o

wavespeed $< c_f$

Fig.5 Smooth wave travelling from right (pressure p_1) to left (pressure p_o). The speed of the whole wave is U. The wavelets in front travel at maximum at velocity c_f.

travel at maximum at a speed c_f given in (1.17) . Since these wavelets are stationary with respect to the wave as a whole, this can at most travel with $U=c_f$. Inserting c_f^2 , from (1.17) for U^2 in (2.9) gives with the aid of (1.5)

$$p_1/p_o \leq 1+ \frac{4\alpha\gamma}{1+\gamma} .$$

(2.10)

For pressure ratio's in the range indicated by (2.10) smooth waves have been observed experimentally [9]. The thickness is, as follows from balancing the nonlinear term $c_o \partial \tilde{p}/\partial x$ with the diffusion term $(c_f^2 - c_o^2)\tau \frac{\partial^2 \tilde{p}}{\partial x^2}$ and taking (2.10) into account, of the order of magnitude

$$\tilde{d} \sim c_o \tau.$$

(2.11)

At pressure ratios larger than (2.10) the wave cannot be completely smooth. What happens provides an interesting possibility of verification of the existence of the two speeds of sound c_o and c_f: A smooth wave of the type just discussed is preceded by a thin shock wave (with thickness of the order indicated in (2.8) which bridges the gap between the pressure p_o and a

pressure p^+, say, such that p_1/p^+ satisfies (2.10) with p^+ in
stead of p_0. In analogy with relaxation shocks in gasdynamics
[12] this may be called a partly dispersed shockwave in contrast
to a fully dispersed wave. In [9] examples of these are shown.
In a shock tube of considerable length the propagation of shock
waves was investigated. In Fig.6 a partly and a completely
dispersed wave are shown taken from the results. The shocks are
as in ordinary shock tubes produced by puncturing a seal separating a high pressure region from a low pressure region. Of
course, for a dispersed or partly dispersed wave to occur the
time elapsed since the puncturing of the seal must exceed τ.
For times $t<\tau$ or, equivalently, distances $x<c_f\tau$, the shock
wave is governed by (2.7) however with c_f instead of c_0. Then
the shock looks like in Fig. 6a.

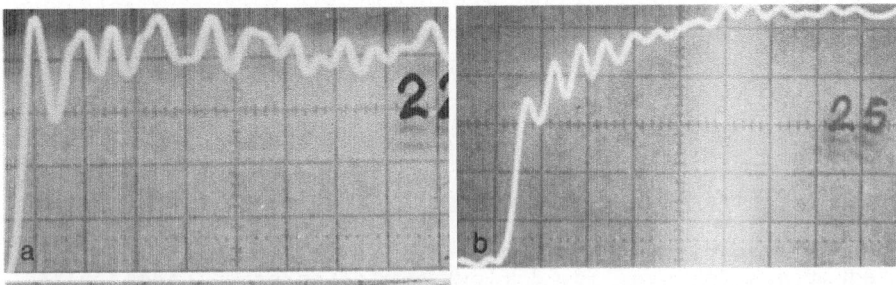

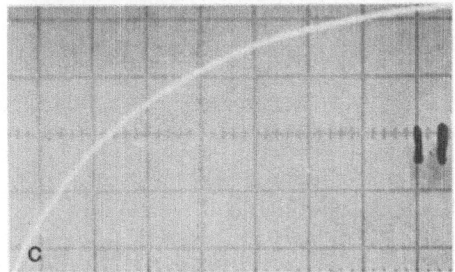

Fig. 6a-c Shock waves in bubbly
liquid (a) Shock wave governed by
[2.7]: balance between nonlinear
steepening and dispersion, (b)
front like in a), at the backside
diffusion by relaxation balances
nonlinear steepening (c) wave,
completely smoothed out by diffusion due to relaxation

3. Conclusion

We have seen that under very restrictive circumstances acoustic
waves in bubbly liquids can be analyzed. For waves of infinitesimal amplitude the restrictions are that all bubbles are spherical with the same radius in undisturbed conditions, that the concentration by volume is small, that the frictional force on a
bubble can be calculated with Levich's model (equation (1.20),
and the added mass is given by (1.16). For practical application
it would be helpful if more were known about the shape which
bubbles assume under the joint influence of surface tension,
viscous and inertia forces. Dispersion and dissipation in mixtures that are not very dilute could be studied if more were
known about effects of hydrodynamic and other interaction between bubbles constituing a bubbly suspension. The techniques
for calculating transport properties in other areas of suspension
technology [16] may be useful here. An example is the calculation [17] of the virtual mass of a bubble in the presence of
other bubbles. From the foregoing sections it follows that,

apart from waves of infinitesimal amplitude, waves of moderate amplitude can be analyzed provided the effects of dispersion and dissipation are small. When this is no longer the case, very little can be said about the propagation of coherent waves because nonlinear bubble dynamics as reviewed in [4] must be coupled to nonlinear conservation equations in the mixture.

References

[1] Van Wijngaarden, L. One dimensional flow of liquids containing small gas bubbles. Ann. Rev. Fluid Mech. 4,369,1972.
[2] Minnaert,M. On musical air bubbles and the sound of running water. Phil.Mag. 16,235, 1933.
[3] Crighton,D.G. & Ffowcs Williams, J.E. Sound generation by turbulent two-phase flow. J.Fluid Mech. 36,585,1969.
[4] Plesset, M.S. & Prosperetti, A. Bubble dynamics and cavitation. Ann. Rev. Fluid Mech. 9,145,1977.
[5] Oldenziel, D.M. Bubble cavitation in relation to liquid quality. Thesis Twente University of Technology, Enschede, 1979.
[6] Voinov, V.V., Voinov, O.V. & Petrov, A.G. Hydrodynamic interaction between bodies in a perfect incompressible fluid and their motion in non uniform streams. Prikl.Math.Mech.37,680, 1973
[7] Van Wijngaarden, L. Some problems in the formulation of the equations for gas/liquid flows. In 'Theoretical and Applied Mechanics' edited by W.T. Koiter. North Holland Publishing Company, Amsterdam 1977.
[8] Levich, V.G. Physicochemical Hydrodynamics Prentice Hall 1962.
[9] Noordzij, L. & Van Wijngaarden, L. Relaxation effects, caused by relative motion, on shock waves in gas-bubble/liquid mixtures, J. Fluid Mech. 66, 115, 1974.
[10] Prosperetti, A. Thermal effects and damping mechanisms in the forced radial oscillations of gas bubbles in liquids. J. Ac. Soc. Am. 61,17,1977.
[11] Whitham, G.B. Linear and Nonlinear Waves. Wiley & Sons, New York, 1974.
[12] Lighthill, M.J. Viscosity effects in sound waves of finite amplitude. In surveys in Mechanics. Edited by G.K. Batchelor and R.M. Davies. Cambridge 1956.
[13] Korteweg, D.J. & De Vries,G. On the change of form of long waves advancing in a rectangular canal and on a new type of long stationary waves. Phil. Mag. 39, 422, 1895.
[14] Miura, R.M. Korteweg-De Vries equation and generalizations. I.A remarkable explicit nonlinear transformation. J.Math. Phys. 9,8,1202,1968.
[15] Karpman, V.I. & Cap, F.F. Non-linear waves in dispersive media. Pergamon 1975.
[16] Kuznetsov, V.V., Nakoryakov, V.E., Pokusaev, B.G. & Shreiber, I.R. Propagation of perturbations in a gas-liquid mixture. J. Fluid Mech. 55, 85, 1978.
[17] Van Wijngaarden, L. Hydrodynamic interaction between gas bubbles in liquid. J. Fluid Mech. 77, 27, 1976.

On the Amplification of Modulated Acoustic Waves in Gas-Liquid Mixtures

F.H. Fenlon

Department of Engineering Sciences and Mechanics and
The Applied Research Laboratory
The Pennsylvania State University
University Park, PA 16802, USA

J.W. Wonn

Advanced Systems Laboratory, Westinghouse Research and Development Center
Pittsburgh, PA 15235, USA

1. Introduction

The problem of acoustic wave propagation in gas-liquid mixtures has been extensively investigated during the past twenty years from many different perspectives. three of which can be briefly summarized as follows:
   (i) WIJNGAARDEN'S [1]-[2] theoretical analysis of solitary wave formation
due to the establishment of a dynamic-steady-state between the nonlinear
waveform distortion incurred by finite-amplitude plane waves and bubble induced dispersivity, as recently confirmed experimentally by KUZNETSOV et al.
[3].
   (ii) Acoustic radiation via cavitation and turbulence as analyzed for
example by CRIGHTON and FFOWCS WILLIAMS [4], LYAMSHEV [5], BOGUSLAVSKII et al.
[6], and by KRASILNIKOV and KUZNETSOV [7].
   (iii) Harmonic and intermodulation frequency generation via acoustic wave
propagation in gas-liquid mixtures as analyzed by ZABOLOTSKAYA and SOLUYAN
[8], SAFAR [9], and ZABOLOTSKAYA [10].

   Although the analytical concepts underlying these areas of investigation
are similar and complementary this paper has been primarily influenced by
the latter since it pertains more directly to the application with which we
are concerned, namely the detection of entrapped gas bubbles in a fluid via
their effect on small-signal acoustic waves. As WELSBY and SAFAR [11] have
shown this effect manifests itself via a considerably more rapid change of
the second-order acoustic properties relative to corresponding changes in the
first-order properties when an acoustic signal propagates in a gas-liquid
mixture. It is therefore possible to determine the presence of bubbles by
monitoring the appearance of nonlinearly generated harmonic and intermodulation frequency components during the propagation of small-signal bifrequency
acoustic waves through the two-phase fluid. During the past decade this phenomenon has led to the development of acoustic threshold detectors for regis-
tering the presence of bubbles in a liquid. Before discussing some of the
diverse applications of this technology however, we will now provide a review
of the underlying theoretical background followed by a brief summary of the
respective roles played by physical mechanisms such as nonlinearity, absorption, and dispersion on the amplification of modulated small signal acoustic
waves in gas-liquid mixtures.

2. Theory

In a gas-liquid mixture denoting the equilibrium density of the liquid by ρ_0,
the small-signal speed-of-sound by c_0, the ambient pressure by P_0 and the gas

specific heat ratio by γ_G, the linear compressibilities C_G and C_L of the gas and liquid phases are respectively given by $C_G = 1/\gamma_G P_0$ and $C_L \simeq 1/\rho_0 c_0^2$. If an acoustical disturbance propagating in the fluid gives rise to a change in volume v' of a bubble relative to its initial volume v_0, the relative change v'/v_0 can be defined as a function of the excess pressure p' by re-expressing the bubble dynamic equation via the method-of-successive approximations, correct to second-order terms as

$$v'/v_0 = L_\tau^{-1} [C_G p' + \beta_G c_G^2 M(p')] \tag{1}$$

where $\beta_G = (\gamma_G + 1)/2$ is the second-order nonlinear coefficient of the gas,

$$M(p') = (L_\tau^{-1} p')^2 + [(L_\tau^{-1} \dot{p}')^2 + (L_\tau^{-1} p'\ddot{p}')^2]/(6\beta_G \omega_0^2) \tag{2}$$

and $L_\tau = 1 + \mu\partial_\tau + \partial_\tau^2$, $\tau = \omega_0 t$ $\tag{3}$

In this notation t is time, ω_0 is the resonant (angular) frequency of a bubble of radius R_0, given approximately (i.e. neglecting surface tension) by $\omega_0 = (3\gamma_G P_0/\rho_0 R_0^2)^{1/2}$, and μ is the bubble damping coefficient given by $\mu = \omega_0 R_0/c_0 + 4\eta/\rho_0 \omega_0 R_0^2$, where η is the coefficient of viscosity of the liquid.

Assuming a uniform distribution of bubble sizes and small bubble concentration n, the volumetric strains of the gas and liquid phases are given respectively by $(V'/V_0)_G = nv'$, and $(V'/V_0)_L = - (C_L p' - \beta_L c_L^2 p'^2)$, where $V'=V-V_0$; $V = 1/\rho$ being the specific volume and $\beta_L = (1 + B_L/2A_L)$ the second-order nonlinear coefficient of the liquid. Summing these partial strains to obtain the total volumetric strain and substituting (1) in the resulting expression, with $\delta = nv_0$ (i.e. the volumetric void fraction), gives

$$V'/V_0 = nv' - (C_L p' - \beta_L C_L p'^2) \tag{4}$$

$$= \delta L_\tau^{-1} [C_G p' + \beta_G c_G^2 M(p')] - (C_L p' - \beta_L C_L^2 p'^2) \tag{4a}$$

$$\simeq (1 + \delta C_G/C_L) p' + (\beta_L C_L^2) (1 + \delta\beta_G C_G^2/\beta_L C_L^2) p'^2 \quad , \quad \omega/\omega_0 << 1 \tag{4b}$$

where it must be emphasized that (4b) only holds when the highest significant frequency of the acoustic field is significantly below bubble resonance. Eq. (4b) is similar to that derived by WELSBY and SAFAR [11]. It shows that whereas the second-order term begins to change significantly when, $\delta\beta_G C_G^2/\beta_L C_L^2 \gtrsim 1$, the first-order term is not affected unless $\delta C_G/C_L \gtrsim 1$. For example in the case of air bubbles in water, where $C_G/C_L \simeq 1.5 \times 10^4$ and

$\beta_L/\beta_G \simeq 3$, it follows that the second-order term begins to be significantly affected for void fractions $\delta \lesssim 10^{-8}$, whereas corresponding changes in the first-order term only occur whenever $\delta \lesssim 10^{-4}$.

Now in order to estimate the magnitude of bubble induced second-order changes in the field of an acoustical disturbance it is necessary to consider the appropriate acoustic wave equation. As shown by WESTERVELT [12] the propagation of acoustic waves in a fluid under isentropic conditions can be described correct to second-order terms by the equation.

$$(1 + \Lambda\partial_t) \, \nabla^2 p' = \partial_t^2 \, (\rho - p'^2/\rho_o c_o^4) \quad , \quad \partial_t = \partial/\partial t \tag{5}$$

where $\Lambda = 2c_o \, \alpha_\omega/\omega^2$ and α_ω is the thermoviscous attenuation coefficient at frequency ω. Since $\rho = 1/V = \rho_o \, (1 - nv') - 2nv'p'_c/{}^2 + p'/c^2 + (B_L/2A_L)p'^2/\rho_o c_o^4$ from (4), correct to second-order terms, substitution of this expression in (5) with the aid of (1) gives the second-order nonlinear wave equation for a gas-liquid mixture.

$$\left\{(1 + \Lambda\partial_t) \, \nabla^2 - c_o^{-2} \, \partial_t^2\right\} \, p' = - \, \rho_o \, \partial_t^2 \, \left\{\beta_L C_L^2 \, p'^2 + \delta L_t^{-1} \, [2C_L C_G p'^2 + \right.$$
$$\left. \beta_G C_G^2 \, M(p'))]\right\} \tag{6}$$

In the frequency-domain, again correct to second-order terms, this becomes

$$(\nabla^2 + \kappa_\omega^2)p'_\omega = k^2 \, F_\omega \left\{ \beta_L C_L p'^2 + \delta \left[\frac{2C_G p'^2 + (\beta_G C_G^2/C_L) \, M(p')}{1 + i\mu\Omega - \Omega^2} \right] \right\} \tag{7}$$

where $p'_\omega = F_\omega(p')$, $F_\omega(\ ) = \displaystyle\int_{-\infty}^{\infty} (\ ) \, e^{i\omega t} \, dt$, $k = \omega/c_o$

and $\kappa_\omega = (k - i\alpha_\omega) \left(1 + \dfrac{\delta C_G/C_L}{1 + i\mu\Omega - \Omega^2}\right)^{\frac{1}{2}}$, $\Omega = \omega/\omega_o$ $\qquad$ (8)

We now consider the propagation in a gas-liquid mixture of a small-signal carrier wave of frequency ω_* which is sinusoidally modulated by a wave of frequency $\omega_-/2$, whose waveform at the radiator is given by the equation $p'(o,t) = 2p_* \sin(\omega_* t) \cos(\omega_- t/2)$, with $\omega_-/\omega_* \ll 1$. In this instance, with the aid of (2) it is clear that at the nonlinearly generated sum and difference frequencies $\omega_+ = (\omega_1 \pm \omega_2)$, where $\omega_1 = (\omega_* + \frac{\omega_-}{2})$ and $\omega_2 = (\omega_* - \frac{\omega_-}{2})$, (7) becomes

$$(\nabla^2 + \kappa_{\omega_+}^2) \, p'_{\omega_+} = \beta_L C_L k_+^2 \, S_{\omega_+} \, F_{\omega_+}(p'^2) \tag{9}$$

143

with $\kappa_{\omega_\pm} = (k_\pm - i\alpha_{\omega_\pm})\left(1 + \dfrac{\delta C_G/C_L}{1 + i\mu\Omega_\pm - \Omega_\pm^2}\right)^{\frac{1}{2}}$ $\qquad\qquad$ (10)

and $S_{\omega_\pm} = \left[1 + \dfrac{2\delta\beta_L^{-1}(C_G/C_L)}{1 + i\mu\Omega_\star - \Omega_\star^2} + \dfrac{\delta(\beta_G/\beta_L)(C_G/C_L)^2(1 - \Omega_\star^2/6\beta_G)}{(1 + i\mu\Omega_\star - \Omega_\star^2)(1 \pm i\mu\Omega_\star - \Omega_\star^2)(1 + i\mu\Omega_\pm - \Omega_\pm^2)}\right]$

$\qquad\qquad$ (11a)

$\cong \left[1 + \delta\left\{2\beta_L^{-1}(C_G/C_L) + (\beta_G/\beta_L)(C_G/C_L)^2\right\}\right]$, for $\Omega_\star, \Omega_\pm \ll 1$ $\quad$ (11b)

where $\Omega_\pm = \omega_\pm/\omega_0$, $\Omega_\star = \omega_\star/\omega_0$

For the case of progressive waves (9) can be expressed in the form of ZABOLOTSKAYA and KHOKHLOV'S [13] paraxial wave equation in the frequency domain, the solution of which for weak axisymmetric finite-amplitude interactions along the beam axis relative to the carrier field can be deduced as shown by FENLON [14] giving

$$p'_{\omega_\pm}(R,0)/p'_{\omega_\star}(R,0) = (\omega_\pm/\omega_\star)^2 (\beta_L C_L p_\star k_\star r_\star/2) S_{\omega_\pm} H_{\omega_\pm}(R) \qquad (12)$$

where $R = r/r_\star$ is the range normalized with respect to the carrier wave Rayleigh distance $r_\star = k_\star a^2/2$, a being the radius of the radiating piston. Since the following discussion is not concerned with the form of the function $H_{\omega_\pm}(R)$ rather than including it here the interested reader can readily derive it from a slight modification of the equations in [14]. Although the preceding analysis has been applied to the case of sum and difference-frequency formation it should be noted that it is equally applicable to the case of harmonic generation by simply putting $\omega_1 = \omega_2$, $\Omega_\star = \omega_1/\omega_0$ and $\Omega_\pm = 2\omega_1/\omega_0$.

We now proceed to consider the physical significance of (10) and (11). From inspection of the former, it is evident that dispersion only begins to play a role for $\delta C_G/C_L \gtrsim 0.1$ (e.g. if $\delta \gtrsim 10^{-3}$ for the case of air bubbles in water at atmospheric pressure). Likewise, it is clear that absorption, which results both from thermoviscous attenuation and bubble damping is only significant for $\alpha_\omega r > 1$ and at frequencies close to bubble resonance. As for nonlinearity, inspection of (11) and (12) reveals that the respective roles of bubble and hydrodynamic nonlinearity are described by the function $S_{\omega_\pm}$ which can be represented heuristically as $S_{\omega_\pm} = 1 + $ (effective nonlinearity of the bubbles/hydrodynamic nonlinearity). Hence, if $|S_{\omega_\pm}|$ is significantly greater than unity the bubble nonlinearity is dominant, and conversely as it approaches unity. Again, from inspection of (11) it is clear that the bubble nonlinearity is dominant (i) as the void fraction δ increases at frequencies $\omega_\star$ and $\omega_\pm$ such that $\Omega_\star, \Omega_\pm < 1$, (ii) as the pressure P_0 decreases causing the compressibility of the gas, C_G, to increase, and (iii) at frequencies ω_* or ω_+ equal to the resonant frequency of the bubbles (i.e. Ω_*, $\Omega_+ = 1$). On the other hand the hydrodynamic nonlinearity begins to predominate (i) as the void fraction δ approaches zero, (ii) at extremely high values of the pressure P_0 when the compressibility of the gas, C_G, has decreased significantly, (iii) at frequencies ω_*, ω_+ significantly above bubble resonance (i.e. Ω_*, $\Omega_+ \gg 1$), and (iv) whenever the driving frequency ω_* is such that the non-linearity of the gas is cancelled by the dynamic response of the bubbles, which according to (11) occurs at $\Omega_* = (6\beta_G)^{\frac{1}{2}}$ or, for example, when $\omega_* = 2.7\,\omega_0$ for the case of air bubbles in water.

The above remarks, which are in agreement with ZABOLOTSKAYA'S [13] plane wave second-harmonic analysis, may be further illustrated by considering the specific case of air bubbles in water at atmospheric pressure for which (11) becomes

$$S_{\omega_{\pm}} = 1 + \frac{\delta(0.86 \times 10^8)\,(1 - 0.14\,\Omega_*^2)}{(1 - \Omega_*^2)^2\,(1 - \Omega_{\pm}^2)} \quad ; \quad \Omega_*,\ \Omega_+ \neq 1$$

If Ω_*, $\Omega_+ \ll 1$ then $S_{\omega_+} = 1 + (0.86 \times 10^8)$, and bubble nonlinearity is dominant for $\delta \geq 10^{-7}$. Again, in the case of the sum-frequency component, ω_+, if Ω_*, $\Omega_+ \gg 1$, then $S_{\omega_+} \sim 1 - \delta \times 10^7/(\Omega_*\Omega_+)^2$, implying that bubble non-linearity is dominant only if $\delta > (\Omega_*\Omega_+)^2 \times 10^{-6}$, which in turn implies that $(\Omega_*\Omega_+)$ should not exceed 100 for most practical experiments. Finally, in the case of the difference-frequency component, ω_-, if $\Omega_* \gg 1$, but $\Omega_- \ll 1$, then $S_{\omega_-} \sim 1 + \delta \times 10^7/\Omega_*^2$, implying that bubble nonlinearity is dominant if $\delta > \Omega_*^2 \times 10^{-6}$, or in general if $\Omega_* < 100$.

Since the decision to monitor the sum or difference-frequency depends primarily upon the anticipated bubble size that one wishes to detect we merely note at this point that in water at atmospheric pressure the resonant frequency of air bubbles is $f_* = \omega_*/2\pi = (3.27 \times 10^3/R_0)$ kHz/μm. Thus, if $R_0 = 10^3$ μm, then $f_* = 3.27$ kHz, so that, in order to detect the presence of such bubbles it would appear preferable to monitor a difference-frequency signal below bubble resonance. An appropriate difference-frequency in this instance would be, for example, 1 kHz generated when a carrier wave (whose frequency, possibly 5 kHz, is above bubble resonance), sinusoidally modulated by a 1/2 kHz signal, undergoes nonlinear self-interaction while propagating through the gas-liquid mixture. On the other hand, for significantly higher values of f_* due to smaller bubbles, both sum and difference-frequency signals can be monitored for dual threshold detection.

It should be noted that although the analytical model which has been presented is based on an assumed uniform distribution of bubble sizes, experiments have shown that the results remain applicable to moderately nonuniform distributions, where in such instances ω_o represents the mean resonant frequency.

Over the past decade considerable effort has gone toward applying these phenomena to solving practical engineering problems. Engineering applications of liquid-gas nonlinear effects have generally attempted to capitalize on two important features: (i) the promise of detecting extremely small void fractions, and (ii) the inherent immunity to reflections from interfaces. The following section highlights some areas where practical applications hold promise. The diversity of the application areas is exemplified by the wide range of liquid media to be addressed: blood, transformer, oil, liquid metal, and mud. In each case the major thrust is to improve the capability for detecting or monitoring populations of microbubbles which can potentially lead to serious problems.

3. Applications

3.1 Diver Embolism Detection

In the late 1960's investigations by WELSBY and SAFAR [11], and by TUCKER and WELSBY [15], demonstrated the potential of exploiting nonlinear ultrasonic effects in the human body to aid in the early detection of gas bubbles associated with decompression sickness (the "bends") in divers. Standard deep-diving practice requires that the diver, returning to the surface, must interrupt his ascent at specified depths to allow the lungs to remove gas accumulated in the bloodstream. Under other circumstances, a diver is returned to the surface more rapidly and immediately undergoes treatment in a decompression chamber. The treatment consists of rapid repressurization, followed by gradual depressurization to permit a safe release of gasses which have accumulated in the tissues.

In both cases, the rate at which the diver is depressurized is carefully controlled to minimize the possibility of injury from the bends. The ability to monitor the diver's body for bubbles during decompression is considered to be an important part of insuring a safe, effective, and efficient decompression schedule.

Nonlinear ultrasonic methods have been experimentally investigated by MARTIN, HUDGENS and WONN [16] for detecting gas emboli in divers' tissues during decompression. Fig. 1 shows a block diagram of a typical detection apparatus. An oscillator and amplifier drive a transmitting transducer (Tx) with a pure sine wave of frequency f_o. The launched waves traverse the test medium and are received by a receiving transducer (Rx). The levels of both the received fundamental (f_o) and second harmonic $(2f_o)$ are detected and recorded on magnetic tape for later playback. Preliminary experiments, using a 10 watts/m^2 transmitted ultrasonic 150 kHz pure tone, demonstrated detection of 2 to 15 micron diameter gas bubbles in a water bath. Further experiments were carried out at 350 watts/m^2 to 500 watts/m^2 on insonified beef parts with electrolytically produced bubbles injected through the venous system. Bubbles as small as 8-10 μm and void fractions of 10^{-7} (volume ratio of bubble to insonified liquid in blood vessel) were detected. The apparatus was then used to monitor human subjects during decompression following simulated dives in a hyperbaric chamber [16]. Fig. 2 shows a sample of bubble indications obtained from a diver's thigh muscle during an experiment.

146

It is evident from inspection of these results that the second harmonic peaks
induced by the presence of bubbles are readily detectable, thus pointing to
the eventual development of automatic 'early warning' ultrasonic threshold
detectors for rapid deep sea diver decompression.

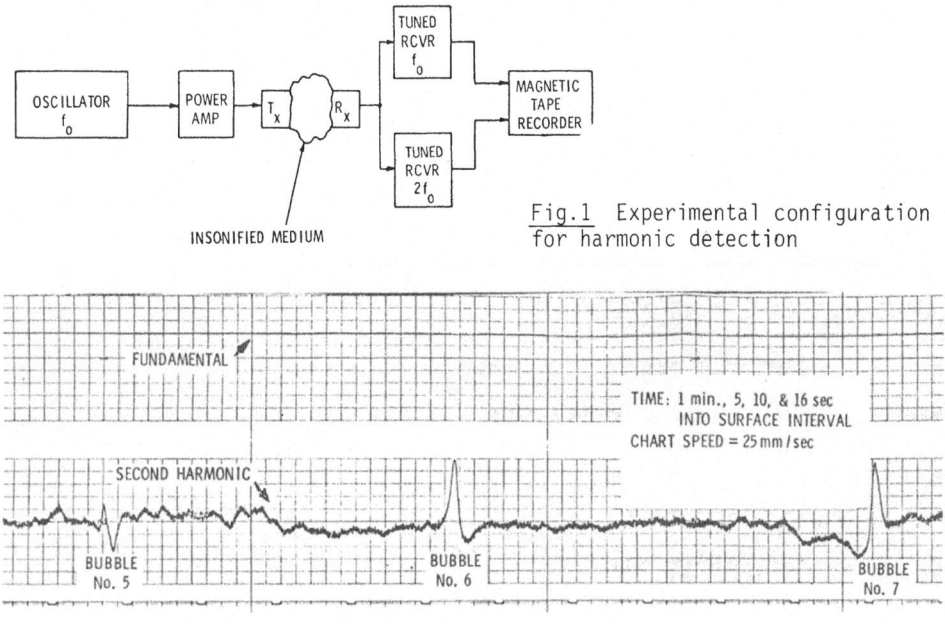

Fig.1 Experimental configuration
for harmonic detection

Fig.2 Second harmonic strip chart recording

3.2 Bubble Detection in Large Power Transformers

Large power distribution transformers are often filled with special oil which
facilitates the dissipation of internal heat and provides important dielectric
properties. It has been found that the onset of transformer deterioration is
often accompanied by the presence of gas bubbles in the oil. Early detection
of this condition permits transformer maintenance or replacement before catastrophic failure occurs, thus avoiding potential hazard and the disruption of
service.

Nonlinear ultrasonic experiments for detecting the presence of gas bubbles
in transformer oil have recently been conducted by NOMM [17] who monitored
both harmonic and difference-frequency signals. In these experiments a controlled stream of uniformly sized bubbles with radii varying from 250 μm to
900 μm were introduced into a 0.75 x 0.28 x 0.3 m^3 glass walled tank filled
with transformer oil. For the case of harmonic detection a 9 kHz signal radiated via a 4 x 10^{-2} m diameter transducer bonded to an outer wall of the
tank by a castor oil film propagated through the bubble stream to a 3 x 10^{-2} m
diameter spherical hydrophone receiver. The latter, in turn was connected
to a high-pass filter capable of rejecting the radiated signal while passing
harmonic frequency components generated via nonlinear interaction with the
bubble stream. As shown in Figs. 3a and 3b, for a transducer drive level of
78 volts no harmonics were detected at levels at least 60 dB below that of

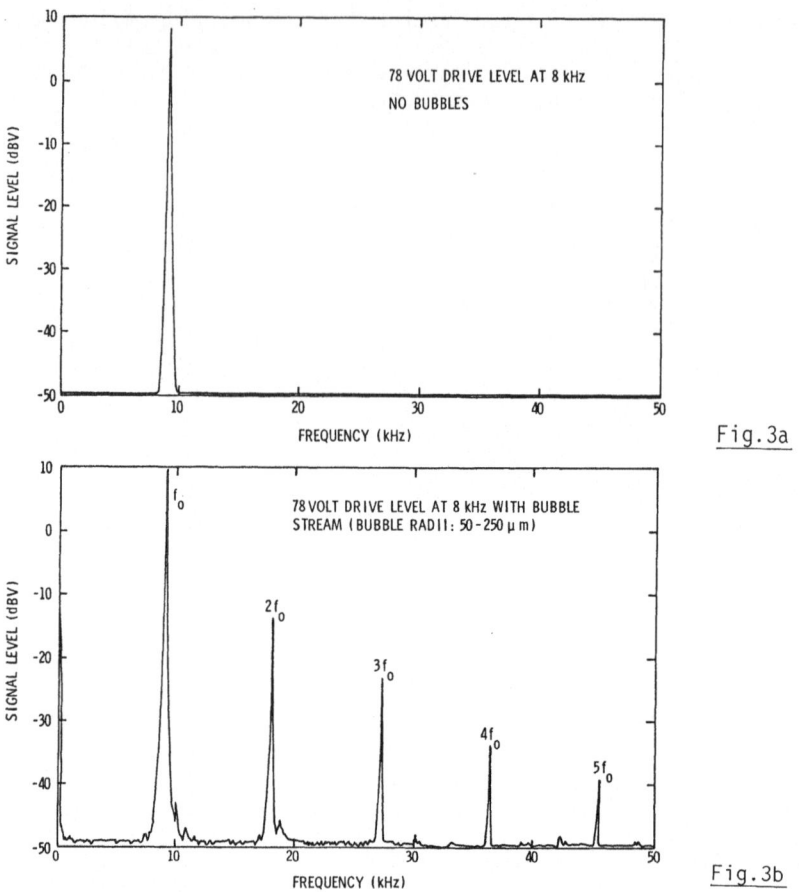

Fig.3a

Fig.3b

Fig.3 (a) Response to a 9 kHz signal in the absence of a bubble stream. (b) Response to a 9 kHz signal in the presence of a bubble stream

the fundamental in the absence of the induced bubble stream, whereas in the presence of the latter a second harmonic level 35 dB lower than that of the fundamental was measured, together with clearly discernable higher harmonics. When the applied voltage was reduced by a factor of five the second harmonic level dropped by ~12 dB in keeping with (12), and the higher harmonics fell below the background noise level.

Again, in another experiment where 9.5 kHz and 15 kHz signals of equal amplitude were simultaneously radiated into the fluid (i.e. equivalent to radiating a 12.25 kHz carrier wave sinusoidally modulated by a 2.75 kHz envelope) and the hydrophone output was low pass filtered to suppress the carrier but permit reception of the 5.5 kHz difference-frequency signal (formed via nonlinear self interaction of the signal in the bubble stream) a difference-frequency response 33 dB below the peak level of the combined carrier level was observed, together with significant harmonic levels as shown in Fig. 4.

148

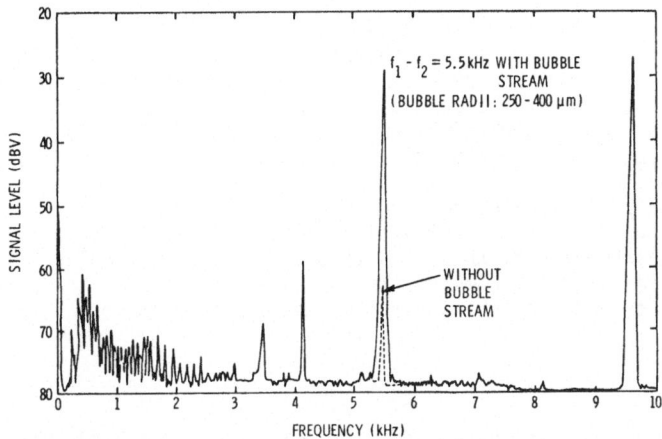

Fig.4 Response to 16 kHz and 9.5 kHz signals in the presense of a bubble stream

It should be noted that although ambiguities in the response arising from the excitation of structural resonances were not a problem in these experiments they may be more significant under practical operating conditions. Hence, the alleviation of such ambiguities via random noise excitation or frequency-swept pulses is currently under investigation.

3.3 Liquid Metal Boiling Detection

Liquid Metal Fast Breeder Reactors (LMFBR) use liquid sodium as the primary heat transport fluid. Under normal operating conditions the circulating liquid metal is free of gas bubbles. Under certain conditions it is possible to achieve undesirable sodium boiling in the vicinity of the fuel bundle. The transition from bubble nucleation to fully developed boiling can proceed quite rapidly due to the high energy density combined with decreased coolant effectiveness caused by the presence of gas voids. It is therefore important to have the earliest possible warning of the presence of microbubbles in the sodium so that remedial action can be taken in time to prevent boiling. Non-linear ultrasonic methods appear to provide the small void fraction sensitivity needed for this application. In liquid sodium the ratio of the vapour to liquid compressibilities $C_G/C_L \sim 3 \times 10^5 \times P_s$, where P_s is the saturation vapour pressure. Thus, for example, at a temperature of 400°C, $C_G/C_L \sim 145$, implying that at ultrasonic frequencies significantly below bubble resonance changes in the second-order properties will begin to be manifested for void fractions $\delta \gtrsim 10^{-4}$. However, at 600°C where $C_G/C_L \sim 10^4$ it should, in principle, be possible to detect void fractions as small as 10^{-8}. Much work remains to be done however, before the applicability of ultrasonic technology in the harsh environment of liquid sodium can be fully implemented.

3.4 Gas "Kick" Warning for Off-Shore Drilling Operations

An important part of conducting off-shore drilling operations is controlling the density of the drilling mud in the long column formed by the drill string and its casing. Failure to maintain sufficient mud density can result in a bouyant mud column which may exert a potentially damaging vertical force on the surface drilling platform. One of the problems associated with maintaining adequate mud density occurs when the drill bit encounters a gas-bearing formation. Released gas can enter the mud column, reducing the local mud density, and producing a condition referred to as a gas "kick". Early detection of this condition is advantageous to maintaining control of the drilling operation. Some experimental work has been conducted indicating that nonlinear ultrasonic equipment located near the bottom of the drill string can provide early warning of a gas kick situation by detecting an abnormally high concentration of gas bubbles in the mud. In this case, a particular advantage of the nonlinear ultrasonic method is its inherent immunity to sonic reflections from sediments and rock chips which are naturally present in the mud.

4. Conclusions

We have reviewed the theory underlying a simplified analytical model for predicting the levels of harmonic, sum, and difference-frequency signals resulting from the self interaction of monotonic and bifrequency ultrasonic waves in gas-liquid mixtures. We have also expressed this model in terms of fundamental dimensionless parameters, namely, the volumetric void fraction δ, the gas-liquid compressibility ratio C_G/C_L, the gas-liquid coefficient of nonlinearity ratio β_G/β_L, and the frequency ratios $\Omega_* = \omega_*/\omega_o$, $\Omega_+ = \omega_+/\omega_o$. Finally, we have briefly discussed four applications of nonlinear ultrasonic technology which are being actively pursued at the present time.

5. References

1. L. Van Wijngaarden, J. Fluid Mech. 33, 465 (1968).
2. L. Noordzij and L. Van Wijngaarden, J. Fluid Mech. 66, 115 (1974).
3. V. V. Kuznetsov, V. E. Nakoryakov, B. G. Pokusaev, and I. R. Shreiber, J. Fluid Mech. 85, 85 (1978).
4. D. G. Crighton and J. E. Ffowcs Williams, J. Fluid Mech. 36, 585 (1969).
5. L. M. Lyamshev, Sov. Phys. Acoust. 15, 494 (1970).
6. Yu. Ya. Boguslavskii, A. I. Ioffe, and K. A. Naugol'nykh, Sov. Phys. Acoust. 16, 17 (1970).
7. V. A. Krasilnikov and V. P. Kuznetsov., Sov. Phys. Acoust. 20, 285 (1974).
8. E. A. Zabolotskaya and S. I. Soluyan, Sov. Phys. Acoust. 18, 396 (1973).
9. M. H. Safar, 'Finite-Amplitude Wave Effects in Fluids', edited by L. Bjorno (IPC Science and Technology Press, Guildford Surry, England, 1973) pp. 174-179.
10. E. A. Zabolotskaya, Sov. Phys. Acoust. 21, 569 (1976).
11. V. G. Welsby and M. H. Safar, Acustica 22, 177 (1969).
12. P. J. Westervelt, 'Nonlinear Acoustics', edited by T. G. Muir, ARL Texas Report No. AD719936, 1969, pp. 165-181.
13. E. A. Zabolotskaya and R. V. Khokhlov, Sov. Phys. Acoust. 15, 33 (1969).
14. F. H. Fenlon, J. Sound Vib. 64, 17 (1979).
15. D. G. Tucker and V. G. Welsby, U.S. Patent No. 3,622,958.
16. F. E. Martin, J. E. Hudgens, and J. W. Wonn, Westinghouse Oceanic Division, Report No. N00014-73-C-0191, May 1973.
17. M. Nomm, Westinghouse Oceanic Division, Memo No. DEM 78-45, March 1978.

Self-Induced Transparency and Frequency Conversion Effects for Acoustic Waves in Water Containing Gas Bubbles

Yu.A. Kobelev, L.A. Ostrovsky, A.M. Sutin

Applied Physics Institute, Academy of Sciences of the USSR
Gorky, USSR

1. Introduction

The presence of gas bubbles exerts a great influence on nonlinear acoustic properties of a liquid. Theoretical estimations [1-3] as well as experimental observations of shock wave and soliton formation in acoustic impulses [1,2] have indicated the increase of the dominant nonlinear parameters by several orders of magnitude due to the presence of bubbles. In most cases, however, the spectra of the acoustic signals lie below the bubble resonance frequencies. At the same time the nonlinear resonance effects become more intense and complicated. In most real situations the bubble sizes are widely distributed and nonlinear effects are most essentially influenced by those bubbles whose resonance frequencies correspond to the spectrum of the acoustic signal. Such resonance nonlinearities have not yet been investigated beyond theoretical calculations of a medium with monoradial bubbles.

This report deals with the results of both theoretical and experimental investigations of nonlinear acoustic wave transformation in water containing bubbles with a wide range of sizes.

The behaviour of an acoustic wave in a bubble-liquid system is governed by the equation [3]

$$\Delta p - \frac{1}{c_0^2} \frac{\partial^2 p}{\partial t^2} = - \rho_0 \frac{\partial^2 V}{\partial t^2} \tag{1}$$

where p is the acoustic pressure, c_0 is the sound velocity in the liquid, ρ_0 is the equilibrium density of the medium, V is the specific volume gas content. Eq. (1) is valid for $R \ll \ell \ll \lambda$ (so that $V \ll 1$), where R is the maximal bubble radius, ℓ is the distance between bubbles and λ is the acoustic wavelength; dissipation and nonlinearity in the liquid phase are neglected.

Another relation between p and V follows from the well-known Rayleigh equation for nonlinear oscillations of a single bubble of volume ω. Taking account of only quadratic nonlinear terms we have [4]

$$\ddot{v} + \omega_0^2 v - \alpha v^2 - \beta(2\ddot{v}v + \dot{v}^2) + \nu\dot{v} = eP \tag{2}$$

151

where

$$\beta = \left(8\pi R_0^3\right)^{-1}, \quad \alpha = 3\beta(\gamma+1)\omega_0^2 , \quad e = 4\pi R_0/\rho_0 , \quad \upsilon = w - w_0 ,$$

and R_0 is the equilibrium bubble radius which is related to its resonance frequency ω_0 by

$$\omega_0 R_0 = \sqrt{3\gamma\, P_0/\rho_0}$$

where P_0 is the equilibrium pressure, γ is the polytropic index of the gas inside the bubble, υ is the loss coefficient (for harmonic oscillations $\upsilon = \omega/Q$, where Q is the quality factor of the 'bubble oscillator').

If the distribution function of bubbles $n(R_0)$ per unit volume is known then one can find the volume gas content

$$V = V_0 + \int \upsilon(R_0)n(R_0)dR_0 \tag{3}$$

Eqs. (1)-(3) form a closed system for p and V. Note, however, that we neglect the bubble-bubble interactions which are apparently responsible for the self-induced transparency effect described below.

2. Difference-Frequency Generation

Let the double-frequency acoustic pressure field with spectral amplitudes $P_1(\omega_1)$ and $P_2(\omega_2)$ act on the bubbles. As follows from (2), the component of volume oscillations of a single bubble at the difference frequency $\Omega = \omega_1 - \omega_2$ due to nonlinearity is described by the expression [4]

$$\upsilon_\Omega = \frac{e^2\left[\alpha-\beta\left(\Omega^2+\omega_1\omega_2\right)\right]P_1 P_2}{\left(\omega_0^2-\omega_1^2-i\omega_1^2/Q_1\right)\left(\omega_0^2-\omega_2^2+i\omega_2^2/Q_2\right)\left(\omega_0^2-\Omega^2-i\Omega^2/Q_\Omega\right)} \tag{4}$$

where $Q_{1,2,\Omega}$ are the corresponding values of Q. Then from integral (3) we find the variation V_Ω of the gas volume at Ω. Substituting this into (1) we obtain an inhomogeneous wave equation for the amplitude of the pressure component at the difference frequency

$$\Delta P_\Omega - \frac{\Omega^2}{c_0^2}\, P_\Omega = -\frac{\varepsilon\Omega^2}{\rho_0 c_0^4}\, P_1 P_2 \tag{5}$$

where

$$\varepsilon(\omega_1,\omega_2) = 2\pi\sqrt{\frac{P_0}{3\gamma P_0}}\int_{R_{min}}^{R_{max}} \frac{\omega_0\left[3(1+\gamma)\omega_0^2-\Omega^2-\omega_1\omega_2\right]n(R_0)dR_0}{\left(\omega_0^2-\omega_1^2-i\omega_1^2/Q_1\right)\left(\omega_0^2-\omega_2^2+i\omega_2^2/Q_2\right)\left(\omega_0^2-\Omega^2-i\Omega^2/Q_\Omega\right)} \tag{6}$$

Eq. (5) is well known in the theory of parametric acoustic arrays (see, for example, [5]). Its solutions have been investigated for many cases by analytical and numerical methods. The bubbles change only the nonlinearity parameter ε which in a homogeneous fluid is equal to 1 + B/2A.

Integral (6) is largely due to bubbles whose radii are close to the resonance ones. Using the assumptions that $\Omega \ll \omega_1$, ω_2 (then $\omega_1 \approx \omega_2 \approx \omega$, $Q_1 \approx Q_2 \approx Q$) and $Q \gg 1$ (6) yields

$$\varepsilon \approx i\pi^2 c_0^4 \left[\frac{(3\gamma+2)n(R_\omega)}{\omega^3(\Omega-i\omega/Q)} - \frac{n(R_\omega)}{\Omega^2\omega^2} \right] \tag{7}$$

If $\Omega \ll \omega/Q$ and $n_\Omega \to 0$ then (7) takes the simple form

$$\varepsilon \approx 3{,}98 \cdot 10^{-2} n(R_\omega) Q \lambda^4 \tag{8}$$

3. Experiment

We observed the generation of difference-frequency signals in a water-filled laboratory tank (40 x 40 x 80 cm³). A sketch of the experimental set-up is shown in Fig. 1.

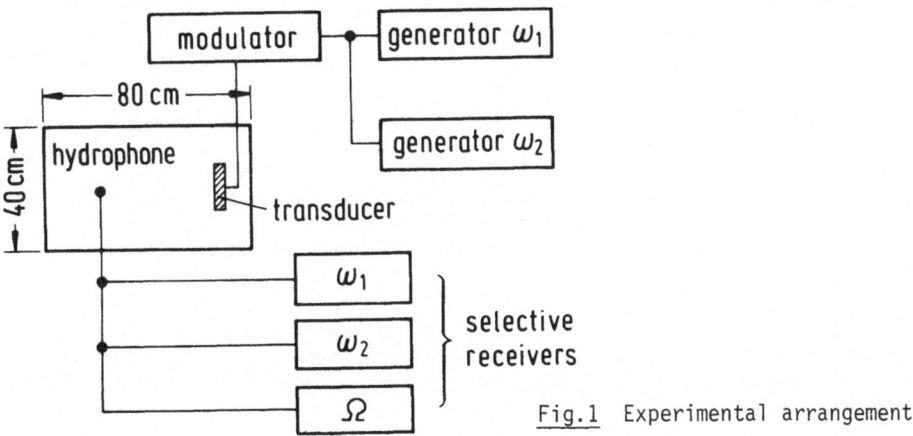

Fig.1 Experimental arrangement

The primary acoustic signal was generated by two sources at 136 and 150 kHz, modulated by single 1 ms pulses and fed to a transducer. After passing through the bubbly water the signal was received by a hydrophone situated at approximately 20 cm from the transducer. The bubbles were created by electrolysis. To investigate the size distribution of bubbles the "linear" attenuation of weak sound was measured in the frequency range 90 - 300 kHz (which corresponds to resonance bubble radii 11 - 36 μm). Using the well-known formula for the decrement of the acoustic wave amplitude α ($\alpha = n(R_0) \cdot 725 \cdot R_0^3$ [6]), which can be directly measured we can calculate the distribution $n(R_0)$. A typical function $n(R_0)$ is given in Fig. 2 (curve 1).

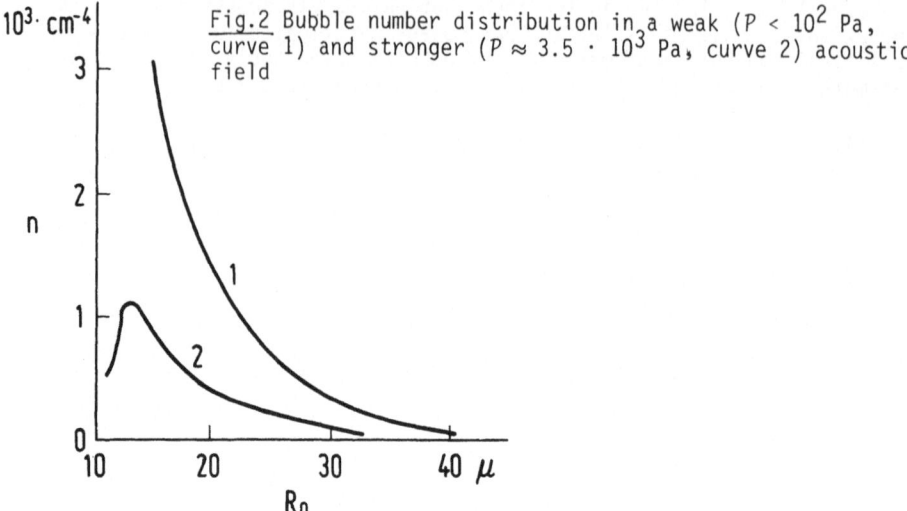

$10^3 \cdot \text{cm}^{-4}$

Fig.2 Bubble number distribution in a weak ($P < 10^2$ Pa, curve 1) and stronger ($P \approx 3.5 \cdot 10^3$ Pa, curve 2) acoustic field

R_0

The measurement of the nonlinear effect from more intense double-frequency waves was performed by filters selecting the signals at 136 and 150 kHz together with the difference-frequency of 14 kHz at the output of the hydrophone. For the amplitude P_Ω on the axis of the acoustic beam one can use the formula known for a parametric array [5].

$$P_\Omega = \frac{\varepsilon\Omega^2\, P_{01}P_{02}\, S}{8\pi\alpha\rho_0 c_0^4\, r} \tag{9}$$

(S is the beam cross-section area, r is the spacing; in our case $\alpha r \gg 1$). It may be shown from (8) and the relation between ε and α that $\varepsilon/\alpha \approx 5.3 \cdot 10^3\, \lambda Q$ so that in this approximation P_Ω is independent of $n(R_\omega)$.

In our experiment $P_{01} \approx P_{02} \approx 8.5 \cdot 10^3$ Pa and $P_\Omega \approx 15$ Pa while (9) gives $P_\Omega \approx 27$ Pa i.e. of the same order of magnitude (some discrepancy may be due to, for example, the reflections from the tank walls). Note that for clean water we should have $P_\Omega = 2 \cdot 10^{-2}$ Pa.

Note also that for the generation of the sum frequency (2ω for $\omega_1 = \omega_2$) both theory and experiment show considerably less effective energy conversion than for the difference-frequency signal.

4. Self-induced Transparency of an Acoustic Beam

This phenomena was investigated with the same set-up as shown in Fig.1 but using a continuous monochromatic signal. The dependence of the acoustic wave decrement α on its pressure amplitude was measured in different conditions. Such a dependence at a frequency of 100 kHz is shown in Fig.3. It is seen that the attenuation of a wave sharply decreases with increase of amplitude P beginning from a value P' as small as 10^3 Pa.

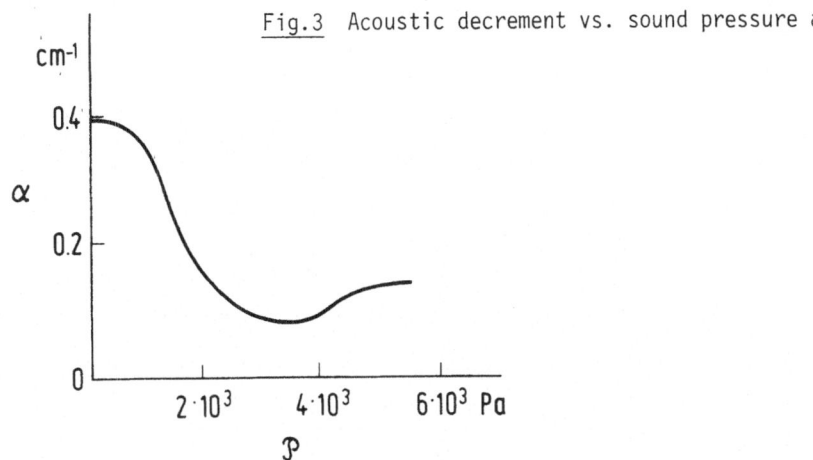

Fig.3 Acoustic decrement vs. sound pressure amplitude

Transient processes are of interest here. It turns out that the decrement α does not decrease immediately after the acoustic field is switched on but several seconds later. The restoration of wave damping after switching off the field took about 10 s. Note that the above-described experiment was made with short impulses expressly to prevent the self-induced transparency changing the bubble distribution function.

To investigate these nonlinear bubble changes we measured the attenuation of a weak probe monochromatic wave (at $f \sim 90 - 300$ kHz) propagating across the main beam. Fig.2 (curve 2) illustrates the distribution of bubbles in an acoustic wave field at a frequency of 130 kHz and an amplitude of $3.5 \cdot 10^3$ Pa. One can see that this field sharply decreases the bubble content in this range.

The explanation of the observed effects lies apparently in bunching and merging of bubbles by the Bjerkness force F_b which arises between any two oscillating bubbles and is known to be [7]

$$F_b = \frac{\rho_0}{4\pi\ell^2} < \dot{v}_1 \dot{v}_2 > \, , \qquad (10)$$

where ℓ is the distance between bubbles and $\dot{v}_i$ is the rate of volume change. It is easy to estimate from (10) the time of confluence of two bubbles. For example, bubbles of $R_1 = 10$ μm and $R_2 \approx 20$ μm at a distance of 1 cm will approach each other in about two seconds and for a pressure amplitude of 10^3 Pa. This agrees by order of magnitude with the time of self-transparency development. The restoration time of the effect is associated probably with the removal of bunched bubbles from the beam due to their rising to the water surface by the buoyancy force. Note that visual observation of comparatively large bubbles revealed both merging and forming of "bunches" of separate bubbles which dispersed after leaving the beam area.

The observed nonlinear effects may presumably be used for diagnostics of bubbles, for the enhancing of parametric array efficiency, etc.

5. References

1 G. A. Ostroumov, C. A. Druzhinin, V. M. Kriachko, A. S. Tokman, in:
  Proc. VI Int. Symp. Nonlinear Acoust., V. A. Akulichev et al. (eds.),
  Moscow 1976, vol. 1, pp 209-219.
2 V. V. Kuznetsov, V. E. Nakoriakov, B. G. Pokusaev, I. R. Schreiber,
  Akust. Zh. 23, 2, 273-278 (1977) [Sov. Phys. - Acoustics 23, 153 (1977)].
3 M. H. Safar, in: *Finite-amplitude wave effects in fluids*, L. Bjørnø
  (ed.) (IPC Press, Guildford 1974), pp 174-179.
4 E. A. Zabolotskaja, S. I. Soluyan, Akust. Zh. 18, 3, 472-474 (1972)
  [Sov. Phys.-Acoustics 18, 396 (1973)].
5 H. O. Berktay, in: *Underwater Acoustics*, vol. 2, V. M. Albers (ed.),
  (Plenum press, New York 1967).
6 The Physics of Sound in the Sea. Summary technical report. U.S. Navy,
  Washington, D.C., 1946.
7 V. F. Kazantsev. Doklady Academii Nauk SSSR 129, 1, 64-67 (1959)
  (In Russian).

Pressure Waves in a Liquid with Gas or Vapour Bubbles

V.E. Nakoryakov, B.G. Pokusaev, and I.R. Shreiber

Institute of Thermophyiscs, Siberian Branch of the USSR, Academy of Sciences
630090 Novosibirsk, USSR

1. Pressure Waves in a Liquid with Gas Bubbles

One of the main peculiarities of the propagation of acoustic disturbances
in a gas-liquid mixture is the low sonic velocity. Numerous experiments with
a low-frequency signal support the validity of the known formula [1]

$$c_0^2 = \gamma P / \rho_1 \varphi \quad . \tag{1}$$

Since the velocities of gas-liquid mixtures in tubes in power and chemical technology are as much as 10-20 m/s and in the near future they will
grow and attain the sonic speed, a necessity arises to discuss the gasdynamics of such flows. Therefore the problems of wave dynamics of such media
become particularly urgent.

The second peculiarity of the above systems is the strong dependence of
the disturbance propagation rate on the frequency [2]

$$1/c^2 = (1 - \varphi)^2/c_1^2 + \omega_0^2/c_0^2(\omega_0^2 - \omega^2) \quad . \tag{2}$$

The dispersion curve constructed according to (2), shows that the rate of
disturbance propagation decreases with the frequency ω, reaching zero at
$\omega = \omega_0$. Then there is a region where sound does not propagate, but beginning
with some frequency the sound speed in the system without dissipation becomes infinite and with the frequency increase it drops to the value in the
liquid, c_1. The main reason of strong sound dispersion in a bubble mixture
consists in the fact that a bubble gains the volume which equals to its
equilibrium at this pressure value not instantaneously, but, due to inertia,
in a certain period of relaxation.

Variations of the bubble radius in a viscous incompressible liquid are
described by the known oscillator equation

$$R \cdot R_{tt} + 3/2R_t^2 + (4\mu/R) \cdot R_t - P(R)/\rho_1 = -P_\infty/\rho_1 \quad . \tag{3}$$

Here R is the bubble radius, P_∞ is the pressure at an infinite distance from the bubble.

Since in a homogeneous model the mixture density is

$$\rho = \rho_1(1 - \varphi) + \rho\varphi \quad ,$$

and the gas void fraction φ for the fixed bubble number is a function of the radius, then assuming that the pressure at an infinite distance from the bubble is the mixture pressure, an equation is obtained relating the mixture pressure and density,

$$P'_\infty = P'_2 + \eta\rho_t + \beta\rho_{tt} \quad , \tag{4}$$

where η, β are the coefficients of effective viscosity and dispersion, respectively.

In combination with the discontinuity and Eulerian equations

$$U_t + UU_x = -\rho^{-1}P_x \tag{5}$$

$$\rho_t + (U\rho)_x = 0 \tag{6}$$

we have a system to describe the nonlinear acoustics of gas-liquid media. In [3,4,5] it has first been shown that in the case of weak dispersion and nonlinearity the behaviour of the above system under adiabatic compression is described by the Burgers-Korteweg-de-Vries equation (BKV)

$$\tilde{P}_\tau + \tilde{P}\tilde{P}_\xi - Re^{-1}P_{\xi\xi} + \sigma^{-2}\tilde{P}_{\xi\xi\xi} = 0 \quad , \tag{7}$$

where $(\Delta P/\Delta P_0) = \tilde{P}$, and τ,ξ are the dimensionless time and coordinate in the reference system with the velocity C_0; $Re = U_0 \ell_0/\eta$ is the Reynolds number, $\sigma = \ell_0 (U_0/\beta)^{\frac{1}{2}}$ is the dispersion parameter, ℓ_0, U_0 are the width and amplitude of the initial disturbance, respectively.

On the basis of numerical solutions in [6] a map of solutions of BKV equations has been constructed (Fig. 1). It is shown that at low σ and high Re (e.g., short signals and low viscosities) the dome-shaped initial signal evolves as a wave packet (Fig. 1c). At high σ the signal breaks up

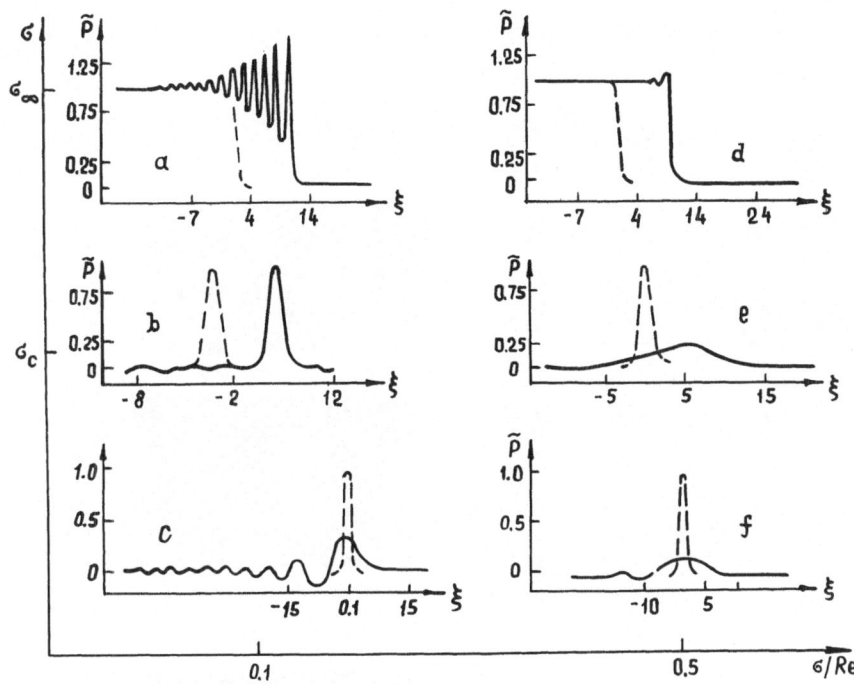

Fig. 1a-f. Map of solutions of BKV equations

into averaged waves-solitons (Fig. 1b) and at infinite σ an oscillatory shock wave propagates in the medium (Fig. 1a). Thus the signal propagation at low viscosities is described by the Korteweg-de-Vries equation

$$\tilde{P}_\tau + \tilde{P}\tilde{P}_\xi = -\sigma^{-2}P_{\xi\xi\xi} \quad . \tag{8}$$

In the case of high viscosities the BKV equation takes the form of the Burgers equation

$$\tilde{P}_\tau + \tilde{P}\tilde{P}_\xi = Re^{-1}\tilde{P}_{\xi\xi} \quad . \tag{9}$$

In this case the signal evolves as shown in the right-hand side of Fig. 1.

Experimental verification of the theory was performed in a shock two-phase tube [7,8]. Gas bubbles of the same radius ranging from 0.05 to 2 mm, uniformly fill the tube. The signal is formed by breaking the diaphragm which separates the high- and low-pressure chambers, and was fixed by wall pressure gauges.

Oscillographic recordings are given in Fig. 2 in direct comparison with the calculated results.

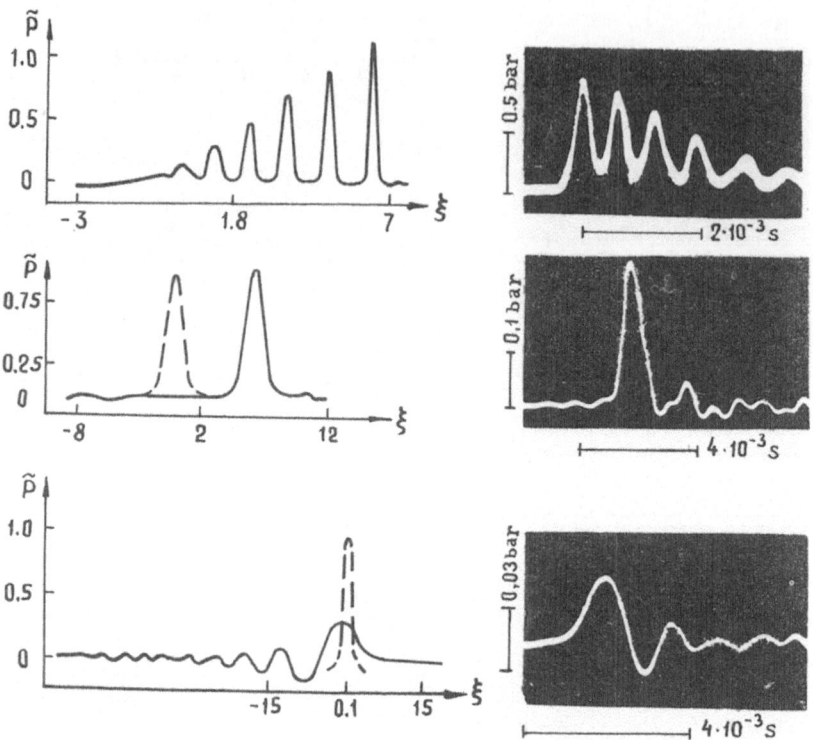

Fig. 2. Comparison of calculated (8) and experimental data at $R_0 = 1.4 \cdot 10^{-3}$m, $\varphi_0 = 0.01$

Recent results show that when one takes into account the liquid compressibility the signal evolution is described by the so-called two-wave equation

$$P_{tt} - C_0^2 P_{xx} + 2\beta \cdot C_1^{-2}(P_{tt} - C_1^2 P_{xx})_{tt} = 2C_0^2(\Delta P/P_0)^2_{xx}$$

$$-2k \cdot C_1^{-2}\rho_0^2(P_t^2)_{tt} - 2\mu_1 \cdot C_1^{-2}(P_{tt} - C_1^2 P_{xx})_t \quad .$$

$$(10)$$

At slow processes, low frequencies and great times of relaxation the equation transforms into the Korteweg-de-Vries one and at high frequencies and small times of relaxation into the Klein-Gordon equation whose possible application was reported in [9].

The most interesting effect here is the appearance of a term in a low-frequency approximation, of the type $\alpha(dP/dx)(d^2P/dx^2)$, which can be responsible for the sharpening of solitons and the increase of signal amplitude observed in the experiments with strong waves.

A numerical calculation has been performed of two-velocity (Fig. 3a-d), Klein-Gordon and Korteweg-de-Vries equations. In the solution of the two-velocity equation it has been shown that the disturbance breaks up into two groups of signals propagating at a sonic speed in the liquid, i.e., a "forerunner" and a signal in the mixture. The frequency of the main signal and the bubble resonance frequency are close, while that of the "forerunner" is by C_1/C_2 times higher. Ibid (Fig. 2a-e) experimental results are shown [10] in a shock tube (refer also to [7]).

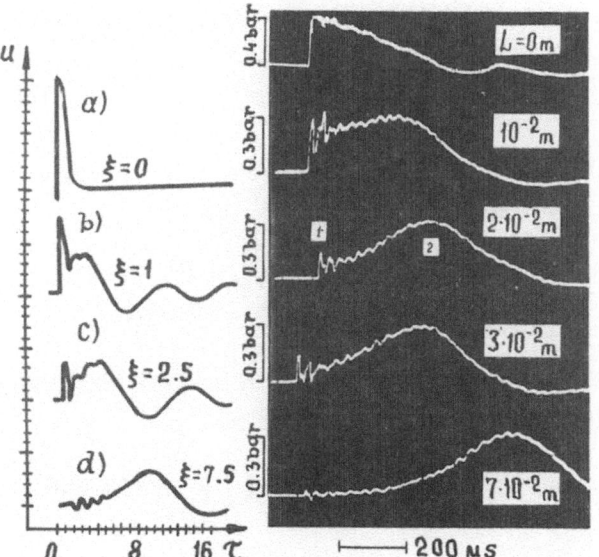

Fig. 3. Comparison of solution (10) with experiment [11] at $R_0 = 1.5 \cdot 10^{-3}$m, $\varphi_0 = 6 \cdot 10^{-3}$ at various distances from the signal entrance into medium

2. Pressure Waves in a Liquid with Vapour Bubbles

For a vapour-liquid mixture on the saturation line and slow processes one can assume a mechanism of sound propagation only due to the phase transition. This mechanism leads to the sound speed described by the Landau relation

$$c_1^2 = Lp^2/\rho_1^2 C_v B^2 T^3 \quad ,$$

161

where L is the latent heat of evaporation, B is the gas constant, T is the temperature.

But the experiments show that in practice sound propagates at a sound speed in a gas-liquid mixture (1).

The theory of sound propagation in a vapour-liquid mixture is based on the following assumptions:

a) the vapour-liquid interface is in thermodynamic equilibrium and on it the Clapairon-Klausius condition is met: $dP_S/dT_S = LP_S/T_S$;

b) the liquid is a thermostat and retains its initial temperature at a sufficient distance from the bubble;

c) the vapour behaviour inside the bubble is adiabatic;

d) the homogeneous model of the mixture is realized.

An energy equation for an individual bubble will be written in the form

$$(P_2)_t + 3\gamma P_2 R^{-1} R_t = 3\gamma P_2 \cdot q_L/R\rho_2 \ . \tag{11}$$

Here P_2 is the pressure in the bubble, q_L is the heat transfer from bubble. The bubble-to-liquid heat transfer is written as

$$q_L = \lambda \left[(T_S - T_0) R_0^{-1} - \int_0^t [1/\sqrt{\pi a}(t - \tau)] \cdot (T_S - T_0)_\tau \cdot d\tau \right] \ , \tag{12}$$

where λ is the coefficient of thermal conductivity. For small disturbances from the above equations one can obtain a single equation

$$P_t + (\alpha C_0/2) \cdot P \cdot P_\xi + \beta \cdot C_0 P_{\xi\xi\xi} = - k\delta P - (kR_0/\sqrt{\pi a}) \int_0^t P_\tau (1/\sqrt{t - \tau}) d\tau \tag{13}$$

and the solution of the linear version is

$$(P - P_0)/(P_H - P_0) = \exp(-kx/C_0) \left[1 - \mathrm{erf}\ \frac{kR_0 x}{4\sqrt{\pi a C_0} \cdot \sqrt{t - x/C_0}} \right] \ , \tag{14}$$

where a is the thermal diffusivity, $k = (\sqrt{\pi C_0} \cdot 3\gamma P_S \cdot \lambda)/R_0^2 L^2 \rho_2^2$, P_H is the disturbance pressure.

The type of solution, signal form and a qualitative comparison with the experiment are represented in Fig. 4.

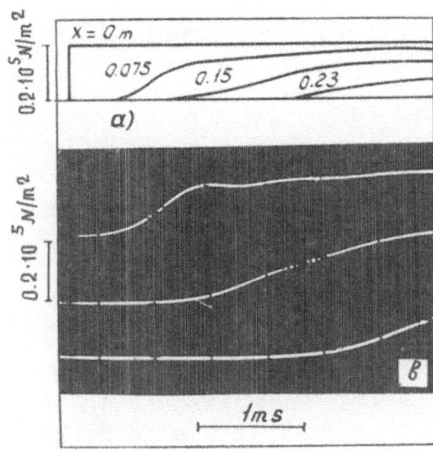

Fig. 4a,b. Pressure wave evolution
along the tube in a liquid with
vapour bubbles, (a) calculation (14),
(b) experiment

References

1. A. Mallock: Proc. Roy. Soc. A*84*, 391 (1910)
2. G.K. Batchelor: Fluid Dynamics Trans. (PWN, Warszaw 1969) p.425
3. A.P. Burdukov, V.E. Nakoryakov, B.G. Pokusaev, V.V. Sobolev, I.R. Shreiber: Chislennye Metody Mekhaniki Sploshnoi Sredy (VTs SO AN SSSR, Novosibirsk 1971) *2*, 5, p.32
4. V.E. Nakoryakov, V.V. Sobolev, I.R. Shreiber: Izv. Akad. Nauk SSR, Mekh. Zh. i Gaza *5*, 71 (1972)
5. L. van Wijngaarden: Ann. Rev. Fluid Mech. *4*, 369 (1972)
6. G.V. Gasenko, V.E. Nakoryakov, I.R. Shreiber: Nonlinear Wave Processes in Two-Phase Media (ITF SO AN SSSR, Novosibirsk 1977) p.17
7. V.E. Nakoryakov, B.G. Pokusaev, I.R. Shreiber, V.V. Kuznetsov, N.V. Malykh: Wave Processes in Two-Phase Systems (ITF SO AN SSSR, Novosibirsk 1975) p.54
8. V.V. Kuznetsov, V.E. Nakoryakov, B.G. Pokusaev, I.R. Shreiber: J. Fluid Mech. *85*, part I, 85 (1978)
9. N.V. Malykh, I.A. Ogorodnikov: Continua Dynamics, vyp.29, Mechanics of Explosion (Novosibirsk 1977) p.143
10. V.V. Kuznetsov, B.G. Pokusaev: Laminar-to-Turbulent Transition of Boundary Layer. Two-Phase Flows (ITF SO AN SSSR, Novosibirsk 1978) p.61

Dynamics of a Liquid with Gas Bubbles During Interaction with Short Large-Amplitude Pulses

N.V. Malykh and I.A. Ogorodnikov

Institute of Thermophysics, Siberian Branch of the USSR Academy of Sciences
630090 Novosibirsk, USSR

We have studied the velocity of propagation and absorption of short (τ =50-200μs) pulses of 10-1000 bar amplitude from the explosion of a small charge in a liquid with gas bubbles at low (φ =10^{-4}-10^{-1}) gas volume concentration. Under consideration is also bubble dynamics in such waves.

2 Experimental Procedure

The velocity and absorption of pulses in a liquid with gas bubbles were studied in a hydroacoustic reservoir and in a shock tube. Pulses were recorded by a miniature hydrophone with frequency-independent response. The velocity and absorption of pulses in a gas-liquid medium were calculated by the signal structure oscillograms, whose typical form is shown in Fig. 1 (a,b), [3],[4].

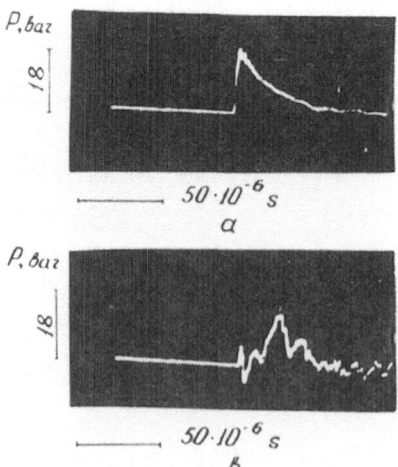

Fig.1 Pulse structure

In the case "b" at the point of observation a high-frequency peak ("forerunner") first appears at C_o velocity. Then the "main part" of the pulse arrives at a lower velocity, which carries the main part of energy. The moment of "forerunner" arrival is taken as a reference point to estimate the velocity of the "gravitational center" of the "main part". The absorption is estimated by the amplitude ratio of the incident signal and of the signal measured in the gas-liquid layer. The amplitude was measured by the peak of the "main part" of signal. Bubble dynamics was studied by high-speed recording.

3 Experimental Results

The degree of pulse damping versus the amplitude of the wave which is incident to the interface of a 0.35 m thick layer with nitrogen bubbles ($R_o = 0.1 \cdot 10^{-3}$m) is shown in Fig.2.

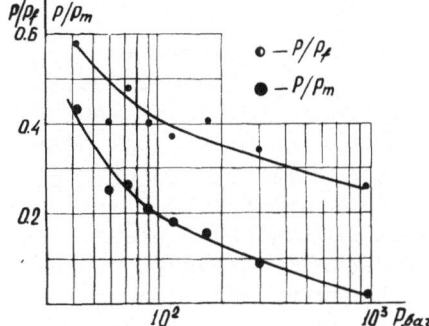

Fig.2 Wave damping vs amplitude

To eliminate the influence of spherical divergence, the pulse amplitude was registered at a fixed point first without bubbles (P_f) and then with bubbles (P). The P/P_f ratio (upper curve) is compared to the P/P_m (bottom curve), which allows for the combined influence of bubbles and spherical divergence (P_m is the incident wave pressure). It should be noted that the damping increases with the amplitude of the incident pulse. A fairly low (by 4 times) damping of the pulse of 970 bar is due to the fact that the given series of experiments was performed for the low ($7 \cdot 10^{-4}$) volume gas void fraction.

The damping of waves versus the distance passed along the medium, x, is shown in Fig.3. It is seen that increasing the P_m amplitude increases the damping degree. It can also be seen that data on the volume gas void fraction cleave. Ibid the curves $\exp(-\eta x)$ with different η are given. For the pulses with P_m of several hundreds bar the experimental points are

165

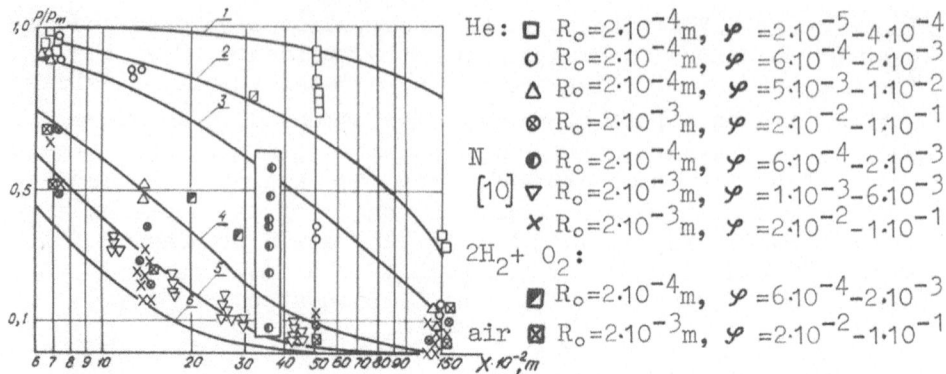

He: □ $R_0 = 2 \cdot 10^{-4}$m, $\varphi = 2 \cdot 10^{-5} - 4 \cdot 10^{-4}$
○ $R_0 = 2 \cdot 10^{-4}$m, $\varphi = 6 \cdot 10^{-4} - 2 \cdot 10^{-3}$
△ $R_0 = 2 \cdot 10^{-4}$m, $\varphi = 5 \cdot 10^{-3} - 1 \cdot 10^{-2}$
⊗ $R_0 = 2 \cdot 10^{-3}$m, $\varphi = 2 \cdot 10^{-2} - 1 \cdot 10^{-1}$

N ◑ $R_0 = 2 \cdot 10^{-4}$m, $\varphi = 6 \cdot 10^{-4} - 2 \cdot 10^{-3}$
[10] ▽ $R_0 = 2 \cdot 10^{-3}$m, $\varphi = 1 \cdot 10^{-3} - 6 \cdot 10^{-3}$
✕ $R_0 = 2 \cdot 10^{-3}$m, $\varphi = 2 \cdot 10^{-2} - 1 \cdot 10^{-1}$

$2H_2 + O_2$:

◪ $R_0 = 2 \cdot 10^{-4}$m, $\varphi = 6 \cdot 10^{-4} - 2 \cdot 10^{-3}$

air ⊠ $R_0 = 2 \cdot 10^{-3}$m, $\varphi = 2 \cdot 10^{-2} - 1 \cdot 10^{-1}$

Fig.3 Wave damping vs distance. In the frame points(P/P_f) are separated obtained values of P_m 42, 59, 71, 89, 117, 171, 300 bar from top to bottom (1 - η^m= 0.01, 2 - 0.05, 3 - 0.1, 4 - 0.3, 5 - 0.5, 6 - 0.8).

grouped near the curve with $\eta \approx 0.5$, i.e. the maximum experimental damping values are much higher than those calculated [2] without allowing for the wave energy losses for increasing the bubble intrinsic energy. It should be noted that despite the great damping the wave is not completely absorbed at distances of about one meter even at such great volume gas void fractions as 1-10%. However, the estimates of energy losses in a wave performed by the experimentally observed variations in the bubble radius show that the wave should completely damp at much lower distances. This can be assigned to the fact that the bubble energy can "return" to the wave due to the bubble radiation in the phase of maximum compression (similar to the radiation of cavitation bubbles).

The values of pulse rates versus the volume gas void fraction are given in Fig.4. In all experiments the rate of a "forerunner" is independent of the volume gas void fractions, incident wave amplitude and gas type and is equal to the sound speed in a liquid without bubbles. The rate of propagation of the "main part" of the pulse significantly depends on the volume gas void fraction and decreases with its increase. Figure 4 also represents the experimental data [5-9] for fairly small P_m values (below 6 bar), therefore, all of them are grouped around curve 5, which describes the equilibrium sound speed C [1].

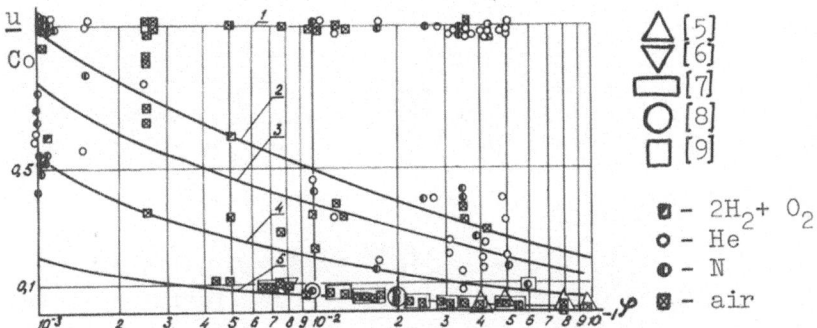

Fig.4 Pulse rates vs volume gas void fraction. 1 - sound
speed in a pure liquid. 2,3,4,5 - calculated for P = 80, 40,
10, 1 bar, respectively [1]

The pulse rate versus the amplitude is shown in Fig.5.

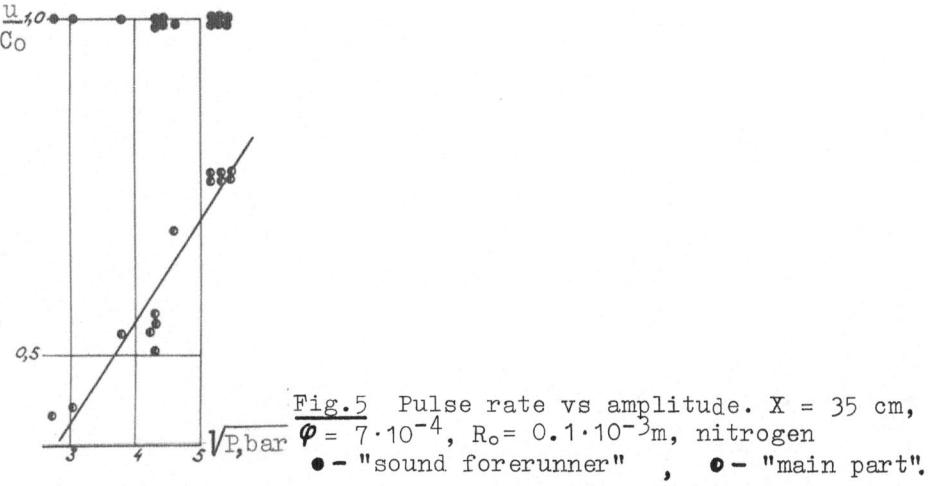

Fig.5 Pulse rate vs amplitude. X = 35 cm,
$\varphi = 7 \cdot 10^{-4}$, $R_0 = 0.1 \cdot 10^{-3}$m, nitrogen
● – "sound forerunner"   ,   ◑ – "main part".

It can be seen that the rate of the "main part" of the pulse
linearly depends on $\sqrt{P}$, and the rate of the "forerunner" is
independent of P_m. In terms of the pulse structure (Fig.1) it
means that with increasing the P_m the "main part" approaches
the "forerunner" and hence its rate attains the sound speed
in a liquid without bubbles C_0.

The analysis of experimental results shows that the propagation of short pulses of great intensity is essentially a non-
steady process. The only constant value which does not depend
on the volume gas void fraction, **wave pressure and type of gases
under study** is the rate of the "forerunner". Its constancy

permits us to use it as a reference point to determine the
rest pulse characteristics which do not remain constant during
the propagation in the medium. The rate of propagation of the
"main part" of the pulse and damping are variable values and
depend on the current amplitude and medium structure values.

4 Bubble Dynamics

Bubbles in the field of a short pulse of large amplitude have
no time to attain parameters which are in equilibrium with respect to the wave. Thus when the liquid with bubbles is affected by an explosion pulse with the front amplitude of 10-1000
bar, the difference in the pulse and bubble pressures can attain the value which is equal to the pulse front amplitude.
This pressure drop greatly accelerates the bubble wall thus
governing its further inertial behaviour. The damping influence
of viscosity, thermal conductivity and acoustic radiation does
not exclude the predominant influence of inertial forces. Therefore the minimum and maximum bubble radii can significantly
differ from the equilibrium. Depending on the relation between
wave time and bubble intrinsic time, the bubble can attain its
extremal sizes in the wave or behind it. Thus under action of
the wave with 20 bar pressure and 200 μs duration, the experimentally observed maximum size of air bubbles in 1250 μs is
3.2 R_o (Fig.6).

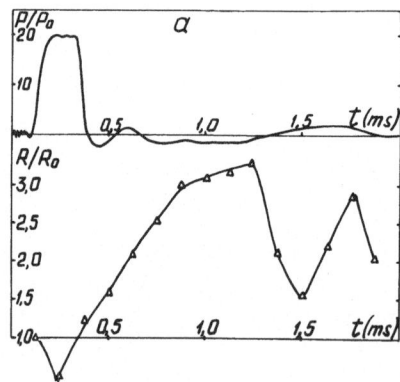

Fig.6 Bubble radius vs time

The dynamics of these nitrogen bubbles in a shock wave is
shown in Fig.7. At a given rate of recording the minimum size
of bubbles is not registered, but the maximum size (3-5 R_o
(Frames 8-9)) is readily determined. The overexpansion behind

the wave is the circumstantial evidence of the nonequilibrium overcompression of bubbles in the wave. One can also emphasize the presence of cumulative jet streams and subsequent crushing of cavitational bubbles which are typical for the dynamics of collapse (Frames 5-12).

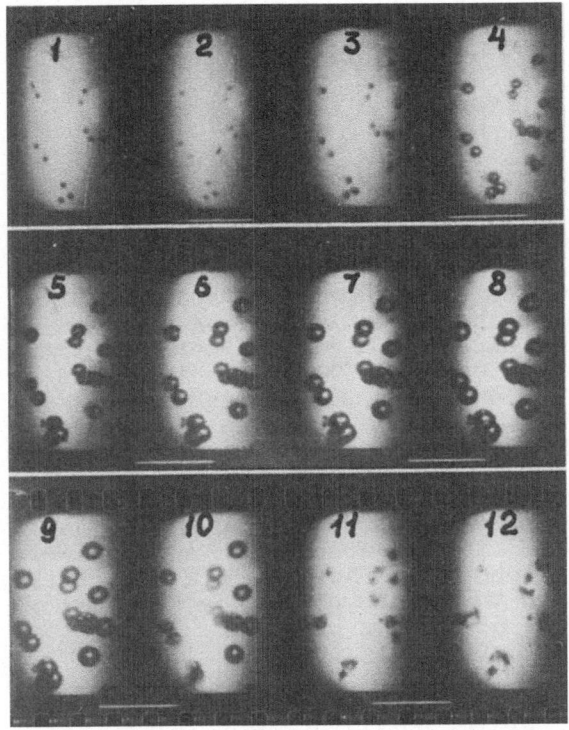

Fig.7 Cinegram of nitrogen bubble dynamics. $R_0=0.35$ 10^{-3}m, P=20 bar, $\tau=200$ 10^{-6}s, wave propagation from right to left, recording speed = 2000 frames/s. Frame 1 - initial bubble size, Frame 2 - moment of wave arrival

References

1 G.K.Batchelor, Fluid Dynamics Trans., (PWN,Warszaw,1969),p.425.
2 L.van Wijngaarden, Ann.Rev.Fluid Mech., 4, 369 (1972).
3 N.V.Malykh, I.A.Ogorodnikov, Continua Dynamics, vyp.29, Mechanics of Explosions (Novosibirsk,1977),p.143.
4 N.V.Malykh, I.A.Ogorodnikov, Proc. 8th Int.Symp.on Nonlinear Acoustics, Paris 1978 (unpublished)
5 L.J.Campbell, A.S.Pitcher, Proc.Roy.Soc., Ser.A243, 1235,534 (1958).
6 F.W.Gibson, J.Acoust.Soc.Amer.,48,5,1195 (1970).
7 L.Noordzij,Unsteady Water Flows with High Velocities (Nauka,Moskva 1973),p.369.
8 V.E.Nakoryakov, B.G.Pokusaev, I.R.Shreiber, V.V.Kuznetsov, N.V.Malykh, Wave Processes in Two-Phase Systems (ITF SO AN SSSR, 1975),p.54.
9 B.E.Gelfand, S.A.Gubin, S.M.Kogarko, E.I.Timofeev, Izv.Akad.Nauk SSSR, Mekh.Zh.i Gaza, 6, 58 (1974).
10 V.V.Kuznetsov, B.G.Pokusaev, Laminat-to-Turbulent Transition of Boundary Layer. Two-Phase Flows (ITF SO AN SSSR, 1978), p.73.

Shock Wave Transformation in Bubbly Liquids

V.K. Kedrinskii

Institute of Hydrodynamics
Siberian Branch of the USSR Academy of Sciences
Novosibirsk 630090, USSR

The experimental results and numerical estimations for the structure of compression waves generated by bubble layers at the absorption of short shock waves are presented.

1.   The experiments were performed using a hydrodynamical shock tube. The steady plane shock wave was formed within a certain section of the tube as a result of underwater explosion of a wire near one of the tube ends. The bubble layer was created with the help of hollow spherical bubbles made of very thin rubber and fastened to a special frame. The possible effect of this frame on the wave structure was easily controlled.

Pressure transducers recording the form and parameters of incident and resulting waves were placed in front of and behind the layer at a constant distance from its front boundary. The other end of the shock tube either was closed (the reflection process in the layer on the solid wall was researched) or had a special device for the absorption of waves allowing the tube to be considered as an infinitely long one. The length 1 of the layer ranged from 1 to 30 cm, the volume gas concentration k_0 from 0.004 to 0.3, the shock wave amplitude from 10 to 30 atm, the duration of its positive phase from 50 to 100 s.

The main problem is to realize in detail the mechanics of the shock wave transformation by the bubble layers. Figs. 1,2 show the pressure oscillograms describing this process. According to the data of the first five oscillograms presented in Fig. 1 obtained for k_0 = 0.08 and 1= 0,1,2,3,12 cm with the layer length increase the shock wave energy absorption becomes more intense as a result of both the radial flow formation and the internal gas energy increase at the bubble collapse. This absorbed energy doesn't disappear and is re-radiated by the layer with some delay as a compression wave [1]. The arising time of its maximum amplitude corresponds to the bubble collapse time. A practically full absorption of the incident shock wave takes place along the length 1_0 depending on the wave parameters and the gas content of the layer. To judge by the oscillogram character, such layer behaves as a single collective bubble.

Instead of the shock wave at $1 > 1_0$, only a track as a precursor before the layer radiation is recorded. It is obvious that the same process of absorption and re-radiation will take place for the radiation of layer l_0 if the layer length increases. Consequently, there will be always such a layer l_1 which absorbs energy of this radiation and re-radiates it with a certain time delay. As is shown in Fig.1 (Oscillogram 6) for a sufficiently great value of l the above-mentioned process can be repetitive.

The resulting radiation is a wave packet of periodical structure with rather constant frequency ν depending on a number of parameters of the medium and incident wave (Fig. 2,

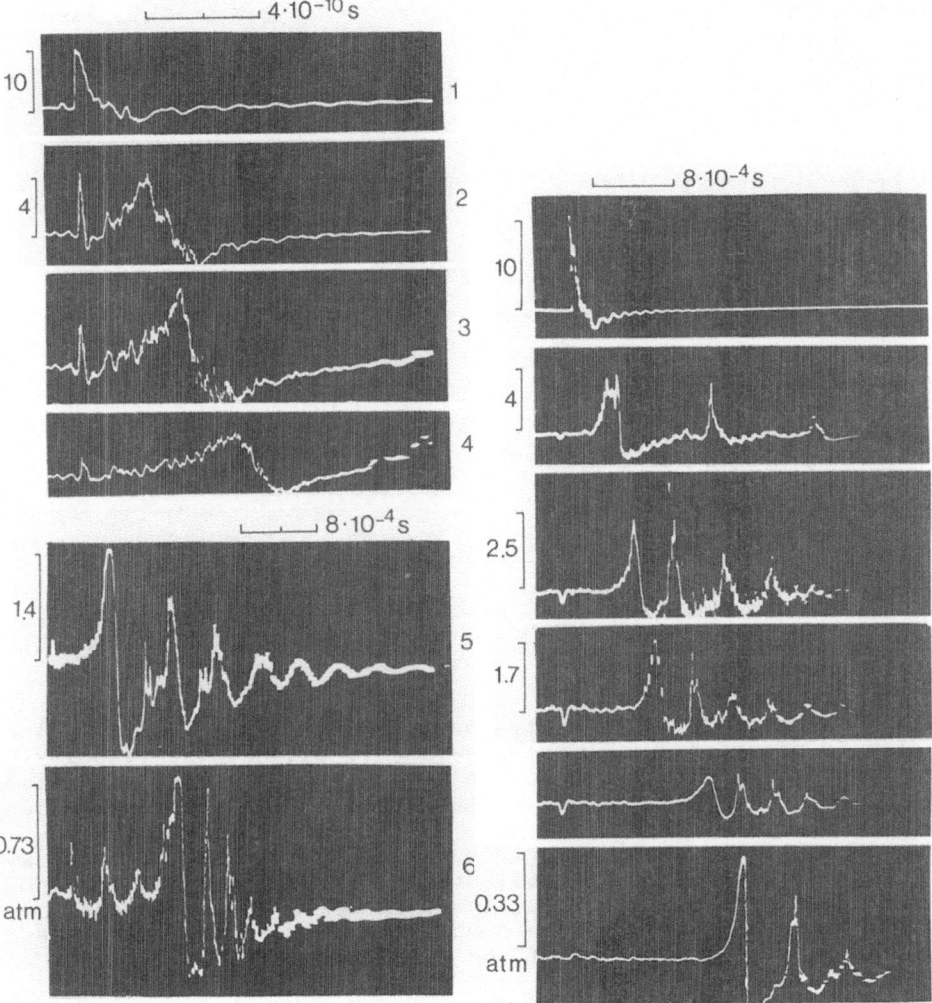

Fig.1. Absorption and re-radiation of shock wave energy with the formation of precursors.

Fig.2. Formation of a wave pocket by a bubble layer.

$k_0 = 0.15$, $1 = 0$, 2, 4, 6, 10, 14 cm, respectively). The experiments show that the dependence $\nu(1)$ has a maximum which is decreased and displaced to the region of small values of 1 with increase in k_0. One of the most interesting results is the delay time τ_0. It is the time interval between the recording of the disturbance propagating with the sound velocity c_0 in a pure liquid and the maximum of the layer radiation. The dependences $\tau(1)$ are presented in Fig.3 for the following values of k_0: 0.04, 0.08, 0.15, 0.3, respectively.

It is considered that the character of $\tau(1)$ dependences usually determines the disturbance propagation velocity in the two-phase medium which can be described by the well-known relationship $c_1^2 = \gamma p_0/\rho_0 k_0$. But in this case the value of $\beta = c_1^2 \cdot k_0$ must be constant for one and the same gas and p_0 and independent of k_0. The data presented in Fig.3 for Curves 1-4 result in the following values of β : 800, 940, 1400, 880 m^2/s^2

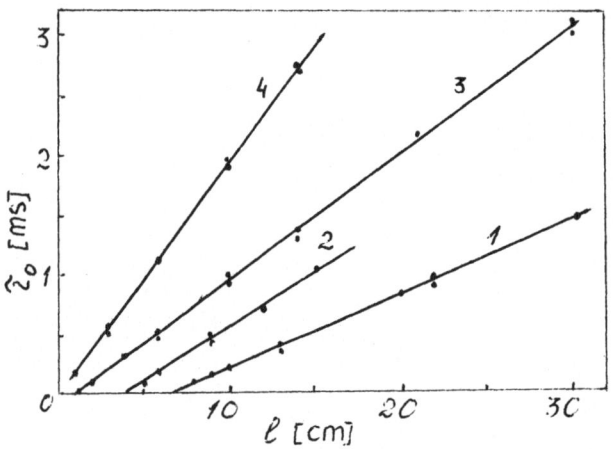

Fig. 3. Maximum delay time of radiation vs length of layer.

The main reason of this contradiction consists in that only the delay is recorded during the experiment, but not the velocity of propagation. And the delay is a result of the late generation of the disturbance, but not a small velocity of its propagation.

2.   The liquid containing gas bubbles has a number of peculiarities: the pressure non-equilibrium in liquid and gas phases causing relaxation processes, the primary influence of gas phase compressibility on the compressibility of the medium as a whole and the influence of the bubble dynamics on the non-linearity of the process.

A one-dimensional motion of a two-phase medium containing gas bubbles of different radius R_i and described by the system of equations [1-3]:

$$\rho_t + (\rho u)_x = 0, \quad \rho u_t + \rho u u_x + p_x = 0, \quad \rho = (\rho_o + \frac{p-p_o}{c_o^2})(1+k)^{-1}$$

$$k = \sum_{i=1}^{N} k_i, \quad k_i = k_{io}(\frac{R_i}{R_{io}})^3, \quad \rho_o(R_i \cdot \ddot{R}_i + \frac{3}{2}\dot{R}_i^2) = p_o(R_i/R_{io})^{-3\gamma} - p \tag{2.1}$$

is considered within the framework of the conditions $L \gg l_i \gg R_i$ for the characteristic dimensions of process L and medium l_i at a constant number of bubbles per unit volume. Linearization (2.1) leads to

$$p_{tt} - c_o^2 p_{xx} - \rho_o c_o^2 \sum_{i=1}^{N} (k_i)_{tt} = 0,$$

$$(k_i)_{tt} + \Omega_i^2 \cdot k_i = -\Omega_i^2 \cdot p/\gamma p_o, \quad \Omega_i^2 = 3\gamma p_o/\rho_o R_{io}^2 \tag{2.2}$$

which allows the dependence to be obtained for the phase velocity of sound c_*

$$c_o^2 \cdot c_*^{-2} = 1 + c_o^2 \cdot c_1^{-2} \cdot \int_0^\infty \frac{k(R)dR}{1 - \omega^2/\Omega^2(R)} \tag{2.3}$$

when $N \to \infty$ and $p = a \cdot \exp[i(\omega t - mx)]$, $k_i = b_i k_{io} \cdot \exp[i(\omega t - mx)]$ at $\int_0^\infty k(R)dR = 1$, where $k(R)$ is the concentration of bubbles. The dependence (2.3) is dispersive, and when $k(R) = a \cdot (R/b)^2/(1 + (R/b)^4)$, it results in

$$c_o^2 \cdot c_*^{-2} = 1 + c_o^2 c_1^{-2}(1 - \omega^2/\Omega^2(b))/(1 + \omega^4/\Omega^4(b)) \tag{2.4}$$

obtained under the condition that the experimental and theoretical data coincide at a certain characteristic point of the dependence. The analysis of (2.4) shows [1,3] that in the case of different size bubbles the theoretical curve is in full agreement with experimental results [4] and its resonance peculiarity disappears.

If the liquid component of the medium is incompressible, the linearization of the system (2.1) leads to (at $\varphi = k/k_o$)

$$p_{xx} + \rho_o k_o \varphi_{tt} = 0,$$

$$\varphi_{tt} = 3\varphi^{1/3}(p_o \varphi^{-\gamma} - p)/\rho_o R_o^2 \tag{2.5}$$

The solution of it, with some additional approximations, has to form

$$p = p_o \cdot \varphi^{-\gamma} + A(\varphi)e^{-\sqrt{\varphi^{1/3}} \cdot \eta} + B(\varphi) \cdot e^{-\sqrt{\varphi^{1/3}} \cdot \eta} \tag{2.6}$$

where $\eta = (3k_o)^{1/2} \cdot x/R_o$. The coefficients A and B are determined for each problem from boundary conditions, and the joint solution of (2.6) and the second equation of (2.5) allows obtaining significant, if though approximate, results. In fact, due to incompressibility of the liquid component, the arising of the pressure jump on the boundary of the two-phase medium when t = 0 changes instantaneously the value $\varphi_{tt}(x)$ and, consequently, the distribution of p along x according to (2.5). This distribution was found to correspond to the experimental results on the change of the shock wave amplitude propagating within the bubble layer l. Figure 4 compares these experimental and numerical results for different k_o (* - 0.004, x - 0.02, 0 - 0.06, Δ - 0.08) and allows the conclusion that the coefficient $\eta = (3k_o)^{1/2} l/R_o$ is a similarity criterion.

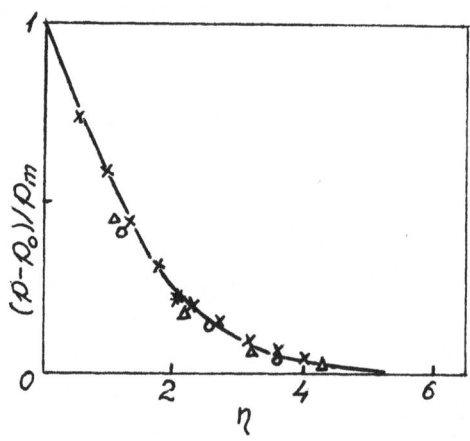

Fig.4. Decreasing of shock wave amplitude in the layer.

The calculation of the transformation process of the shock wave by the bubble layer (Fig.5) corresponding to the experimental conditions for the oscillogram 1-4 presented in Fig.1 and its rather good agreement with the experiment confirm the two-phase model reality. If the pressure on the boundary of the two-phase half-space with k_o = 0.002, R_o = 0.4 cm decreases in accordance with the shock wave form of a maximum amplitude of 20 atm and a duration of 10^{-4} s, the pressure wave profile evolution takes place in such a way as is shown in Fig.6. The numbers at the curves correspond to times of 140, 250, 330, 440, 580 μs . The signal seems to propagate in a medium with some decreasing group velocity: 364, 250, 182 and 143 m/s . The theoretical value is 264 m/s. This comparison shows that the notion of the disturbance velocity introduced here is conditional. The experimental and theoretical data for the delay of the maximum disturbance τ have been obtained on the background of the sound velocity in an acoustic (c_o) or incompressible ($c_o = \infty$) liquid and have the same nature.

The analogy between the wave-bubble interaction in the case of short shock waves and long shock waves was shown in [3] . At constant pressure on the boundary of the two-

phase half-space absorption of the front part of the wave and
re-radiation of its energy as decreasing pulsations modulating
the wave profile take place.

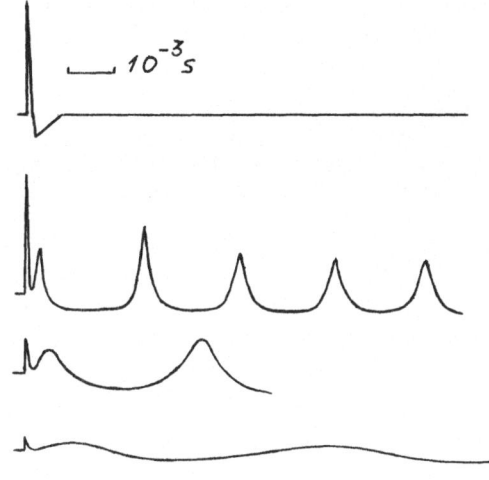

Fig.5. Calculation of the
mechanics of shock wave
transformation in bubble
layers.

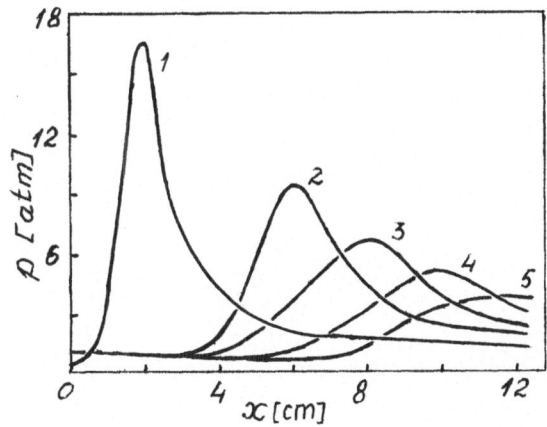

Fig.6. Propagation of comp -
ression wave in two-phase
half-space.

REFERENCES

1.   Kedrinskii, V.K., The disturbance propagation in a liquid
containing gas bubbles (Rasprostranenie vozmuschenii v zhidkosti soderzhaschei puzyr'ki gaza), Prikladnaya mekhanika i
tekhnicheskaya fizika, No 4, 1968.

2.   Iordansky, S.V., On equations of liquid motion containing
gas bubbles (Ob uravneniiah dvizheniia zhidkosti soderzhaschei

puzyr'ki gaza), Prikladnaya mekhanika i tekhnicheskaya fizika,
No 3, 1960.

3.   Kedrinskii, V.K., Cavity dynamics and waves.   In: Unsta-
tionary hydrodynamics problems (Dinamika polosti i volny.
Nestatsionarnye problemy gidrodinamiki), vyp.38, Institute of
Hydrodynamics, Siberian Branch of the USSR Academy of Sciences,
Novosibirsk 1979.

4.   Fox , F., Kerly, S., Larson, G., Phase velocity measurements and sound absorption in liquid containing air bubbles
(Izmerenie fazovoi skorosti i pogloschenie zvuka v vode soderzhaschei vozdushnye puzyr'ki), In: Problems of Modern Physics,
No 8, 1956.

Relaxation Effects in the Propagation of Underwater Shock Waves

T.N. Fedoseeva, F.E. Fridman, V.N. Goldberg, and I.G. Zarnitsina

Radiophysical Research Institute
Gorky 603600, USSR

1. Introduction

As is known, when a spherical shock wave propagates in the ocean the mutual action of nonlinear and dissipative effects leads to smoothing of the front and to a transformation into a linear acoustic wave [1,2]. A detailed investigation of this process is of interest. Here we consider the solution of the nonlinear Burgers' equation describing the propagation of a spherical shock wave in a homogeneous dissipative and relaxing medium [3]:

$$\frac{\partial u}{\partial z} - \beta_1 v \frac{\partial v}{\partial \tau} - \beta_2 e^z \frac{\partial^2 v}{\partial \tau^2} - \beta_3 e^z \frac{\partial}{\partial \tau} \int_{-\infty}^{\tau} \frac{\partial v}{\partial \tau'} e^{-\frac{\tau-\tau'}{a}} \, d\tau' = 0 . \tag{1}$$

The following symbols are introduced:

$$v = ur/u_m r_0 , \qquad z = \ln(r/r_0) , \qquad \beta_1 = \varepsilon r_0 u_m / c^2 T_M ,$$

$$\beta_2 = br_0/2c^3 pT_m^2 , \qquad \beta_3 = mr_0/2cT_m , \qquad \tau = (t-r/c)/T_m ,$$

u is the velocity of particles in a wave, r is the distance, t is the time, u_m and T_m are the characteristic amplitude and the duration of the initial disturbance, r_0 is the boundary condition coordinate value, b is the viscosity coefficient, m is the relaxation coefficient, $a = f_*^{-1} T_m^{-1}$, f_* is the relaxation frequency, c is the sound velocity, p is the medium density, ε is the nonlinearity parameter. Eq. (1) is the modified Burgers' equation; the spheric divergence of a wave is manifested in it by the exponential increase of the viscosity and relaxation coefficients. It is a fact that an increase in terms describing dissipative effects leads to the transformation of a shock wave into a linear acoustic wave. However, the distance over which such transformation occurs may be sufficiently large if the coefficients β_2 and β_3 are small. This takes place in the propagation of acoustic signals from underwater explosive sound sources.[1] That is why the boundary condition for $r = r_0$ is written in a form which is sufficiently accurate to describe a leading shock wave from an explosive source

$$v(z = 0,\tau) = (1-\tau)e^{-\tau} \qquad \tau > 0 \tag{2}$$

---

[1] We give characteristic parameters. For a TNT charge of 1 g weight $r_0 \simeq 0.1$ m, $\beta_2 \simeq \beta_3 \simeq 10^{-5}$, $a \simeq 1$, $\beta_1 \sim 1$.

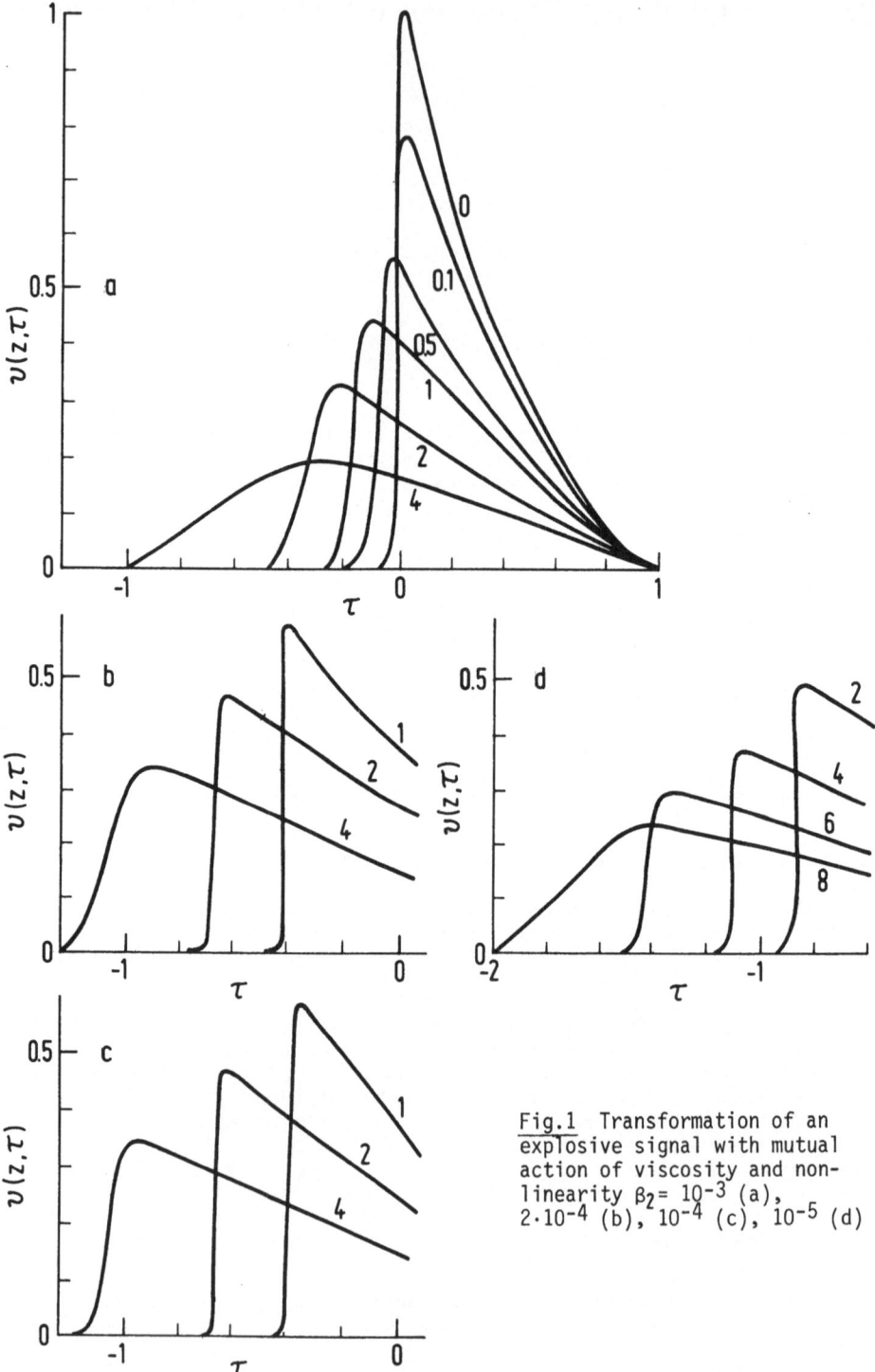

Fig.1 Transformation of an explosive signal with mutual action of viscosity and non-linearity $\beta_2 = 10^{-3}$ (a), $2 \cdot 10^{-4}$ (b), 10^{-4} (c), 10^{-5} (d)

We do not introduce coefficients in the boundary conditions (2) since in equation (1) we have three coefficients β_1, β_2 and β_3 for the direct analysis of the influence of different factors. Obviously, a transformation of the coordinate z allows preserving only two coefficients in (1) (a coefficient will then appear in boundary condition (2)).

2. Shock Wave Propagation Without Relaxation

Eq. (1) with boundary condition (2) is solved numerically. We now present the results of the calculation. When relaxation is absent ($\beta_3 = 0$) the dissipation is defined by the viscosity of sea water only. Then (1) describes the mutual action of nonlinearity, viscosity and geometric divergence of a wave. The mutual competition between viscosity and nonlinearity leads to the formation of a stationary shock wave. However, in a spherical wave (in contrast to a plane one) the stationary shock wave no longer exists beyond a certain distance. The process of transformation of the shock wave into a stationary wave and then into a linear acoustic wave is considered in detail in Fig.1a. The numbers near the curves on the diagrams give the

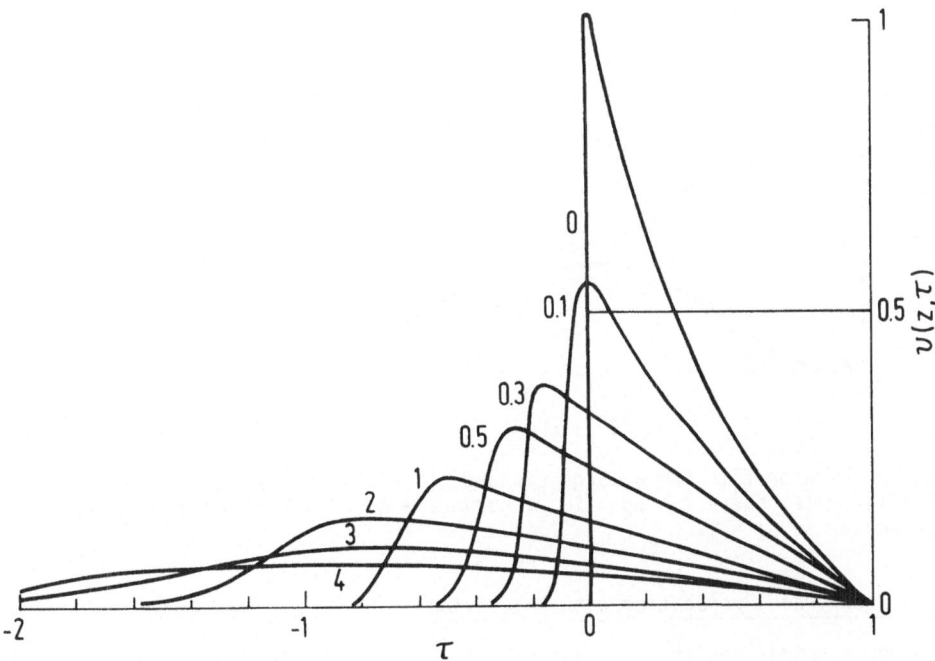

Fig.2 Transformation of an explosive wave profile for $\beta_1 = 1$, $\beta_2 = 10^{-2}$, $\beta_3 = 0$. In contrast to the diagram in Fig.1a the viscosity here is increased 10 times and the nonlinearity is increased 5 times. This is equivalent to an effective increase in viscosity of a factor of two which leads to more rapid smoothing of the shock wave.

179

value of the dimensionless coordinate z. The decrease of viscosity increases the region of existence of a stationary shock wave, since a weak smoothing of the discontinuity (due to viscosity) may be restrained by the nonlinearity for a sufficiently long distance (despite of its decrease due to spherical divergence of the signal). Figs.1a-1d illustrate the process of waveform variation when the dimensionless viscosity coefficient decreases from $\beta_2 = 10^{-3}$ to $\beta_2 = 10^{-5}$. When the nonlinearity parameter β_1 increases, the process of front smoothing occurs more slowly and the front remains shocklike for large distances. A similar increase in the nonlinearity parameter and viscosity coefficient produces negligible change in the character of motion of a shock wave. Fig.2 gives results of a calculation of the wave profile with increased values of the viscosity coefficient and the nonlinearity parameter (compared to with the calculation given in Fig.1). Since the viscosity is increased more than the nonlinearity, the smoothing of the front occurs more quickly.

The role of viscosity is especially marked at large distances. With a constant nonlinearity, (Fig.3) smaller values of the viscosity coefficient lead to preservation of the shock front. At large distances it is not difficult to follow the transformation of a wave having a shock front into a linear wave with nearly symmetric profile (Fig.4).

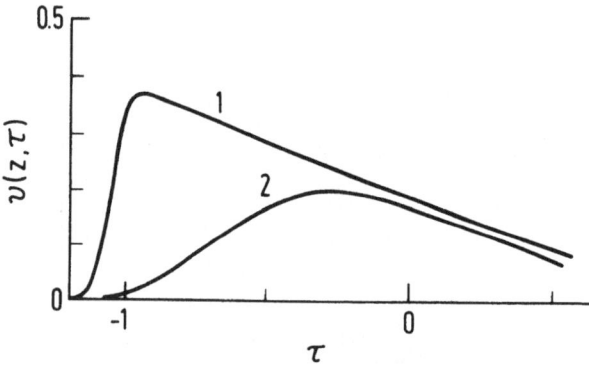

Fig.3 Wave profile for a large distance ($z = 4$) with different viscosity coefficients: curve 1 - $\beta_2 = 10^{-3}$, curve 2 - $\beta_2 = 10^{-4}$. The nonlinearity parameter is $\beta_1 = 1$.

3. Shock Wave Propagation with Relaxation

We now consider results of a solution of equation (1) in the presence of a relaxation integral. The presence of relaxation causes an additional viscosity and dispersion which is associated with a change in time scale (this scale is the time of relaxation a). The dispersion results in an additional distortion of the wave profile. Thus, low frequency dispersion changes the sound velocity and, in a coordinate system moving at velocity c, the impulse is shifted as a whole due to the velocity increase (Fig.5). The role of relaxation in propagation of an acoustic pulse is seen from comparison of the curves in Fig.5. At small distances z = 1 the relaxation leads to a further

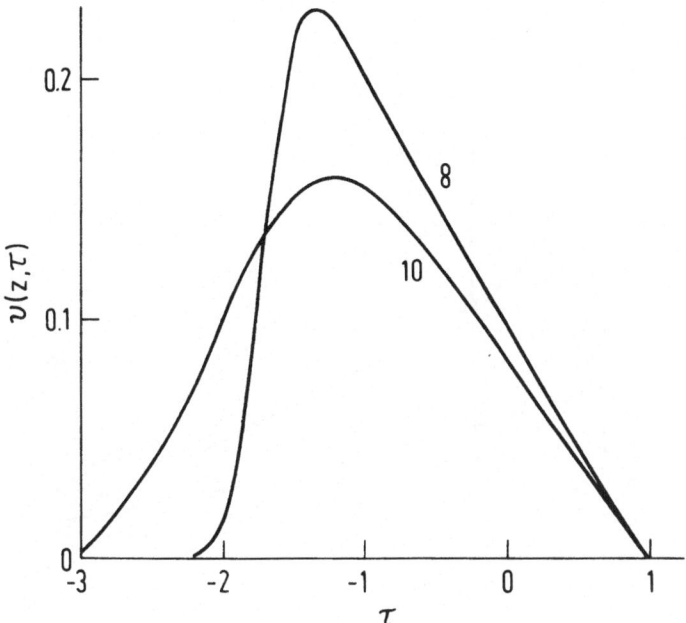

Fig.4 Wave profile at different distances for viscosity coefficient $\beta_2 = 10^{-5}$. The nonlinearity parameter is $\beta_1 = 1$.

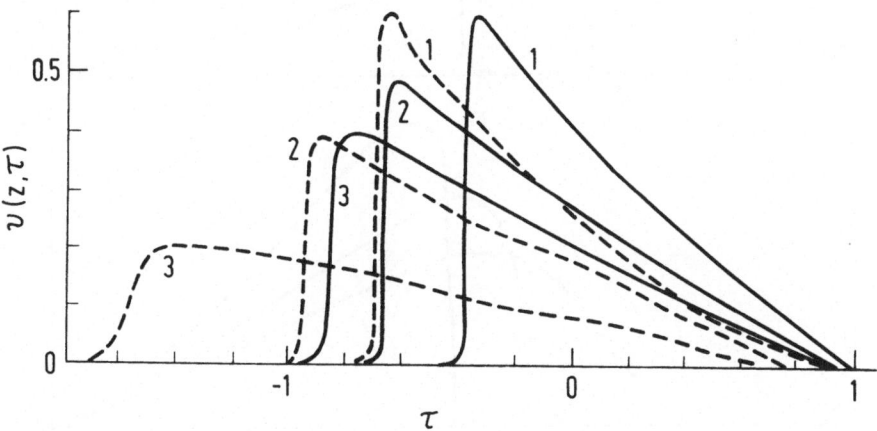

Fig.5 A comparison of an explosive signal in a viscous medium (solid curves) and in a medium with viscosity and relaxation (dashed curves). In both cases the nonlinearity parameter is $\beta_1 = 1$. The viscosity coefficient is $\beta_2 = 10^{-4}$. The relaxation coefficient is $\beta_3 = 5 \cdot 10^{-2}$, the relaxation time is $a = 1$.

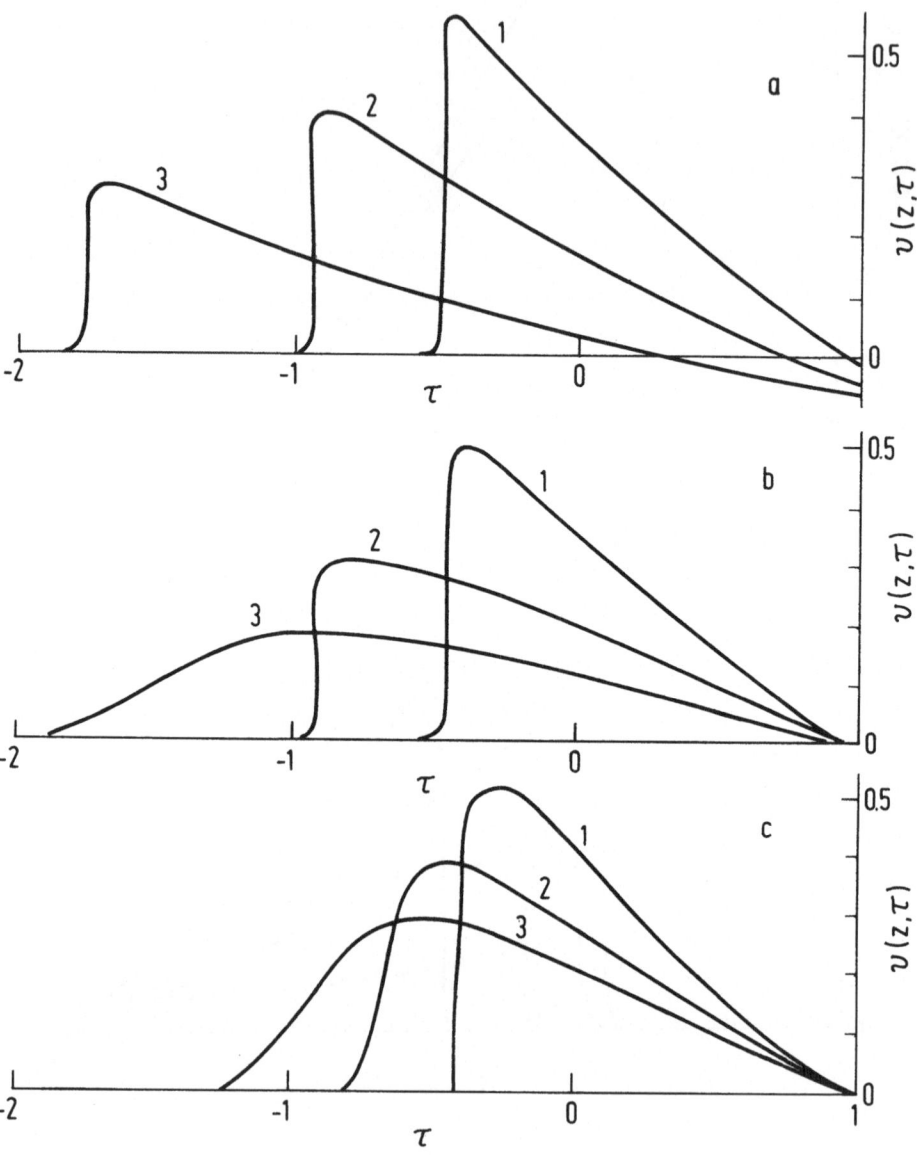

Fig.6 Pulse form with different time of relaxation for $\beta_1 = 1$, $\beta_2 = 10^{-4}$, $\beta_3 = 5 \cdot 10^{-2}$. a = 5(a), 0.5(b), 0.05(c)

shift of the wave front only. At the large distance z = 2, due to dissipation specified by the relaxation integral, and since the wave is a shock the amplitude is smaller. At a still larger distance z = 3 a greater difference is seen between relaxing and nonrelaxing media. The presence of relaxation results in strong damping of the field and in smoothing of the shock wave front. Note that the duration of a pulse in a medium with relaxation is essentially higher than in a viscous medium. We consider now the role of the relaxation time when a strong acoustic wave propagates in a medium with relaxation (Fig.6). With large relaxation time low frequency dispersion and nonlinearity is of great importance (Fig.6a). If the time decreases from a = 5 to a = 0.5 the dispersion decreases and dissipation increases. Then the pulse form becomes smoother with smaller amplitude and smaller shift of the pulse in a moving system of coordinates (Fig.6b). For a very small relaxation time, a = 0.05, the low frequency dispersion is practically completely absent. In this case, the relaxation is the reason for occurrence of peculiar "rounded" acoustic signals. At a smaller relaxation time the front width becomes larger but the amplitude decreases more slowly (Fig.6c). The variety of forms of acoustic signals in the relaxing medium is defined by the relation between the relaxation time and the pulse duration. If these values are comparable the relaxation effect is maximal: the pulse is shifted and sharply changes its form. With small relaxation time the dispersion manifests itself only in the narrow region close to the discontinuity. Thus numerical calculations show that the large increase of an explosive wave observed in experiment (see, for example [4]) may be associated with the mutual action of nonlinear and relaxation effects.

4. References

1 K.A. Naugol'nykh, Akust. Zh. 18, 579-583 (1972) [Sov. Phys.-Acoustics 18, 475 (1973)]
2 V.E. Fridman, *Wave motion*, 1979 (in press)
3 O.V. Rudenko, S.I. Soluyan, *Theoretical foundations of nonlinear acoustics*, (Nauka Press, Moscow 1975), Translation: (Consultants Bureau, New York 1977)
4 A.B. Arons, J. Acoust. Soc. Am. 26, 343-346 (1954)

Part III

Bubble Spectrometry

Acoustical Bubble Spectrometry at Sea

H. Medwin

Physics and Chemistry Department, Naval Postgraduate School
Monterey, CA 93940, USA

Introduction

We've come a long way since the early 1960s when a prominent oceanographer
wrote a memorandum titled "Do Invisible Microbubbles Exist in the Sea?" Our
first bubble count at sea used an optical technique [1]. But the optical
scattering cross section is at most equal to the geometrical cross section.
Therefore, photographic techniques are restricted to small volumes. Furthermore, the optical identification of bubbles in a sea laden with detritus requires at least proof of sphericity. For small bubbles this proof is additionally constrained by the Rayleigh criterion for resolution.

On the other hand, acoustical bubble spectrometry takes advantage of the
facts that: (1) the bubble resonance frequency is inversely proportional to
its radius; (2) at resonance the scattering and extinction cross sections of
a bubble are about 1000 times greater than its geometrical cross section;
and (3) a bubbly medium is dispersive.

There are three types of measurements from which we have determined in-
situ bubble densities as a function of radius for well-defined volumes of
the order of one cubic meter: scatter, excess attenuation, and sound velocity, measured as functions of frequency. It is my purpose to summarize
briefly the theory that permits bubble numbers and their fluctuations to be
inferred from acoustical measurements and to review the techniques that we
have used at sea.

Attenuation and Backscatter

For acoustical spectrometry, it is adequate to consider the motion [2] of a
single gas bubble of radius, a, in the low frequency regime $ka \ll 1$. The
differential equation describing the forced displacement ξ due to an incident
plane wave of amplitude P_P is then

$$m\ddot{\xi} + R_M\dot{\xi} + s\xi = 4\pi a^2 P_p e^{i\omega t} \tag{1}$$

where $m = 4\pi a^3 \rho$ and $s = 12\pi \gamma P_A b\beta$ $\tag{2}$

The resonance frequency is $f_R = (2\pi)^{-1}(s/m)^{\frac{1}{2}} = (2\pi a)^{-1}(3\gamma P_A b\beta/\rho)^{\frac{1}{2}}$ $\tag{3}$
The mechanical damping constant R_M represents energy losses caused by sound
reradiation, shear viscosity, and thermal conductivity. In terms of the
commonly-used damping constant at resonance $\delta_R = 1/Q$ we have

$$\delta_R = R_M/(\omega_R m) \tag{4}$$

Similarly, we define the component damping constants due to reradiation,
viscosity and thermal conductivity at resonance

$$\delta_R = \delta_{Rr} + \delta_{RV} + \delta_{Rt} \tag{5}$$

where $\delta_{Rr} = k_R a_R$, $\qquad \delta_{RV} = 4\mu/(\rho\omega_R a_R^2)$, $\qquad \delta_{Rt} = (d/b)]_{f_R}$ $\qquad\qquad$ (6)

The parameters b, d, β (section A6.1.1,[2]) are functions of the physical constants of the bubble gas and liquid; (d,b) take account of non-adiabatic processes and (β) includes surface tension. The shear viscosity is μ, the ratio of specific heats is γ and P_A is the ambient pressure.

The large response of a bubble near resonance is due to δ_R being of order 0.1 and is measured by the acoustic cross sections for scatter σ_s, absorption σ_a, and extinction $\sigma_e = \sigma_s + \sigma_a$ which are

$$\sigma_s = 4\pi a^2/\{[f_R/f)^2-1]^2 + \delta^2\}; \quad \sigma_e = (\delta_R/\delta_{Rr})\,\sigma_s; \quad \sigma_a = [(\delta_{Rt} + \delta_{RV})/\delta_{Rr}]\sigma_s \quad (7)$$

When there are bubbles of different sizes, the number per unit volume depends on the radius increment. We define the bubble density for increment da (commonly one micron) as

n(a)da = (number of bubbles of radius between a and a + da)/volume $\qquad$ (8)

The extinction cross section per unit volume S_e for sound traversing a random mixture of non-interacting bubbles is calculated by integrating,

$$S_e = \int_0^\infty \sigma_e n(a)da = \int_0^\infty \frac{4\pi a^2 (\delta/\delta_r)n(a)da}{[(f_R/f)^2-1]^2 + \delta^2} \qquad [m^{-1}] \qquad (9)$$

The attenuation due to bubbles α_b is calculated by integrating

$$dI = - I_p S_e dx \qquad (10)$$

to obtain $\alpha_b = (10/x) \log_{10} (I/I_0) = 4.34 S_e$ dB/m $\qquad\qquad$ (11)

Since backscatter from bubbles is omnidirectional, the cross section for backscatter is $\sigma_{bs} = \sigma_s/4\pi$ and the volume backscattering coefficient is the backscattering cross section per unit volume s_v, calculated from

$$s_v = \frac{1}{4\pi} \int_0^\infty \sigma_s n(a)da \qquad (12)$$

In order to use (11) and (12) to determine bubble densities n(a)da by attenuation and backscattering experiments it is necessary, first, to perform the integration of (9). Past practice has assumed that only bubbles at or very near resonance are important and that δ and δ_r are constant in the integration. This procedure [3], yields

$$\alpha_b = 8.68\pi^2 (a_R^3/\delta_{Rr})n(a_R) \qquad (13)$$

from which the number of bubbles of radius a_R is determinable by a measurement of excess attenuation.

There is another approach [4]. Measurements of bubble densities at sea and in the laboratory can be fitted by a power law

$$n(a) \sim a^{-q} \text{ where } q > 0 . \qquad (14)$$

All reported measurements yield 2 < q < 5. When q is known this makes the numerical integration (9)(12) feasible with no assumptions about constancy of δ. For example, we have found that the attenuation rate calculated by numerical integration, for a 50-kHz sound traversing a mixture of air bubbles in water in which q = 4, is larger than (13) by the factor 1.67.

This theory was first used to calculate bubble numbers in a pulse-echo experiment at sea [1]. The location was the NOSC Oceanographic Tower, San Diego. The simple equipment consisted of a 24 cm mylar piston source = receiver facing a 30 cm diameter reflector at a separation of 76 cm. The echo patterns were calibrated by using a tank of clear water as the no-bubble reference for attenuation. It was self-calibrated for backscatter by using the rigid reflector as a reference. Pulse-echo patterns, such as in Fig. 1 which show attenuating echos and volume backscatter between the echos, were used to calculate bubble densities from (11) and (12). Simultaneous determinations by the two techniques agreed within a factor of approximately 1 to 5. Decreasing numbers were found at greater depths, and there were fewer small bubbles (<60μ) at night then in the daytime. Empirical laws are in [1].

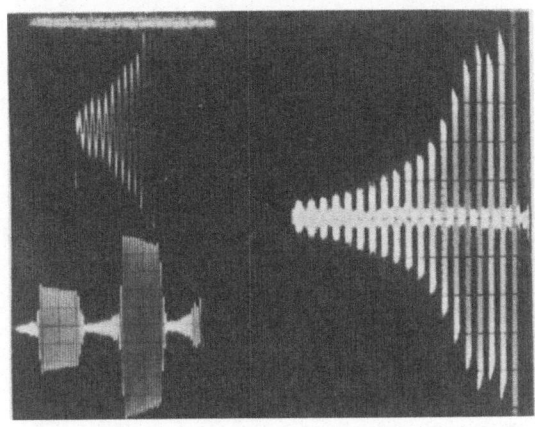

Fig.1 Oscillograms of pulse echo patterns at sea [1]. Left: 200 kHz signal showing 19 echoes. Right (top): two echoes of 30 kHz signal with backscatter between them; (middle) backscatter alone; (bottom) noise alone.

More recently [5] we have used digital analysis of short range CW transmission runs to simultaneously measure excess attenuation at 64 frequencies. In this configuration a harmonic-rich source of fundamental 2.5 kHz faces a coaxial line of point hydrophones at separations 1,2,3,4, and 5m. The 5m long rig containing source and hydrophones is lowered from a ship. Attenuation and phase velocity propagating past two hydrophones are calculated at sea by simultaneously A/D converting the two signals at the sampling rate 320,000/sec with 12 bit resolution. This is followed by dual FFTs of the two time series and then computer calculation of attenuation and phase shift for the frequencies 2.5 to 160 kHz. The minimum attenuation, as measured at much greater depths, is conservatively assumed to be the bubble-free value. In laboratory studies we obtain highly reproducible results at 0.05 dB/m resolution. This corresponds to a bubble density resolution of approximately 0.5 bubble/m^3 for bubbles of radius 300 μ (one micron radius increment) and 200/m^3 for bubbles of radius 20 μ. Since typical bubble densities that we found in coastal waters range from order 10/m^3 of the largest bubbles to 10^4 or 10^5/m^3 of the smallest bubbles, we have a high level of confidence in our results. At sea, of course, there is great variability.

These experiments [5] have provided additional conclusions about ocean bubbles: in coastal waters of depths less than 40m bubble densities within a few meters of the surface are about 10 times greater in the summer months than in the winter, suggesting that the bubbles are biological in origin (Fig. 2); these ambient bubble densities appear to decrease due to Bernoulli pumping with increasing wind speed up to about 10 knots and the densities increase due to wave-breaking processes only when the wind speed is greater than this. A few meters under a sea surface slick or under windrows there is

189

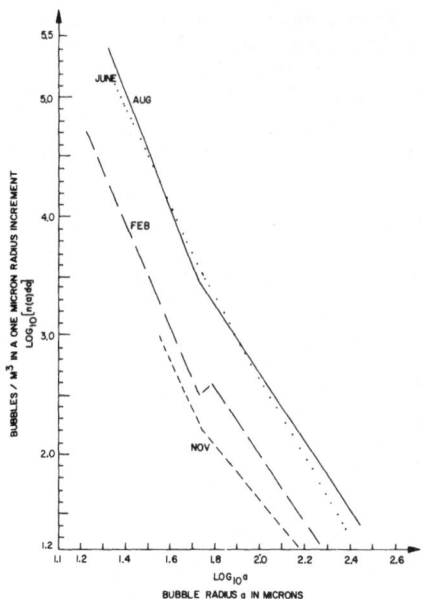

Fig.2 Some seasonal values of bubble densities in coastal waters of depth 40 m or less [5].. Experiment depths 3 to 10 m; wind speeds less than 5 m/s.

an order of magnitude increase in bubble densities compared to usual ocean surfaces, even in deep water locations. The number of near surface bubbles of radius over 60 µ increases in coastal waters from one hour before sunset to a peak at about three hours after sunset; a heavy fog at daybreak "clouds" the expected reverse behavior.

Dispersion and Phase Fluctuations

The total fraction of gas in bubble form independent of bubble radius, i.e., the relative volume of bubble gas to liquid, U, comes from a separate measurement of the low frequency asymptote of the sound velocity dispersion.

The dependence of the sound speed on the compressibility and density is given by

$$c^2 = (\rho\kappa)^{-1} \tag{15}$$

The compressibility is made up of a part due to the bubble-free water κ_0 and a complex part due to the bubbles themselves κ_1

$$\kappa = \kappa_0 + \kappa_1 \tag{16}$$

$$\text{where} \qquad \kappa_0 = (\rho c_0^2)^{-1} \tag{17}$$

The complex compressibility due to the bubbles (section A6.2,[2]) is

$$\kappa_1 = \frac{NS^2}{m\omega^2[(-1 + \omega_R^2/\omega^2) + i\, R_M/(\omega m)]} \tag{18}$$

where N is the number of bubbles per unit volume. For the present, we assume all bubbles are of radius a. To simplify, we use (4) and define the frequency ratio

$$Y = (f_R/f) = (\omega_R/\omega) \tag{19}$$

The speed in bubbly water is the real part of the square root of (15) using (16) through (19),

$$Re\{c\} = c_0 \{1 - [(3U\gamma^2)(\gamma^2 - 1)]/[2a^2 k_R^2(\gamma^2 - 1)^2 + \delta^2]\} \qquad (20)$$

where $k_R = \omega_R/c_0$ is the value of k_0 at resonance, and $U = N(4/3)\pi a^3$, Fig. 3.

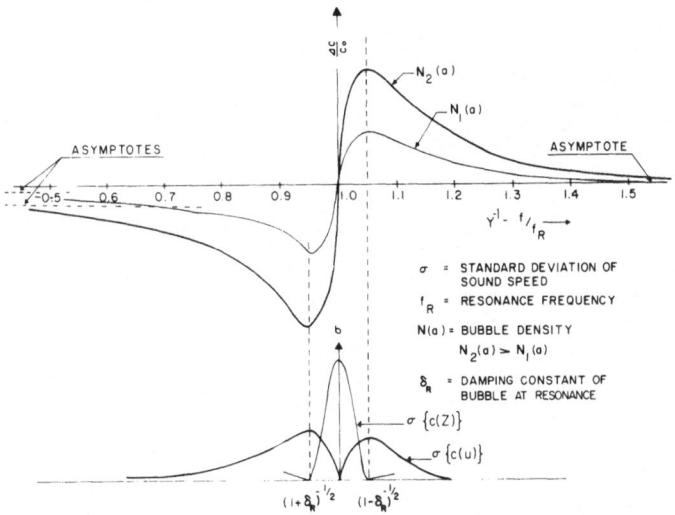

Fig.3 Fractional change of sound speed as a function of frequency near resonance (upper graph) and standard deviation of speed fluctuations (lower graph) due to changing resonance frequency, $\sigma\{c(z)\}$, and due to changing, bubble fractions, $\sigma\{c(u)\}$, at sea [8].

The behavior that we are interested in occurs at low frequency, $f \ll f_R$, where the speed is

$$c_{\ell f} = c_0 [1 - 3U/(2a^2 k_R^2)] \qquad (21)$$

independent of the particular bubble radius. Since ak_R is a constant for a given gas at a given depth, so long as the frequency condition $f \ll f_R$ is fulfilled a measurement of $c_{\ell f}$, combined with knowledge of c_0 which is a function of temperature, salinity and depth, will yield the total bubble fraction U. The generalization to a spectrum of bubbles is trivial.

Within several meters of the ocean surface there may be enough ambient bubbles to create a dispersive acoustic index of refraction. Furthermore, the variability of the ocean over times of order seconds causes fluctuations in the sound phase measured in-situ over a path of the order of meters [8]. Three major sources of speed fluctuations due to bubbles have been identified: (a) Change of phase velocity due to change of total volume fraction of bubbles. From (20), at frequencies well below resonance, the standard deviation (σ) of speed is proportional to that of U

$$\sigma\{c(U)\} = 3c_0/(2a^2 k_R^2) \, \sigma\{U\} \qquad f \ll f_R \qquad (22)$$

(b) For a predominant bubble radius, change of number of bubbles will cause peak fluctuations at the frequencies of the speed maximum and minimum given by $Y = (1 \pm \delta_R/2)$ [6]. From (20) we find the standard deviation

$$\sigma\{c[u(a)]\} = [3c_0(1 \pm \delta_R)/(4a^2 k_R^2 \delta_R)] \, \sigma\{u(a)]\} \, .$$

191

The (+) sign, giving the larger fluctuation occurs at the frequency below resonance (Fig. 3). (c) For a predominant bubble radius in a distribution, a change of ambient pressure will change the resonance frequency (3) and thereby cause a variation in speed. This effect is also shown in Fig. 3, where it is identified as producing an rms change $\sigma\{c(z)\}$. The magnitude of the effect can be readily derived: Consider the region near resonance where

$$Y = f_R/f \simeq 1 + \Delta, \qquad \text{where } \Delta \ll 1. \qquad (24)$$

For the slow rate of pressure change accompanying the ocean wave system,

$$\Delta \simeq (h/12) \exp(-K_0 Z)/(1 + Z/10) \qquad (25)$$

where Z is experiment depth, h is wave height and K_0 is the wavenumber at surface peak frequency all in MKS units. Then

$$\sigma\{c(Z)\} = \{Bu(a)da \, c_0/(k_R a)^2[1 + (Z/10)]\delta_R^2\} \, \sigma\{h\} \, \exp(-2K_0 Z) , \qquad (26)$$

where $B = 0.083 \, M^{-1}$, $\sigma\{c(Z)\}$ is the standard deviation of fluctuations in sound speed due to perturbations of bubble resonance frequency, and $\sigma\{h\}$ is the rms height of the surface in meters.

When $f \ll f_R$ for the great majority of bubbles we find evidence of the mechanism described by (22). In this case the source of the variation is random and the spectral density of the phase fluctuations are nearly Gaussian (Fig. 4).

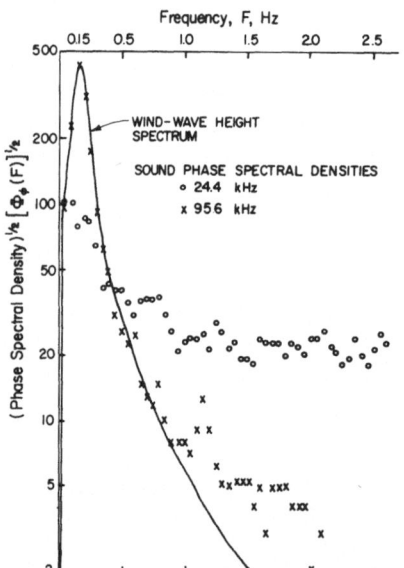

Fig.4 Spectral density of phase fluctuations at 3m depth at sea [7]. Signal at 24.4 kHz is below the resonance frequency of most bubbles. At 95.6 kHz there is resonance for a predominant bubble radius and the sound phase changes are highly correlated with the ocean wave height. The solid line is the Pierson-Moskovitz wind wave height spectrum.

When sound irradiates a bubbly region with a predominant radius at its resonance frequency the mechanism of (25) is apparent; then the sound phase is very highly correlated with the ocean surface height and the phase spectral density follows the Pierson-Moskovitz spectrum of a wind-driven sea (Fig. 4). As a further diagnostic tool, the variation of the phase spectral density with depth has been used to determine the depth dependence of these predominant bubbles within a few meters of the ocean surface [7].

References

1. Herman Medwin, J. Geoph. Res. 75-3, 599-611 (1970).
2. C.S. Clay and H. Medwin, "Acoustical Oceanography", 544 pp, Wiley, New York, 1977.   Sections 6.3, 6.4, A6.
3. "Acoustic Theory of Bubbles", in Physics of Sound in the Sea, edited by R. Wildt, N.D.R.C. Summary Technical Report Div. 6, Chap. 28, Vol. 8, Washington, DC (1946).
4. Herman Medwin, J. Acoust. Soc. Am. 62, No. 4, 1041-44 (1977).
5. Herman Medwin, J. Geoph. Res. 82-6, 971-976 (1977).
6. Peter C.C. Wang and Herman Medwin, Quart. J. Appl. Math 31, 411-425 (1975).
7. Herman Medwin et al., J. Geoph. Res. 80-3,405-413 (1975).
8. Herman Medwin, J. Acoust. Soc. Am. 56-4, 1100-1104 (1974).

Acoustical Scattering from Near-Surface Bubble Layers

P.A. Crowther

Marconi Space and Defense Systems Limited
Frimley, Surrey, England

Abstract

A critical examination is made of the inhomogeneities present in the upper
ocean owing to wind-wave induced bubble layers, with reference to the acoustic backscatter and absorption in the over 1-100 kHz band resulting in
critical situations involving near horizontal propagation, such as that of
a surface duct. Comparison between bubble scattering and interface scattering is made for backscatter and forward scatter situations, and regimes of
relative dominance are indicated. Approximate estimates of wind-wave
bubble distribution laws are made based on experimental acoustic backscatter
and dimensional arguments.

1. Introduction

1.1 General

Experimental data linking surface derived bubbles with acoustic backscattering [1,2,3], and speculation on the rôle of bubbles as attenuators of duct
propagated sound in the ocean [4,5] have been available for many years.
The purpose of this paper is to re-examine quantitatively the hypothesis of
scattering by surface derived bubbles in the light of available information,
and attempt to estimate laws for the generation and dispersion of surface
derived bubbles.

To do this, because of the lack of direct data on bubble distributions,
we have to appeal to indirect data - principally acoustic backscattering
from the sea surface as a function of frequency, windspeed and angle of
incidence. This we believe to be due to 2 parts - (a) sea interface scatter,
and (b) scatter from near-surface bubbles, which are respectively discussed
in Sections 2, 3 below and compared critically against experimental data in
Section 4.

1.2 Data Sources for Surface Backscatter

Measurements have been made of acoustic backscatter from the sea surface by
a variety of experiments together covering a frequency range from < 1 kHz
to 60 kHz, in sea states due to windspeeds from $\gtrsim$ 1 m/s to $\sim$ 18 m/s, and
using either directional array sources [1,2,6,7], or explosive sources and
recorded returns filtered over selected frequency bands [8-11].

A general picture of what happens may be formed by reference to Figs. 1-5,
taken from a few of these sources. In general, backscatter is found (a) to
decrease with incidence angle at a rate dependent on other parameters, (b)

to increase with frequency, and (c) most markedly to increase with sea state and windspeed.

The scale of windspeed and frequency dependence is much greater than has been found for backscattering of electromagnetic waves from (above) the sea surface, as reported by GUINARD and DALY, for a comparable wavelength range and even larger sea-state range [12]. This weak dependence in the electro-magnetic case is consistent with expectations based on surface wave scatter-ing theory and knowledge of surface wave spectra, discussed in Section 2, and strongly suggests the agency of an additional mechanism for acoustic scatter, which is not active electromagnetically, and which increases with sea state and windspeed. It is the contention of this paper that near surface bubbles fill this rôle.

1.3 Notation

The following general notation is used:

θ = incidence angle

$\underline{k}$ = $\{\underline{\mu}, \lambda\}$ = scattered wave vector ($\underline{\mu}$ = horizontal component, λ = vertical component)

k = $|\underline{k}|$ = radian wave number

$Z(\underline{x})$ = spectral covariance function of surface deflection, ζ

$\underline{x}$ = 2D separation vector

σ^2 = variance of ζ

$Y(\underline{x})$ = $\sigma^2 - Z(\underline{x})$

M_{SI} = surface scattering strength due to interface roughness

M_{SB} = surface scattering strength due to bubble layer scatter

f, f_R = frequency, resonant frequency

U = windspeed

g = gravity

R, R_f = bubble radius, resonant radius

$V(R)$ = bubble terminal ascent velocity in still water

z = depth

MKS units are used throughout in all equations involving dimensional quantities.

2. Brief Review of Theoretical Model for Rough Interface Scattering

We use a model for surface scattering, employing a stochastic adaption of Lord Rayleigh's supposition that the scattering of a plane wave may be

represented by series (or integral) of plane wave scattered components [13].
The resulting approximate expression for surface back scattering strength
from this model is

$$M_{SI} = (\lambda/2\pi)^2 \int \exp\left[-4\lambda^2\ Y(\underline{x})-2i\ \underline{\mu}.\underline{x}\right]d^2x \tag{1}$$

It is worth pointing out that this differs from the conventional Kirchhoff
approximation solution radically in near grazing geometry, [14] but this is
equivalent to the solution obtained from the Helmholtz integral by an improvement to Kirchhoff's approximation [15].

Numerical evaluation of (1) by the present author shews that in practice,
for the sea surface spectra and Y ($\underline{x}$) functions actually encountered, the
scattering strength at all but near specular geometries does not greatly
differ from that derived by MARSH, on the basis of a power series expansion
in terms of the Rayleigh roughness parameter, [16] - namely:

$$M_{SI} \approx 4\lambda^2 E_2(2\underline{\mu})\quad (\theta \gtrsim 30^o) \tag{2}$$

where

$$E_2(\nu) = (1/2\pi)^2 \int Z(\underline{x})\ \exp(i\underline{\nu}.\underline{x})d^2x, \tag{3}$$

the 2D surface power spectrum.

We may obtain quite accurate data on the expected level of E_2 from use of
the empirical PIERSON-MOSKOWITZ spectrum for the 'fully developed' sea,
expressible as a non-directional spectrum by integration over azimuth, ψ, as

$$E_1(\nu) = \int_0^{2\pi} E_2(\nu,\psi)d\psi = \frac{1}{2}\ a\nu^{-3}\ \exp\ (-bg^2/\nu^2 u^4), \tag{4}$$

where a = $8.1.10^{-3}$, b = 0.74. Over the range of frequencies and windspeeds
with which we are most concerned, the wavenumber involved is usually sufficiently high that we are well into the saturation region of the spectrum,
where

$$E_t(\nu) \sim \frac{1}{2}\ a\nu^{-3}; \tag{5}$$

and, ignoring the rather small azimuth angle dependence

$$E_2(\nu) \sim (a/2\pi).\nu^{-4}. \tag{6}$$

Putting (6) into (2) gives, essentially, MARSH'S formula

$$M_{SI} \sim 2_{10}^{-4}\ cot^4\theta\quad (\theta \gtrsim 30^o) \tag{7}$$

In summary, for interface backscattering away from specular, we expect
that frequency and windspeed dependence should be weak, and that the
scattering should fall off violently as grazing incidence is approached
($\theta \to 90^o$). That none of these statements is always true occasions the
investigation of bubble scattering mechanisms; but before proceeding to
the main topic, it is worth pointing out that attempts have been made to
improve the approximation inherent in (2,7) by considering the surface as
split into twin structures - a long scale structure consisting of swell

196

waves, on waves, on which is carried the small scale structure responsible for the (diffractive) backscattering represented by (2,7), the idea being to modify (2) by considering the local scattering as modulated by the slope structure carried by the long scale structure. In numerical terms, however, this slope modulation effect is still very small compared with the frequency and sea state dependence encountered. For example, if we study Bachmann's composite surface model [17] in detail, we see little backscatter variation due to the long scale wave slope, and most due to effectively an assumed variation in 'a' in (4,6) with windspeed, which is quite inconsistent with present accumulated knowledge on surface wave statistics.

3. Bubble Scattering Model

3.1 Resumé of Acoustic Properties of Bubbles

The following is an outline of basic bubble mechanics: for a more detailed review reference is made to Medwin [18]. Individual bubbles act as resonant scatters, with a resonant frequency, near the surface, of $f_R \sim 3.2/R$ [MKS]. The resonance is damped by a loss tangent, δ, comprising re-radiation, δ_r, and dissipation, δ_d, given approximately by

$$\delta = \delta_r + \delta_d,$$

$$\delta_r = 0.013, \quad \delta_d = 3.8_{10}{}^{-4} f_R^{\frac{1}{2}} \qquad (f \lesssim 10^5 \text{ Hz}) \tag{8}$$

The bubble scattering cross section is given by

$$\sigma_s = 4\pi R^2 / \left[((f_R/f)^2 - 1)^2 + \delta^2 \right] \tag{9}$$

Bubble layers may be defined by a spectrum, $N(R,z)$, being the density of number of bubbles per volume per interval of radius, at depth z. A bubble at resonance is an isotropic scatterer; and the volume scattering strength (target strength per unit volume) is

$$M_v(z) = (1/4\pi) \int_0^\infty N(R,z)\sigma_s(R,f)dR \approx \frac{\pi}{2} R_f^3 N(R_f,z)/\delta \tag{10}$$

where $R_f = 3.2/f$ (MKS) = resonant radius, and where the approximation is made of ignoring all but near-resonant bubbles in obtaining the approximation on the right in (10). Also of interest is the amplitude absorption coefficient in a bubble layer, $\alpha(z)$. This is related to $M_v(z)$ as

$$\alpha(z) = 2\pi(\delta_d/\delta_r). \quad M_v(z) = \pi^2 R_f^3 N(R_f,z). \quad \delta_d/(\delta.\delta_r). \tag{11}$$

3.2 Bubble Distribution Laws

Bubble density spectra have been obtained by BLANCHARD and WOODCOCK, using water extracted from breaking waves, [19] and by MEDWIN and co-workers, [20,21] working with natural bubbles probably of biochemical origin at depths 3-15 m. Both of these quite different methods gave rise to a density following roughly $N(R) \alpha R^{-4}$ (scale free) laws, measurements ranging over $R \sim 8.10^{-5}$ to 3.10^{-4}m, (BLANCHARD and WOODCOCK) and $R \sim 1.6. 10^{-5}$ to 10^{-4}m. (MEDWIN). Bubble densities derived from these sources do not relate closely to the surface scattering problem, being too high to explain observed backscattering at other than the highest sea states at appropriate frequency;

neither do the spectra cover frequencies $\lesssim$ 12 kHz. We must conclude that these spectra hold only locally for the inshore conditions concerned - not for the open ocean.

We have to consider some model of the generation and dispersion of surface derived bubbles. Supposing bubbles to be generated very near the surface from breaking waves, spray and capillary waves, we then assume them carried downwards by vertical turbulence. Ignorning change of radius with pressure, gas absorption and other effects in the full equation [22]; we postulate a simplified statistical transport equation as

$$\frac{\partial}{\partial t} N(R,z) - \kappa \frac{\partial^2}{\partial z^2} N(R,z) - V \frac{\partial}{\partial z} N(R,z) = S(R). \; \delta(z-\varepsilon) \tag{12}$$

where κ = turbulent diffusion constant, S(R) = superficial source function at depth ε, V(R) = bubble terminal velocity. This has a steady state solution as $\varepsilon \to o$:

$$N(R,z) = (S(R)/V(R)). \; \exp(-x/D), \tag{13}$$

where the e-fold layer depth is

$$D = \kappa/V(R) \tag{14}$$

Bubble densities based on this conjectural and simplified model require 3 basic inputs for computation: (a) S(R), (b) V(R), and (c) κ. None of these is completely understood. Total ignorance of S(R) is matched by only very rough ideas of κ; MONIN, for example suggests $\kappa \sim 10$ cm²/sec, but does not qualify this with any indication of sea state or depth dependence [23]. V(R) is known to vary from a viscous drag regime for $R \lesssim 100$ µm to separated flow and fully turbulent regimes [19,22,24]. For the present, we approximate V(R) by:

$$V(R) = Min \left\{ \left[0.3 \, R^{-\frac{1}{4}} + 6.10^{-4} \, R^{-1} \right]^{-2}, \; 0.25 \right\} \; (R \stackrel{<}{\sim} 7.10^{-3} m) \, [MKS] \tag{15}$$

This is consistent with calculations and supporting data, BLANCHARD and WOODCOCK [19], and experimental observations of DATTA et al [24].

We may estimate the source function dependence on windspeed, U, and R from a dimensional argument; if we assume that S(R) $\propto$ W, the surface rate of wind-wave working (power/area) and is additionally a function of surface tension, τ, radius, R, and water density, ρ, then the only permissible form is

$$S(R) \propto W. \; \tau^{-1} \, R^{-3}, \tag{16}$$

The law W $\propto U^3$ is expected for a scale modelled sea, so that the surface bubble density is expected to go as

$$N(R) = N(R,0) \propto U^3 \, \tau^{-1} \, R^{-3}/V(R). \tag{17}$$

The layer thickness, D, will also depend on windspeed via the dependence of κ on U. This is hard to predict, but a crude dimensional argument for a fully developed sea would run as follows:-

$$\kappa = \kappa(U, \rho, g) \propto U^3/g \quad ? \tag{18}$$

so that

$$D \propto U^3/V(R). \tag{19}$$

Since the dimensional arguments are very 'strethed', we take (13,14,16-18) only as rough indications of likely windspeed, size and depth dependences, allowing some departure on the basis of acoustic results.

3.3   Scattering from a Bubble Layer

Backscattering due to the bubble layer is modified by (a) absorption of sound in the layer and (b) reflection from the surface. This may be modelled for a general bubble depth dependence by considering the propagation via direct and (pressure-release) surface reflected paths from the far field to each resonant bubble and vice-versa. By the ray acoustics approximation, this gives:

$$M_{SB} = M_o \int_0^\infty p(z) \left[\exp(-\beta F(z)-i\lambda z) -\exp(-\beta[2-F(z)] + i\lambda z)\right]^4 dz, \qquad (20)$$

where

$$M_o = \pi B.D./(2.\delta); \quad B = R_f^3 N(R_f); \quad D = 1/p(o);$$

$$\beta = \int \alpha(z)dz/\cos\theta = \pi^2 \delta_d B.D./(\delta_r.\delta.\cos\theta) = 1\text{-way amplitude absorption}$$

exponent for a ray at angle θ reaching the surface;

F(z) = cumulative bubble distribution function, such that

$F(0) = 1$, $F(\infty) = 0$; $p(z) = -dF/dz = N(R_f,z)/\int_0^\infty N(R_f,z')dz'$ = probability density function.

Eq (20) transforms to

$$M_{SB} = M_o \int_0^1 \left[\exp(-2\beta F)-2 \exp(-2\beta)\cos(2\lambda z)+\exp(-2\beta(2-F))\right]^2 dF. \qquad (21)$$

By expanding, we reduce this to:

$$\begin{aligned}
M_{SB} = M_o \Big\{ &4 \exp(-4\beta) + \left[1-\exp(-8\beta)\right]/(4\beta) - \\
&-4 \exp(-2\beta) \int_0^1 \left[\exp(-2\beta F) + \exp(-2\beta \overline{2-F})\right] \cos(2\lambda z)dF \\
&+2 \exp(-4\beta) \int_0^1 \cos(4\lambda z)dF \Big\}
\end{aligned} \qquad (22)$$

This equation is still general with regard to depth distribution.

There are several limiting cases, as follows:-

Case (1): Layers thick on a wavelength scale. The oscillatory terms in the integrals render them insignificant, and we approximate to:

$$M_{SB} \sim M_o \{4 \exp(-4\beta) + \left[1-\exp(-8\beta)\right]/(4\beta)\} \qquad (\lambda D >> \tfrac{1}{4}) \qquad (23)$$

Case (2): Layers thick on an absorption scale, where

$$M_{SB} \sim M_o/4\beta = \delta_r.\cos\theta/(8\pi\delta_d). \quad (\beta >> \tfrac{1}{4}) \qquad (24)$$

This predicts an upper limit to M_{SB} theoretically, no matter how many bubbles are generated: the limit is independent on the details of the bubble distribution, but depends only on the fairly well understood parameters δ_r and δ_d.

Case (3): Layers thin on an absorption scale do not have a universal closed form behaviour, independent of depth distribution; but if we assume an exponential, $F(z) = \exp(-z/D)$, and transform (22) back to integrals over z, we obtain:

$$M_{SB} \sim M_o \left[6 - 8/(1+4\gamma^2) + 2/(1+16\gamma^2)\right], (\beta << \tfrac{1}{4}). \qquad (25)$$

where $\gamma = \lambda D$. Of particular interest is the limiting case where the layer is thin both in γ and in β. Here

$$M_{SB} \sim 384.M_0 \gamma^4, \quad (\beta << \tfrac{1}{4}, \; \gamma << \tfrac{1}{4}) \tag{26}$$

predicting that a <u>thin weak</u> surface bubble layer should scatter asymptotically as $Cos^4\theta$. This is the same law as the interface scattering as $\theta \rightarrow 90^\circ$, and explains several experimental results at low frequency.

Thus, contrarily to what is sometimes assumed, we see that a bubble layer can produce a variable backscattering strength vs. angle dependence, and exhibits a saturation effect at high levels following a $Cos\theta$ Lambert Law. At low frequencies and sea states, it resembles interface scattering angle dependence at near-grazing incidence.

4. Analysis of Experimental Backscatter Data

The scattering formulae and dimensional arguments qualitatively explain observed frequency and sea state dependence of backscattering. Some qualitative comparison follows. At 60 kHz, the saturation prediction is compared with experiment in Figs (1,2). The saturation level can be seen' to coincide with experimental estimates at $\theta \gtrsim 25^\circ$ to within a few dB. At lower frequencies, saturation is not reached for sea states encountered experimentally — coming to within ~ 10 dB below saturation at $U = 18$ m/s at ~ 3 kHz in Fig (4), for example. The low layer thickness limit (25,26) is in Fig (3), however, where backscattering at 3 kHz and 6 m/s has been fitted by the addition of interface and bubble scatter, the latter dominating at $\theta \gtrsim 60^\circ$. The effect of the $Cos^4\theta$ law predicted for a thin layer can clearly be seen.

Fig (4) shows how 3 kHz data can be fitted at higher sea states by progressively higher density and layer thickness, carrying the angle dependence from $Cos^4\theta$ to $Cos\theta$. We can also understand the contrast between windspeed dependence at high and low frequency by comparing Figs 2,5. At 60 kHz, we have a rapid increase in scattering with U, leading to saturation, through the ease of transporting the small bubbles involved. At 3 kHz, there is little initial increase with U, the backscatter being dominated by M_{SI}, owing to the difficulty of transporting large bubbles from the surface; but beyond ~ 5 m/s, a rapid increase with U occurs, as M_{SB} begins to dominate.

Similar analysis is possible for other frequencies, although data is inferior through the use of omnidirectional equipment and broadband filtering. Some 160 selected data points from sources referred to in Section 1 have been fitted by a simple universal formula of the type suggested in Section 3. The fit was made on the basis of least squares decibel error in M_S, taken as

$$M_S = \exp(-4\beta) M_{SI} + M_{SB}, \quad (\text{theory}) \tag{29}$$

with M_{SB} evaluated from (22) by a series expansion

$$M_{SB} = M_0 \left\{ 4 e^{-4\beta} + (1-e^{-8\beta})/4\beta + 2e^{-4\beta}/(1+16\gamma^2) \right.$$
$$\left. + \sum_{n=0}^{\infty} [(-1)^n + 4 e^{-4\beta}] \; [1 + \overline{2n+1}^2 \; \gamma^2]^{-1} \; (2\beta)^n/n! \right\}. \tag{30}$$

Over the frequency range 1-60 kHz, this gave a best fit close to:

$$N(R) = 2.3 \, 10^{-10} \, U^3 \, R^{-3}/V(R) \qquad\qquad ? \qquad\qquad (31)$$

$$D = 3.3 \, 10^{-3} \, U/V(R) + 50.R \qquad\qquad ? \qquad\qquad \mathbf{[MKS]} \qquad (32)$$

at which point the error standard deviation is 3.5 dB. Whilst this standard deviation is encouragingly low compared with the 40 dB range of the data, it is still true that individual experiments can be fitted better by selected N,D combinations deviating considerably from (31,32). The N.D product is likely to be better supported than (31,32) separately. Eqns. (31,32) correspond fairly closely to theoretical conjectures for U, R, V(R) dependence, except that the windspeed dependence is weaker than predicted, through (32). The small second term in (32), which improves the fit, may represent a mean source depth for initial creation of the bubble.

5. Bubble Loss in Surface Reflection

We finally consider the bubble contribution to propagation loss. Referring to Section 3 above, we expect to find an amplitude reflectivity for a bubble layer under a flat surface of order $\exp(-4\beta)$, compared with an interface (coherent) amplitude reflectivity of $\exp(-2\lambda^2\sigma^2)$. For the Pierson-Moskowitz sea, $\sigma^2 = 2.8 \, 10^{-5} \, U^4$ (MKS). Using the bubble densities estimated above, it is possible, for a given sea state, to compute both loss terms and compare. This has been done in a specimen case tabulated below.

Table: Interface and bubble reflection loss comparison

Frequency: 3.5 kHz; $U = 8.5$ m/s; $\sigma = 0.4$ m; $R_f = 9.1 \, .10^{-4}$m; $\beta = 7.10^{-5}/\cos\theta$								
Grazing angle: $90^\circ-\theta$:	0.5°	1°	1.5°	2°	2.5°	3°	4°	6°
$\exp(-2\lambda^2\sigma^2) \equiv$ dB loss	0.05	0.18	0.41	0.73	1.14	1.64	2.9	6.6
$\exp(-4\beta) \equiv$ dB loss	0.28	0.14	0.09	0.07	0.06	0.05	.04	.02

This comparison is based on conditions in surface duct propagation experiments analysed in terms of virtual modes and interface reflection loss by BARNARD and DEAVENPORT [24]. The virtual modes correspond to ray grazing angles up to $\sim 2.7^\circ$. It can be seen that we expect surface roughness loss to predominate over the bubble loss for grazing angles $\gtrsim 1^\circ$, and that for the lowest mode, where the ray angle is about 1.4°, the estimated bubble loss is smaller but of similar order to that due to the surface loss. Angular dependence of the two mechanisms are, however, quite different. Thus while for weak ducts bubble loss might be of similar order to interface loss, in stronger channels, e.g. deep sound channel paths involving surface reflections where grazing angles of order $\gtrsim 5^\circ$ are typically encountered, surface bubble loss will be comparatively negligible.

To generalize this example note that since the bubble loss goes according to $N(R).D(R) \propto U^4$, and the surface loss according to $\lambda^2 \propto U^4$, there is not likely to be much transition of mechanism as a function of windspeed. The frequency, f, dependence of bubble loss will go roughly in proportion to $R^3 N(R) \, D(R) \propto f^2$, whilst the interface loss will also go as $\lambda^2 \propto f^2$, so that whilst bubble scattering can dominate interface scattering at high U or high f, no such dominance is expected for propagation loss since dependences are expected to be similar for both mechanisms.

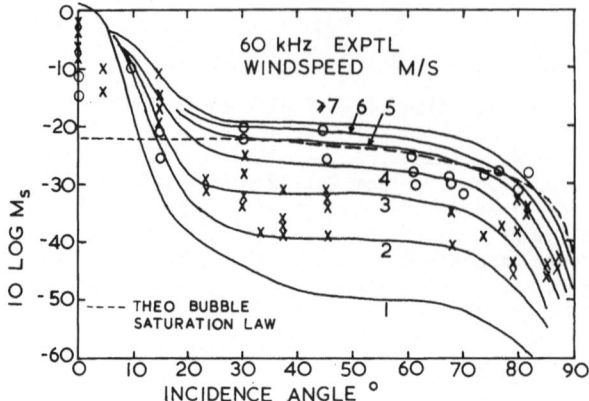

Fig 1 Experimental 60 kHz backscatter. – from GARRISON et al [2]; points from URICK and HOOVER [I], O at 7-8 m/s wind, X at 4-5 m/s wind

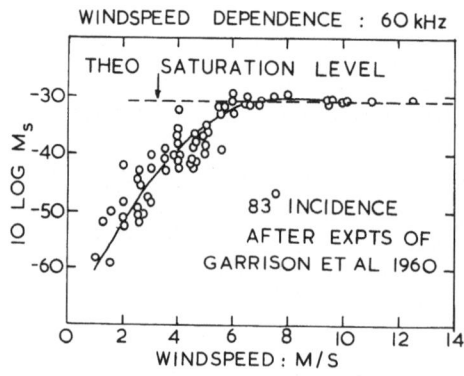

Fig 2 60 kHz data from [2]. The solid curve is empirical. Note approach to theoretical saturation.

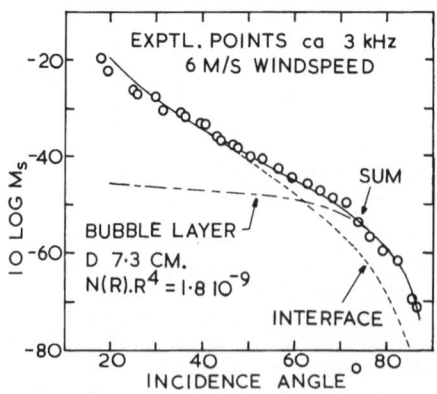

Fig 3 Backscattering at moderate windspeed. MSDS 1968 expts (points). Curves for M_{SI}, M_{SB} and their sum. The layer required is very weak, equivalent to only 0.01 resonant bubbles/metre^2 superficial density.

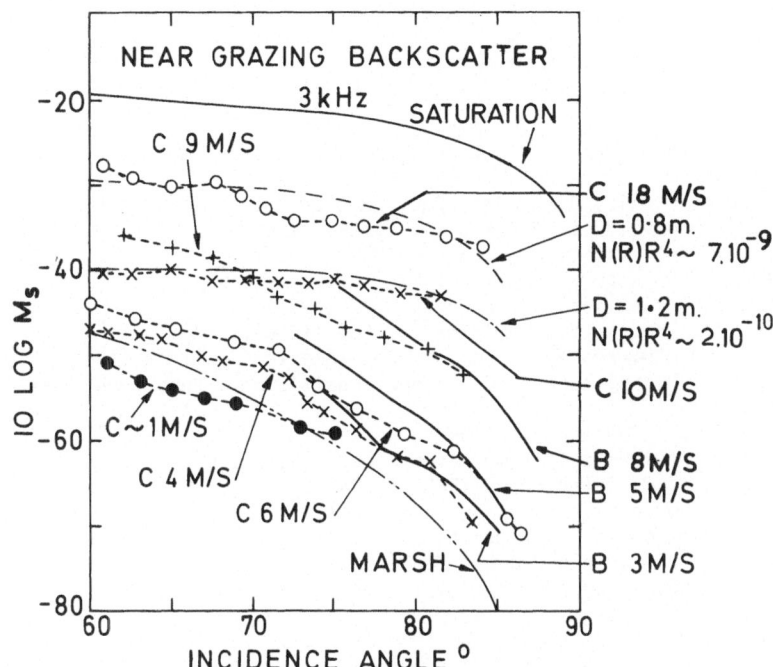

Fig 4 3 kHz backscatter vs. angle for sea states 1-6. Data 'B' are from BACHMANN [7]; Data 'C' are from MSDS expts [6]. See text for theoretical curves.

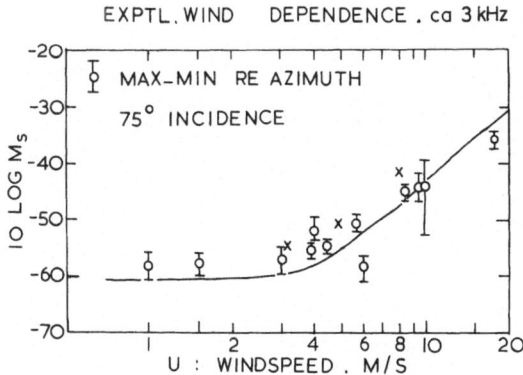

Fig 5 Specimen 3 kHz windspeed dependence. Points experimental – 0 from [6], X from [7]. The curve is the general theoretical fit via (31,32).

Acknowledgement

This work has been carried out with the support of the Procurement Executive, Ministry of Defence.

References

1. R.J. Urick and R.M. Hoover, J. Acoust. Soc. America 28, 1038-42 (1956).
2. G.R. Garrison, S.R. Murphy and D.S. Potter, J. Acoustic. Soc. America 32, 104-11 (1960).
3. C.S. Clay and H. Medwin, J. Acoust. Soc. America 36, 2131-34 (1964).
4. M. Shulkin, J. Acoust. Soc. America 44, 1152-4 (1968).
5. M. Shulkin, J. Acoust. Soc. America 45, 1054-5 (1969).
6. P.A. Crowther, MSDS Reports 1969 and 1972 (unpublished).
7. W. Bachmann, J. Acoust. Soc. America 54, 712 (1973).
8. R.M. Richter, J. Acoust. Soc. America 36, 864 (1964).
9. R.P. Chapman and H.D. Scott, J. Acoust. Soc. America 36, 1735 (1964).
10. J.R. Brown, J.A. Scrimger and R.G. Turner, Pacific Nav. Lab. Tech. Memo 66-8 (1966).
11. M.V. Brown and R.A. Saenger, J. Acoust. Soc. America 52, 944-960 (1972). This paper also reviews data from references |8,9,10|.
12. N.W. Guinard and J.C. Daley Proc. IEEE 58, 543-50 (1970).
13. Lord Rayleigh, The theory of sound, Dover Edition Vol. 2, p.89 (1945).
14. See e.g. P. Beckmann and A. Spizzichino "The scattering of electro-magnetic waves from rough surfaces' MacMillan : New York (1963); L. Fortuin, J. Acoust. Soc. America 47, 1209-28 (Review) and references cited therein.
15. H. Trinkaus , Saclantcen Memorandum SM-15 (1973).
16. H.W. Marsh, J. Acoust. Soc. America. 33,330 (1961).
17. W. Bachmann, J. Acoust. Soc. America 54, 712-6 (1973).
18. H. Medwin, Ultrasonics 15, 7-14 (1977) (Review).
19. D.C. Blanchard and A.H. Woodcock, Tellus 9, 145-58 (1957).
20. H. Medwin, J. Geophys. Res. 75, 599-611(1970).Ibid, J. Acoust. Soc. America 56, 1100-4 (1974).
21. T.B. Huffmann and D.L. Zveare 'Sound Speed Dispersion and Inferred Microbubbles in the Upper Ocean', M.S. Thesis, U.S. Naval Postgraduate School, Monterey (1974).
22. G. Garreston, J. Fluid Mech. 59,187-206, (1973).
23. A.S. Monin, U.M. Kamentovich, V.G. Kort 'The Variability of the Oceans', Eng. Trans., J. Wiley (1977). Chapter 3.3; also A.S. Monin, Sov. Phys. Usp. (U.S.A. Transl.) 16, 121-131 (1973) (Review).
24. R.L. Datta D.H. Napier and D.M. Newitt, Trans. Institution of Chem. Engrs. 23, 14-16, (1950).
25. G.R. Barnard and R.L. Deavenport, J. Acoust. Soc. America 63, 709-714 (1978).

Density of Air-Bubbles Below the Sea Surface, Theory and Experiments

Ir.P. Schippers

Physics Laboratory TNO
The Hague, The Netherlands

1. Introduction

Air-bubbles are usually found in the sea up to a depth of some meters below the surface (bubble radii usually ranging from 1 mm to 10 μm).
   The bubble density may be determined by measuring the acoustic back-scattering from the bubbles. Sound is sent straight upwards from the bottom (North Sea, water depth 18 m). The scattered signal is received at the same place (see Fig. 1). In this paper a formula will be derived in order to calculate the bubble density from the scattered sound level. Some experimental results will finally be presented for one frequency, corresponding to a certain class of bubble radii.

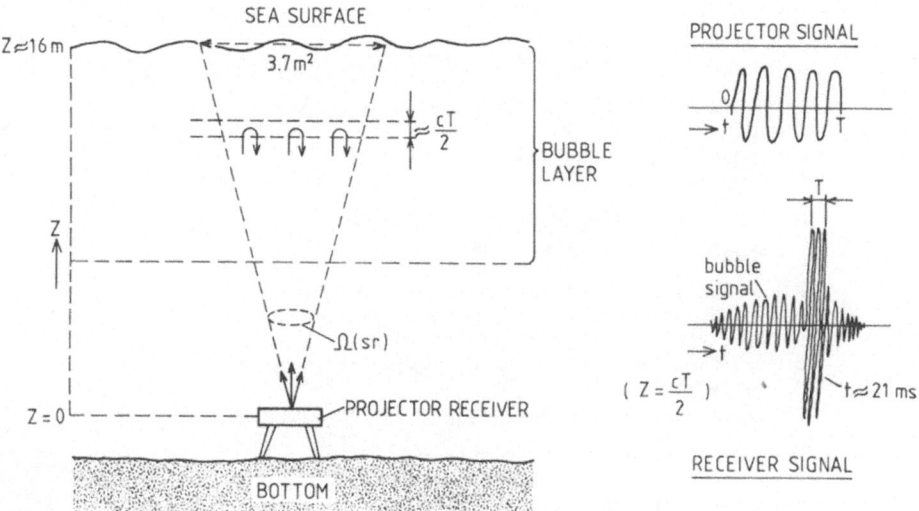

Fig. 1 Measuring system

2. Theory

2.1 Scattering by One Bubble

The scattering of sound by only one bubble (with radius R) is considered first. The wave solution for the total particle speed potential (in the water) can be derived from the linearized form of the Navier-Stokes equation. (This equation contains the shear- and volume-viscosity η and κ.) The wave field obtained is the sum of a plane wave (the incident wave) and a spherical wave (the scattered wave).

Suppose that a CW-signal with radial frequency ω and of unit pressure falls on the bubble. The particle speed vector $\underline{v}$,

$$\underline{v} = V(\omega,r)\, e^{i\omega t}\, \underline{e}_r$$

(with $\underline{e}_r$ = unit radius vector from the bubble centre)
and the stress vector $\underline{s}$,

$$\underline{s} = S(\omega,r)\, e^{i\omega t}\, \underline{e}_r$$

(acting on a small plane with normal $-\underline{e}_r$)
of the scattered spherical wave can be calculated at a distance r from the bubble centre.

$$V(\omega,r)\, e^{i\omega t} = -\frac{R}{r^2}\cdot\frac{\delta+ib}{\omega\rho N}\left(1+\frac{\omega r}{c}\right)e^{i(\omega t - kr)} \tag{1}$$

$$S(\omega,r)\, e^{i\omega t} = -\frac{R}{r}\cdot\frac{\delta+ib}{\omega\rho N}\left(\frac{4\eta}{r^2}+i\omega\rho\right)e^{i(\omega t - kr)} \tag{2}$$

with: R = bubble radius
   δ = damping constant
      theoretical value: $\dfrac{\omega R}{c}+\dfrac{4\eta}{\omega\rho R^2}$, roughly 0.018 at 50 kHz
      measured value   : roughly 0.31 at 50 kHz
   ρ = density of sea water
   c = sound speed (sea water)
   $b = \dfrac{\omega_r^2}{\omega^2}-1$
   ω_r = resonance frequency of the bubble
      equal to: $\dfrac{1}{R}\sqrt{\dfrac{3\alpha P_{go}}{\rho}-\dfrac{2\sigma}{R}}$
   P_{go} = equilibrium gas pressure of the bubble
   α = polytropic constant ($1 \leqslant \alpha \leqslant C_p/C_v = 1.4$)
      α is roughly equal to 1.4 at 50 kHz.
   σ = surface tension of the air-water interface
   $N = b^2 + \delta^2$
   k = wave number (complex), equal to: $\dfrac{\omega}{c}\left[1-i\dfrac{\omega}{K}\left(\dfrac{\kappa}{2}+\dfrac{2}{3}\eta\right)\right]$
   K = compression modulus of sea-water

Suppose that the incident wave is a CW-pulse. The pressure spectrum of the pulse near the bubble $X_{cwp}(\omega)$, is chosen as real. The speed and stress vector $\underline{v}$ and $\underline{s}$ are written as Fourier-integrals using (1) and (2):

$$\underline{v}(r,t) = \text{Re}\left\{\frac{1}{2\pi}\int_{-\infty}^{\infty} X_{cwp}(\omega)\, V(\omega,r)\, e^{i\omega t}\, d\omega\right\}\underline{e}_r$$

$$\underline{s}(r,t) = \text{Re}\left\{\frac{1}{2\pi}\int_{-\infty}^{\infty} X_{cwp}(\omega)\, S(\omega,r)\, e^{i\omega t}\, d\omega\right\}\underline{e}_r$$

The power W_s scattered by the bubble follows from the speed vector $\underline{v}$ and the stress vector $\underline{s}$ of the spherical wave at a large distance r from the bubble:

$$W_s = \lim_{r \to \infty} 4\pi r^2 \; \underline{v}\,(r,t) \cdot \underline{s}\,(r,t) \tag{3}$$

The total scattered energy $E_{s,r}$ of one bubble due to the CW-pulse is calculated by integrating the power W_s of (3) over t, resulting in:

$$E_{s,R} = \frac{1}{\rho c} \int_{-\infty}^{\infty} \frac{R^2 \; X_{cwp}^2\,(\omega)}{b^2 + \delta^2} \, d\omega \tag{4}$$

$$= \frac{1}{4\pi \rho c} \int_{-\infty}^{\infty} \sigma_{s,R} \; X_{cwp}^2 \, d\omega$$

with: $\sigma_{s,R}$ = scattering cross-section of a bubble of radius R

2.2 Scattering by a Quantity of Bubbles

The quantity of bubbles in a volume of water is described by the bubble density distribution function $n(R)$ [i.e. a quantity of $n(R)dR$ bubbles is present per m^3 with radii between the limits R and $R + dR$.] The scattered energy ΔE_s of the bubbles in a small volume ΔV with density distribution $n(R)$ is calculated from (4):

$$\Delta E_s = \int_{R=0}^{\infty} E_{s,R} \; n(R) \, dR \; \Delta V$$

$$= \frac{1}{\rho c} \int_{-\infty}^{\infty} X_{cwp}^2\,(\omega) \int_{R=0}^{\infty} \frac{n(R) \; R^2}{b^2 + \delta^2} \, dR \; d\omega \; \Delta V \tag{5a}$$

$$= \frac{1}{\rho c} \int_{-\infty}^{\infty} X_{cwp}^2\,(\omega) \; J(\omega) \, d\omega \; \Delta V \tag{5b}$$

The above integral $J(\omega)$ over R can be integrated numerically if $n(R)$ were known, or is taken as a constant.
This has not been investigated yet but $J(\omega)$ is approximated, resulting in:

$$J(\omega) \rightleftharpoons \frac{\pi \; R_r^3 \; n(R_r)}{2 \delta_r}$$

with R_r = resonant bubble radius
equal to (about): $\dfrac{1}{\omega}\sqrt{\dfrac{3\alpha \; P_{go}}{\rho}}$
δ_r = resonance value of δ, equal to: $\delta(\omega, R_r)$

The spectrum X_{cwp} in (5b) can be calculated for the CW-pulse of length T and carrier-frequency ω_p. Then the value of the scattered energy ΔE_s by the small volume ΔV can be calculated by numerical integration of (5b). However ΔE_s is approximated in this paper for the case that the CW-pulse possesses a very narrow frequency band (T is large).

$$\Delta E_s = \frac{\pi^2 \; R_r^3 \; n(R_r)}{\delta_r} \; E_i \; \Delta V \tag{6}$$

with: E_i = incident energy per unit area
equal to: $\dfrac{1}{2\pi \rho c} \int_{-\infty}^{\infty} X_{cwp}^2\,(\omega) \, d\omega$

The energy scattered by the bubbles in the volume below the sea surface (see fig. 1) is the integral of $\frac{\Delta E_s}{\Delta V}$ (eq.6; $\Delta V \to 0$) over this volume. However the time spreading of the energy contributions due to the thickness of the volume in the vertical direction has to be accounted for. Moreover $n(R)$ depends on the depth. The value of $\frac{\Delta E_s}{\Delta V}$ is a stochastic variable because the number and the positions of the bubbles in ΔV are stochastic variables. Integration of $\frac{dE_s}{dV}$ results in the following mean intensity (power per m^2) near the hydrophone:

$$I(t) = \frac{c \, \Omega}{8 \, \pi} \cdot \frac{dE_s}{dV} \tag{7}$$

with : Ω = solid angle of projector or receiver $[sr]$

The response of a bubble (amplitude of the scattered spherical wave) decreases during a time τ,

$$\tau = \frac{2}{\omega \, \delta}$$

to a fraction e^{-1} of its original value if no incident signal is present anymore. At 50 kHz we find:

$\tau = 20.5 \, \mu s$

If the pulse length T is of the order of 1 ms we conclude from the above that the insonified layer thickness is with good approximation ($\tau \ll T$) equal to:

$$\Delta Z = \frac{cT}{2} \quad \text{(depth resolution value)} \tag{8}$$

We express $\frac{dE_s}{dV}$ (7) in E_i using approximation (6).
The resulting ratio I/E_i is then expressed in the source level SL at the projector and the received level RL at the hydrophone (SL and RL expressed in dB re 1μPa . m and dB re 1μPa respectively). The result for the bubble distribution is:

$$n(R_r) = \frac{8 \, \delta_r \, z^2}{\pi \, \Omega \, c \, T \, R_r^3} \cdot 10^{\frac{RL - SL}{10}} \tag{9}$$

with Z = distance from the bubble layer to the receiver, equal to: $\frac{ct}{2}$
$R_r = \frac{1}{\omega_p} \sqrt{\frac{3 \alpha P_{g0}}{\rho}}$
and the approximations used to obtain (6).

3. Measurements

Measurements have been executed in the North Sea near the measuring platform Noordwijk (see Fig. 1). Fig. 2 shows some of the received signals at the output of a logarithmic rectifier. These signals have been recorded on analog tape and were digitalized and analyzed in the laboratory. Table 1 gives measured maximum bubble density distributions; eq. (9) was applied to approximately 600 transmissions at 50 kHz. The mean distributions appeared to be a factor 2 to 3 lower than the maximum values for the given mean depths. The depth resolution of (8) for the CW-pulse of 50 kHz and T = 1 ms equals 0.75 m.

The resonant bubble radius R_r of (9) and the scattering cross-section $\sigma_{S,R}$ of (4) for one resonating bubble have been calculated for the depths given in table 1. For the calculation of R_r the surface tension σ is neglected (see [1]) and α is taken as 1.4.

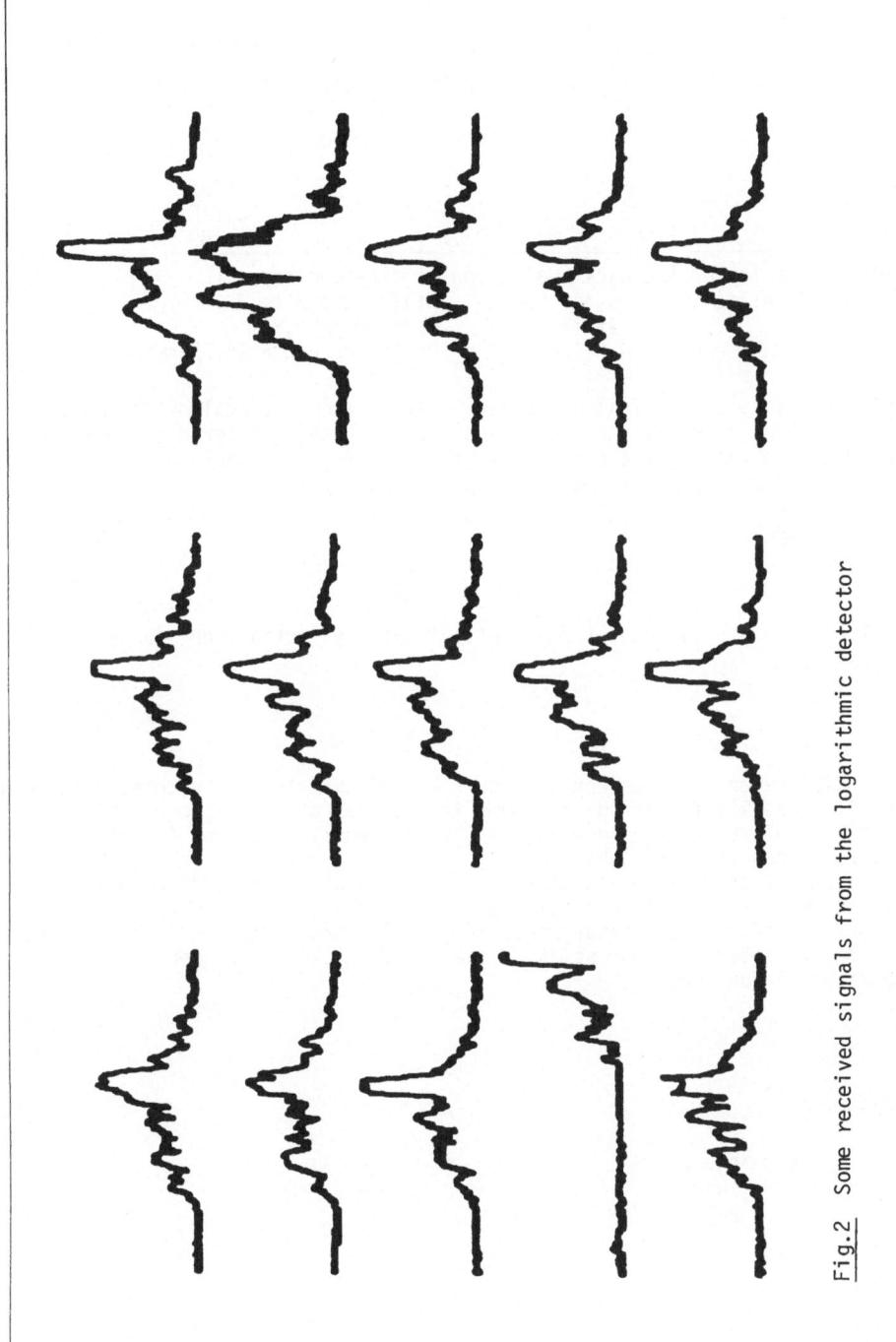

Fig.2 Some received signals from the logarithmic detector

209

Depth [m]	Max. bubble density distribution function $n(R)$ $[m^{-3} \cdot \mu m^{-1}]$	At resonance:	
		bubble radius R_r $[\mu m]$	scattering cross-section $\sigma_{Sr,R}$ $[mm^2]$
0.75	9430	66.8	0.584
1.50	9500	69.1	0.625
2.25	2790	71.4	0.666
3.00	688	73.5	0.707
3.75	309	75.6	0.748
4.50	<1	77.7	0.789

Table 1   Results of the back-scattering of air-bubbles
Conditions: Frequency 50 kHz   Wind speed 3 to 4 B
Sea State 1   Water depth 18 m
Wave height 0.5 m Projector depth 15.75 m

It is interesting to calculate the limits of the interval of the bubble radii which give the main contributions to the total scattered energy. We conclude from (5a) that the total relative interval of bubble radii is the sum of the 3 dB-widths of the CW-pulse spectrum X^2_{Cwp},

$$\frac{\Delta \omega}{\omega} = \frac{2 \pi}{\omega T}$$

$$= 0.02$$

and the resonance maximum of the scattering cross-section for one bubble $\sigma_{S,R}$

$$\frac{\Delta \omega}{\omega} = \delta$$

$$= 0.31$$

So bubbles resonating between 42 and 57 kHz contribute to the total scattered signal. These limits include a volume fraction of air equal to roughly $1.4 * 10^{-7}$ of resonating bubbles, assuming a density $n(R_r)$ of $4 * 10^3$ $m^{-3} \cdot \mu m^{-1}$ at 1.50 m deep (compare table 1).
The experiments (with sea state 1 only) indicated that effects due to bubbles were present using frequencies from 23 kHz up to 70 kHz. It is obvious therefore that measurements in this frequency range will be required in order to obtain an impression of the shape of the bubble density distribution function.

References

[1] R. Wildt.   Physics of Sound in the Sea, Part IV
Summ. of Techn. Rep. of Division 6, NDRC
Vol. 8, Washington, D.C. (1946)

Acoustic Measurements of the Gas Bubble Spectrum in Water

A. Løvik

Electronics Research Laboratory
N-7034 Trondheim - NTH, Norway

1. Introduction

The present paper presents our first result from acoustic measurements of
the bubble spectrum at the sea as well as in a cavitation tunnel. The measurements at sea were done from the research vessel G.O. Sars in two periods in
1978 in the Barents Sea and the North Sea. This work was initiated by the
Marine Institute of Norway, from a desire to measure the attenuation of
acoustic waves through the bubble layer under the vessel in rough weather.
This extra attenuation is important when using echo-integration for estimation of biomass.

A joint project between ELAB and The Ship Research Institute aims at measuring and modelling the cavitation noise from propellers. Since the cavitation process strongly depends on the bubble or nuclei spectrum in the water
some measurements of the nuclei spectrum have been made through this project.

Measurements of the bubble spectrum in both sea water and in cavitation
tunnels have been performed before by several workers using various techniques [1-4]. In this paper the acoustic only determined bubble spectra are
compared with data from MEDWIN [3] and GAVRILOV [1], using acoustics and with
data from Keller [2] using optical backscattering.

2. Methods of Measurement

Two measuring methods have been applied to estimate the bubble spectrum. The
methods are based on resonant backscattering of acoustic waves or resonant
absorbtion. The theoretical background is given in several papers and books
[3,4] . The main results are repeated here.

The bubble is described by its backscattering σ_b, extinction σ_e, and absorption cross section σ_a, where:

$$\sigma_e = \sigma_a + \sigma_b \tag{1}$$

The extinction cross section for a bubble with radius a and resonance frequency f_0 is written as:

$$\sigma_e = \frac{2ac\delta}{f_0 \left[\left(\left(\frac{f_0}{f}\right)^2 - 1 \right)^2 + \delta \right]} \tag{2}$$

as function of frequency f.

The backscattering cross-section is:

$$\sigma_b = \frac{4\pi a^2}{\left[\left(\frac{f_0}{f}\right)^2 - 1\right]^2 + \delta^2} \tag{3}$$

where

$$f_0 = \frac{1}{2\pi a}\sqrt{\frac{3\gamma P}{\rho}} \tag{4}$$

and
δ: effective damping constant
γ: ratio of specific heats of the gas
P: hydrostatic pressure
ρ: density of the water
c: speed of sound in water

The attenuation α of an acoustic wave through a medium containing bubbles with radius density distribution $n(a)$ is given as:

$$\alpha(f) = 4,34 \int_0^\infty n(a)\sigma_e(a,f)da \ , \ [dB/m] \tag{5}$$

If it is assumed that only bubbles of near resonant size will make a large contribution to the integral and that the bubble size distribution is a slowly varying function of bubble radius the integral may be approximated by:

$$\alpha(f_0) = 4,34 \ n(a_0)a_0 \ \delta \ \sigma_e(a_0)$$
$$\alpha(f_0) = 4,34 \ \sigma_e(a_0)N(a_0) \tag{6}$$

where a_0 is the resonant bubble radius, and $N(a_0)$ is the number of near resonant bubbles. The bubble density $n(a_0)$ is often given as the number of resonant bubbles per m^3 in a radius interval of 1 μm, and is found as $n(a_0) = N(a_0)/a_0\delta$, where a_0 is taken in [μm].

Thus by measuring the acoustic attenuation over a distance r, the mean number of resonant bubbles may be found by the use of (6). In the cavitation tunnel this is done over a distance of r = 3 m, and the instrumentation is shown in Fig. 1.

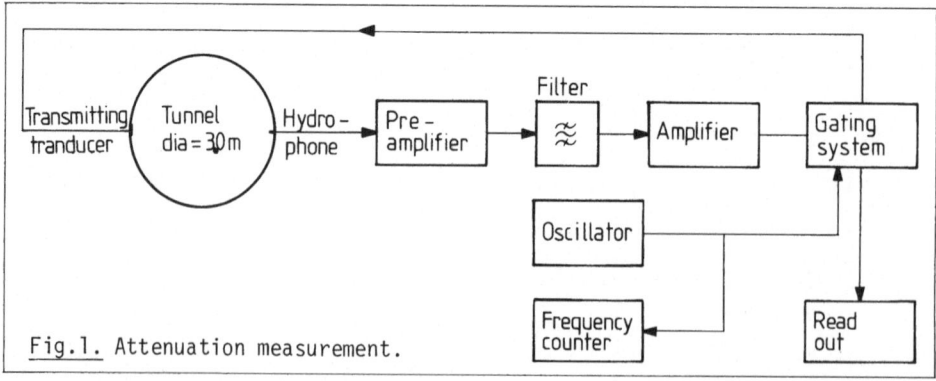

Fig.1. Attenuation measurement.

The frequency may be changed from 20 kHz up to 250 kHz, this is done by using two or three transmitting transducers. The equipment is calibrated both in a laboratory test tank and in the tunnel with degassed, not flowing water.

The backscattering measurement done at sea uses three echosounders at 12, 38 and 120 kHz. The received reverberation intensity I_r from a volume V in the far field of the transducer is:

$$I_r = \frac{I_o}{r^4} s_v \int_V b(\theta,\phi)b'(\theta,\phi)dv \tag{7}$$

where

s_v is the volume backscattering strength
I^o is the transmitted intensity referred to 1 m
$b(\theta,\phi), b'(\theta,\phi)$ is the transmitting and receiving directivity

The trandducers are mounted on a gyrostabilized platform and we may assume that $b = b'$. Transmitting pulses with duration τ, it may be shown that the volume backscattering strength in both the near and the far field is :

$$s_v = 2I_r \cdot r^2 / I_o c\tau\psi \tag{8}$$

where
$$\psi = \int_{4\pi} b^2(\theta,\phi)d\Omega \tag{9}$$

Using first order scattering theory the backscattering from a population of bubbles is:

$$s_v = \frac{1}{4\pi}\int_a \sigma_s(a)n(a)da \tag{10}$$

Eq (10) is again simplified by the assumption that only near resonant bubbles give a large contribution to the integral.
It is thus observed that measurements of the reverberation I_r on a calibrated echosounder give information on the resonant bubble density $n(a)$ through (8), (9) and (10).
The first order estimate of the number of resonant bubbles may be improved by taking into account the extra attenuation caused by the bubbles. This approximation is useful when the scatterer are small and when they are effectively absorbing the energy. This is the fact for resonant gas bubbles having resonance frequencies higher than 10 kHz.
In this case the reverberation intensity is written as:

$$I_r = \frac{I_o}{r^4} \cdot s_v \cdot \frac{c\tau}{2} \cdot r^2 \psi \cdot e^{-2\int_0^r \alpha(r)dr} \tag{11}$$

This gives a differential equation for the number of resonant bubbles N(r):

$$N'-N\frac{N'_o}{N_o} - 2N^2\sigma_e = 0 \tag{12}$$

where $N_o(r)$ is the first order approximation to the number of resonant bubbles:

$$N_o(r) = \frac{8\pi r^2}{c\tau\psi} \cdot \frac{I_r}{I_o} \tag{13}$$

213

Equation (12) is a Bernoulli equation which may be solved exactly. A special and illustrative solution is obtained by assuming that:

$$N_o(r) \sim N_o e^{-\kappa r}$$

(14)

and that the extinction cross section is constant. The solution is then:

$$N(x) = \frac{n_o e^{-\kappa x}}{1 - \frac{2\sigma_e n_o}{\kappa} (1+e^{-\kappa x})}$$

(15)

Figure 2 shows the principal solution, indicating that neglection of the extra attenuation gives an underestimate of the number of bubbles.
The echosounders were standard scientific sounders having a 3 dB main lobe varying from 4,5⁰ to 8⁰.
All the signals were recorded on magnetic tape and processed in the laboratory afterwards. The echosounders were calibrated during the first measuring period in June.

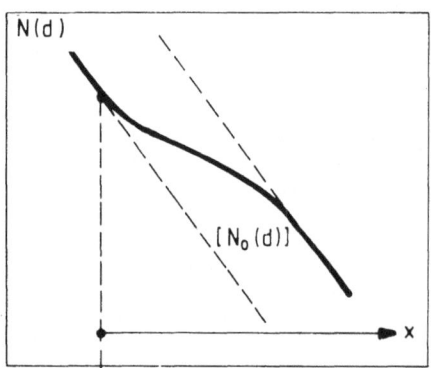

Fig. 2. Number of bubbles versus depth.

3. Results and Discussion

The measurements performed in the cavitation tunnel should illustrate the variation in bubble density with varying total gas content in the water.
The total gas content is measured with a Van Slyke apparatus and gives the sum of free and absorbed gas in the water. The result is given in percentage of the saturated value.
Figure 3 shows the measured bubble spectra and the integrated bubble volume as a function of the total gas content.
It is seen that the form of the bubble spectrum is not changed significantly by increasing the total gas content. Further it is found that below 60-65% total gas content, the total bubble volume increases slowly, while above this percentage the volume increases rapidly. This fact may indicate that addition of gas when total gas content is low is either absorbed or filled into existing cavities. At least the high frequency noise from a cavitating propeller is strongly influenced by the increase in total gas content up to 65% of saturation, while the influence is less pronounced above. The high frequency power level is given mostly by bubble collapse, so this may be taken as an indication of gas being filled into the cavities.

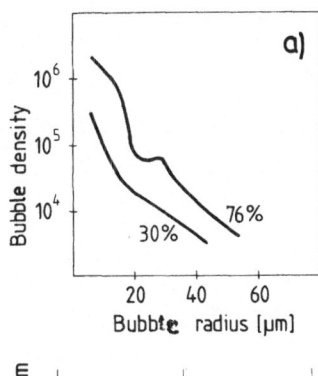

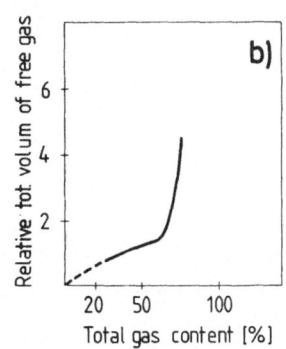

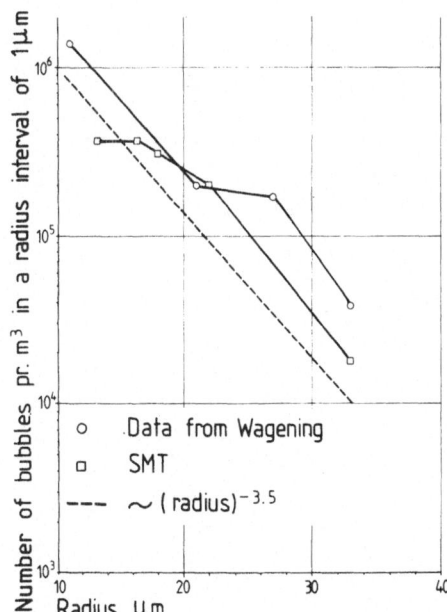

Fig.3 Measured bubble spectra for total gas contents of 30% and 76% (a). Integrated bubble volume (b) given relative to the value of the integral of 30%.

Fig.4 Measured bubble spectra in two different cavitation tunnels.

In Fig. 4 a comparison is given between bubble densities measured in Wageningen with a light scattering technique, the acoustic attenuation method and a line indicating GAVRILOV's result, $n(a) \sim a^{-3.5}$. It is seen that both measurements agree fairly well with this asymptote over the limited range of bubble radii.

The results from the backscattering measurements are shown in Fig.5 as the number of resonant bubbles in a radius interval of 1 μm for the three different frequencies 12, 38 and 120 kHz. The bubble density is shown from a depth of 5 m which is the mean depth of the transducers in a calm sea. The wind velocity is the mean velocity over 10 minutes. With a mean velocity of 45 knots the maximum was measured to 75 - 80 knots.

The data from the first few meters are not significant, they are more or less dominated by ringing in the transducer and the filters. At greater depths all the curves approach a constant density. This level is given by the noise. It is found that all curves have about the same behaviour in the midrange where:

$$n(d) = n_0 \, e^{-1,73d} \tag{16}$$

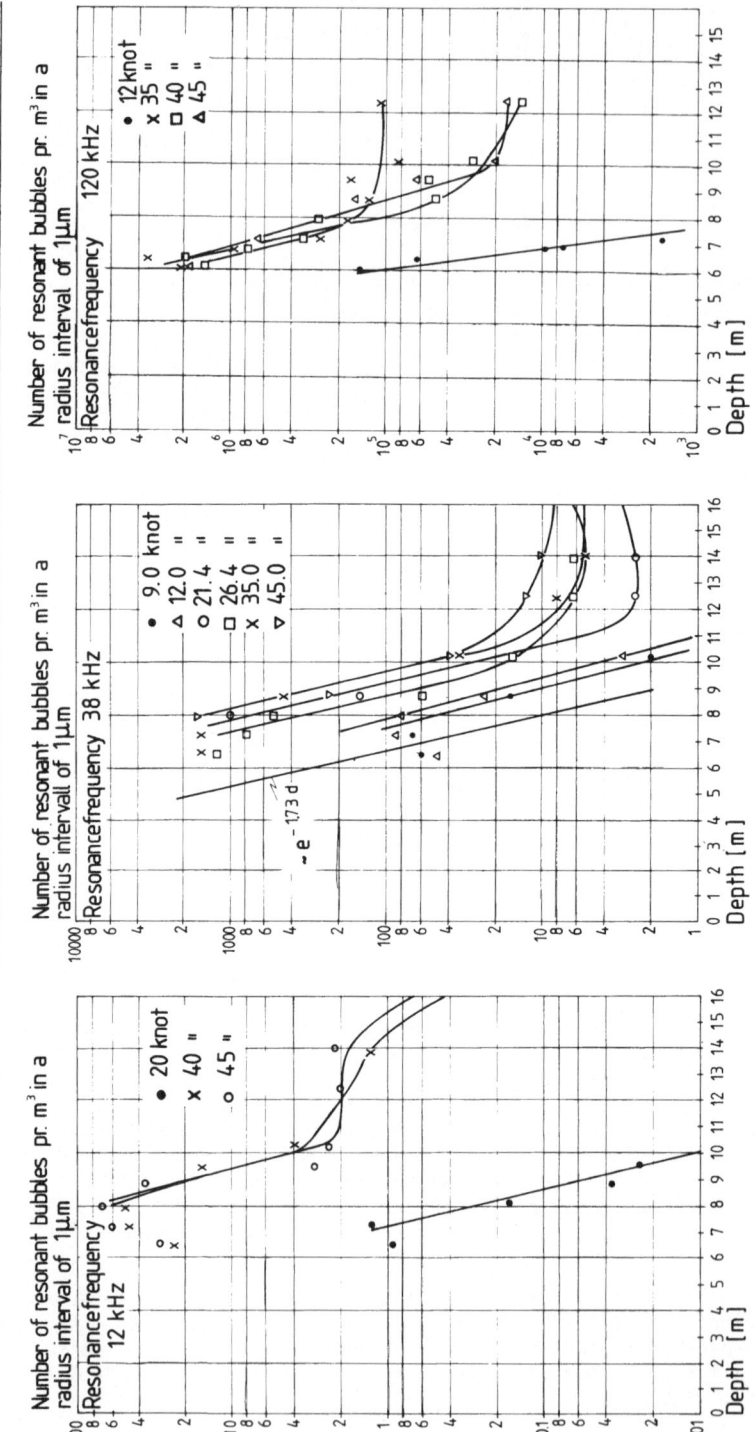

Fig.5 The number of resonant bubbles as function of depth at 12, 38 and 120 kHz with the mean wind velocity as parameter

216

The depth is a nominal depth as all measurements are referred to 5 meter as transducer depth.

Looking at the number of bubbles it is seen to increase with increasing frequency and wind velocity. In average the density decreases as the radius raised to minus 3.8, $R^{-3.8}$. This is in fairly good agreement with GAVRILOV's: -3,5.

The power law dependence between 12 kHz and 38 kHz is - 2,6, while it is -4,2 between 38 kHz and 120 kHz. Averaged over the depth interval the corresponding bubble radii are 380 µm (12 kHz) 120 µm (38 kHz) and 49 µm (120 kHz). MEDWIN has observed a change in the power law for radii between $50 - 80$ µm. He suggests that the number of bubbles vary as R^{-4} below and as R^{-2} above. The discrepancy between the observations is not great and may be caused by the few measuring frequencies used in this work.

The number of bubbles found in this and MEDWINs work are in good agreement. For instance MEDWIN finds at 40 kHz and wind velocities of 4 to 6 knots, sea state 1, at a depth of 7-8 meters, a bubble density of 100 per m^3. This corresponds fairly well to the measured result at 38 kHz at 9 knots. The same agreement is also found at higher frequencies.

The number of bubbles is found to increase with the wind velocity, and Fig. 6 shows the bubble density at 8 meter resonant at 38 kHz.

It is seen that the density increases rapidly with the wind velocity. Two measured points are above the line, 21.4, and 35 knots. In these two cases the sea was rougher than expected at that wind velocity. This indicates that the bubble density also depends on the roughness of the sea.

The extra attenuation caused by the bubbles is estimated by integration of the near resonant absorption over the depth. A bubble-density as shown in Fig. 7 is assumed.

Performing the integration from 5 meters to infinity we obtain the results shown in Fig. 8. In this figure the total mean attenuation is twice the one way attenuation.

The results show that the high frequency echosounders are more exposed to attenuation caused by gas bubbles.

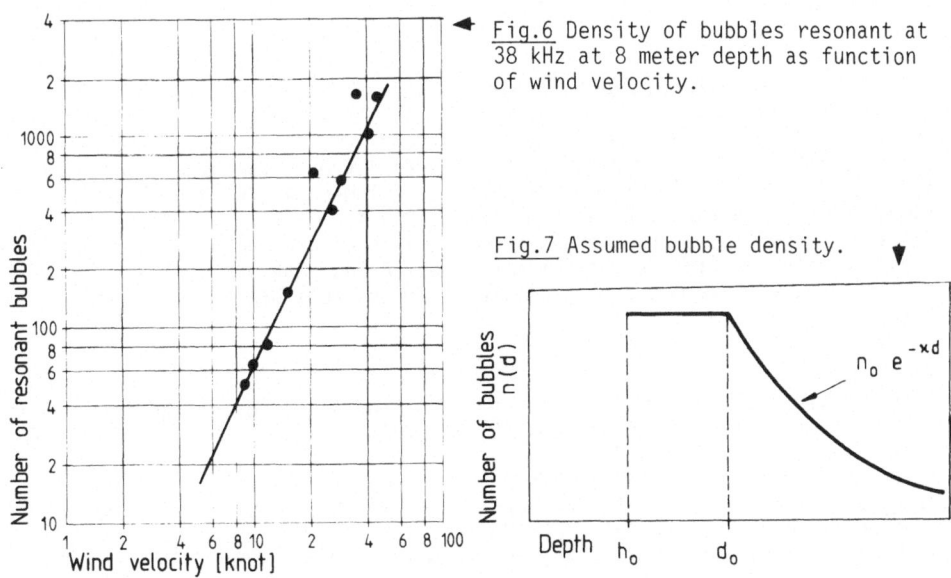

Fig.6 Density of bubbles resonant at 38 kHz at 8 meter depth as function of wind velocity.

Fig.7 Assumed bubble density.

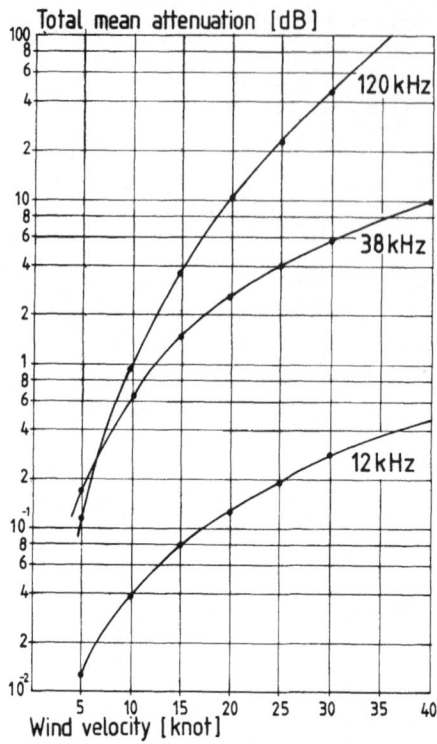

Fig. 8. Total mean attenuation caused by gas bubbles as function of wind velocity.

The 38 kHz echosounder is the main instrument used for acoustic biomass estimation in Norway. From the results here we find that wind velocities from
the range 15-20 knots may cause large errors in the biomass estimate.
It is interesting to note that the shape of the attenuation curves in Fig. 8
are similar.

Conclusions

The acoustic measurements of bubble densities in water indicate that the
number of bubbles is a continuous function of the total gas content in the
water, the wind velocity and the sea roughness.
The measurements indicate that the bubble density n(R) varies as radius raised
to minus 4.2, $n(R) \sim R^{-4.2}$ for small bubbles, $R<100\mu m$, and as $R^{-2.6}$ for larger
bubbles. The mean variation is found to be $R^{-3.8}$.
The attenuation caused by the gas bubbles may preclude the use of hydroacoustic shipborne instruments at high frequencies. The use of lower frequencies, here 12 kHz, may give significantly more reliable results in rough
weather.

References

1. L.R.Gavrilov:Free gas content of a liquid and acoustical techniques for its
   measurement. Sov. Phys. Ac. Vol 15, 3, 1970.
2. A.P.Keller,R.E.A.Arndt:Free gas content effects on cavitation inception and
   noise in a free shear flow. IAHR Sym. Grenoble 1976.
3. H.Medwin:In situ acoustic measurements of bubble populations in coastal
   ocean waters. J.Geoph.Res. Vol 75 No 3 1970.
4. Clarence S.Clay & H.Medwin:Acoustical oceanography. Wiley 1977.

Determination of Bubble Size Spectra by Digital Processing of Holograms

G. Haussmann

Drittes Physikalisches Institut, Universität Göttingen, Bürgerstr. 42-44
D-3400 Göttingen, Fed. Rep. of Germany

Abstract

Bubble fields in water which are recorded on holograms with the help of a pulsed ruby laser are automatically analyzed under control of a computer. The evaluation takes place directly at the real image of the hologram.

1. Introduction

In order to have a better comparison between numerically calculated bubble oscillations and measured cavitation noise spectra knowledge of the bubble size distribution in acoustically produced cavitation bubble fields is of great interest. It turns out that it is possible to resolve this problem by recording the bubble fields on holograms with the help of a pulsed ruby laser [1] and to analyze the particle distribution by computer processing of the three dimensional real image of the hologram. Because of the large variation in size of the particles in cavitation bubble fields the application of off-axis-holography is advantageous [2]. The development of a computerized analysis scheme is necessary due to the large number of bubbles involved.

2. The Experimental Setup

Computer processing of three dimensional pictures involves a great many data and accompanying difficulties in data handling. The problems arising from this can be avoided by using the real image of the hologram itself as a kind of "analogue picture storage". This analogue storage process however cannot give as good data consistency as a digital computer storage process. Figure 1 shows a block diagram of the experimental setup used for the automatic analysis of three dimensional pictures. As a coherent light source for the reconstruction of the hologram we use an ion laser with a radiation power of about 600 mW at a wavelength of 647.1 nm. The input device to the computer is an image dissector camera (optical data digitizer, EMR-Schlumberger), which allows addressing 4096 x 4096 points with random access. The camera is mounted on a translation table and can be moved under the control of a computer (Honeywell H 632) through the different planes of depth of the three dimensional picture. In this configuration it is possible to determine the Z coordinates of each particle from the position of the translation table. X and Y coordinates are calculated from the two dimensional intensity distribution on the camera sensor.

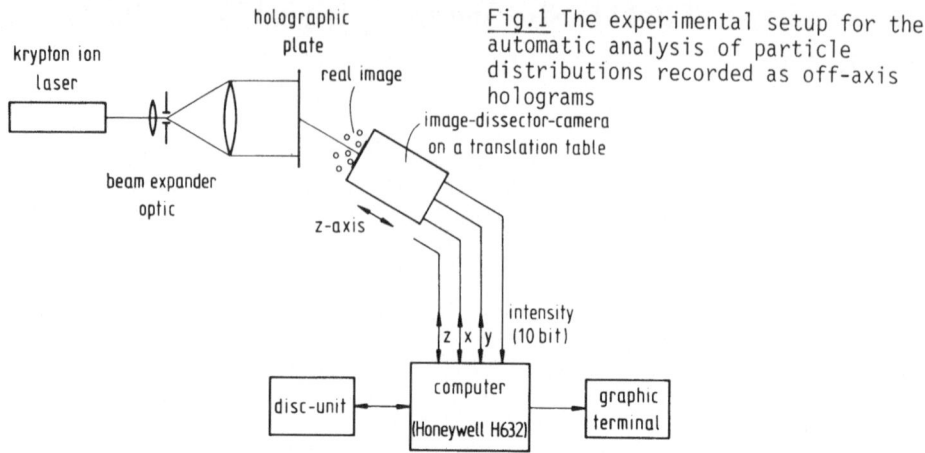

krypton ion
laser

beam expander
optic

holographic
plate

real image

z-axis

image-dissector-camera
on a translation table

intensity
(10 bit)

z x y

disc-unit

computer
(Honeywell H632)

graphic
terminal

Fig.1 The experimental setup for the automatic analysis of particle distributions recorded as off-axis holograms

3. Automatic Focusing on Single Particles

To solve the problem of finding and focusing on particles in the different planes of depth of the three dimensional picture without using special hardware processors we applied methods from the field of digital picture processing to particle analysis. The sharpness of edge was selected as a criterion for the Z position of an object.

The application of local gradient operators to edge detection problems in hologram reconstructions is difficult because extremely high contrast speckle noise is present throughout the whole picture and is strongly amplified by the gradient operator. Without special techniques for the suppression of this noise the plane of depth of maximal edge sharpness cannot be determined (refer to Fig. 2a and 3b).

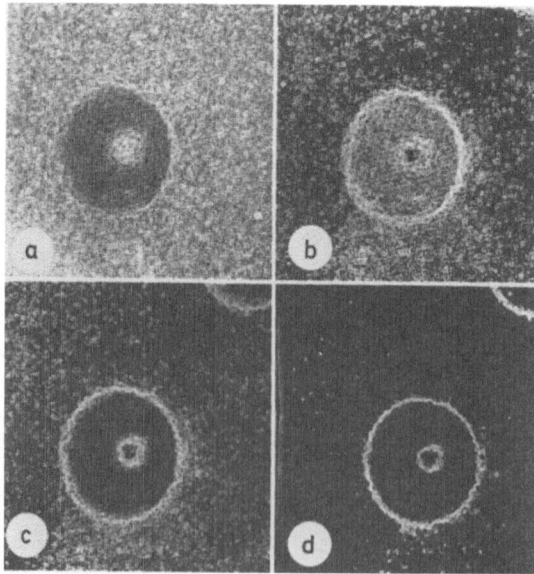

Fig.2 Gradient filtered picture of a focused bubble in water
a) without noise suppression
b) weighted with a linear function of input intensity
c) weighted with a triangular function of intensity
d) weighted with a Gaussian function of intensity

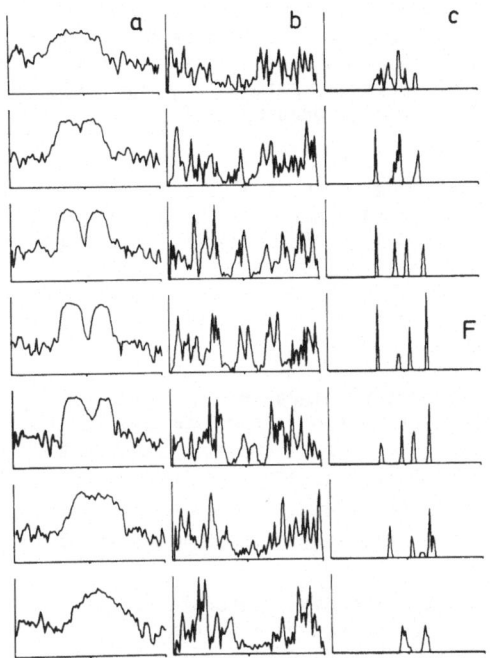

Fig.3 Connection between the input picture and the gradient-filtered picture in different planes of depth (F = focus)

a) cross section of a bubble (the intensity is inverted)
b) gradient picture without noise suppression
c) intensity weighted gradient picture (with a Gaussian weighting function)

The problem can be overcome by matched weighting of the gradient picture with a weighting function derived from the intensity histogram of the input picture (refer to Fig. 2b-d and 3c). This weighting function can be interpreted to describe the probability that a picture point with a certain intensity value belongs to the object contour [3]. Different functions were tested and, it turns out, a Gaussian function with maximum between the average object intensity and the average noise intensity is the best description of the probability density. After this intensity weighting the separation of edge points from noise points by simple thresholding becomes possible.

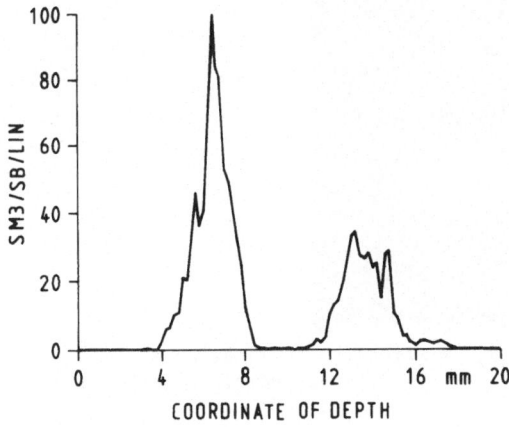

Fig.4 Focusing parameter computed from the intensity-weighted gradient picture as a function of the coordinate of depth.
The local maxima indicate the positions of three bubbles in different planes of depth.

The connection among the shape of a bubble in different planes of depth (refer to Fig. 3a), the gradient-filtered picture without noise suppression (Fig. 3b) and after intensity-weighting with a Gaussian weighting function (Fig. 3c) demonstrates, that the plane of depth of maximal edge sharpness can be identified with the one of maximal average gradient values.

A focusing parameter of sufficient selectivity is calculated from the processed gradient picture [4]. It is assigned to the analyzed image area, not to single particles. The local maxima of this parameter as a function of the Z coordinate indicate the exact Z positions of the objects (refer to Fig. 4).

4. Determination of Geometric Features

If the computer has steered the image dissector camera into the plane of depth of a local parameter maximum, image processing algorithms are applied to the two dimensional image area in order to compute the most important geometric features of the focused particles such as area, perimeter, and the three spatial coordinates. In order to do this

- local thresholds are determined from the intensity histogram to separate object points from the background,
- the picture is scanned, the number of points of each separate object is counted and assigned to the particle as a measure of its area,
- the perimeter (number of border points) and the spatial coordinates are determined,
- holes in the bubbles are filled
- particles that are too small are deleted.

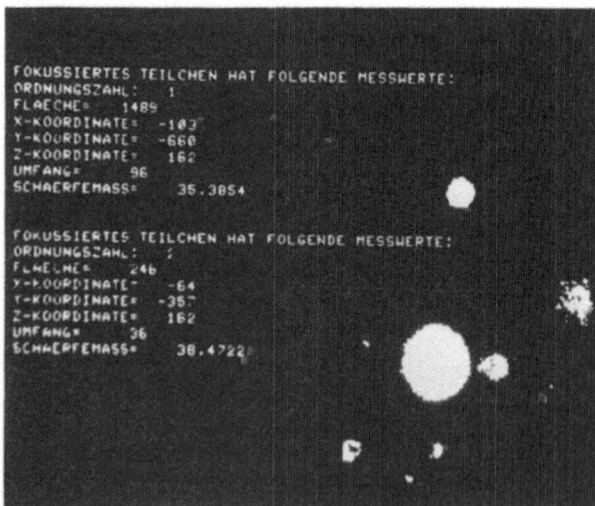

Fig.5 Display from the computer program after the application of image processing algorithms to the image area to compute geometric features of the focused bubbles

If there is more then one bubble in the image area, the computer calculates a measure of sharpness for each particle to separate focused particles from blurred ones. The display from the computerprogram on a graphic terminal is shown in Fig. 5. The segmented particles are shown in a binary picture together with their geometric features. Two of the particles in the image area in Fig. 5 are accepted by the computer and their measured values are stored, three bubbles are rejected and separately focused, and the rest is deleted because of insufficient size.

5. Application Test

The automatic analysis scheme was tested with holograms of air bubbles in water which were recorded with a pulsed ruby laser [5]. Before starting the evaluation the framed picture area in Fig. 6a was subdivided into 25 subpictures of equal size with overlapping regions and each of these subpictures was analyzed one after the other in the Z dimension. Measurements of particles which were detected and stored twice or more are reduced to a single measurement with the maximal sharpness number. The result of the experiment in Fig. 6b shows that the computer has found nearly all bubbles in the tested liquid volume. Z coordinates are not displayed but are also available for further study.

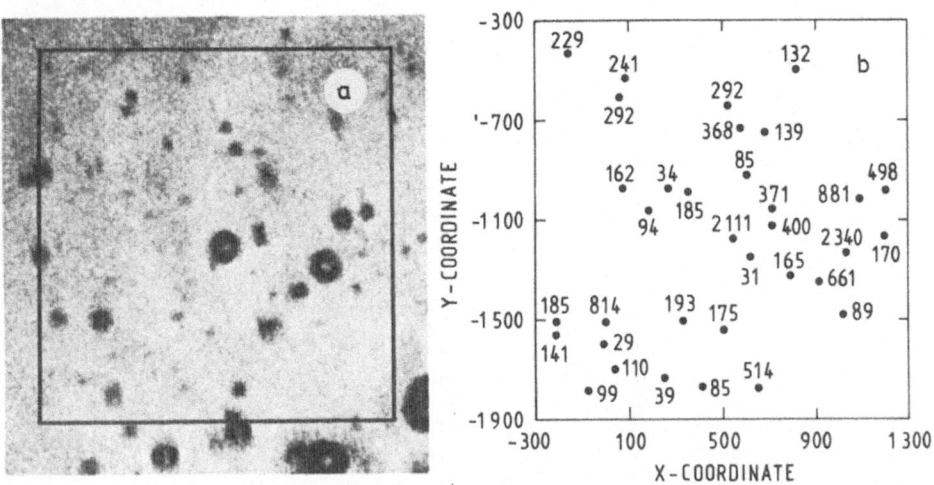

Fig.6 Result of a test analysis of air bubbles recorded on a hologram with the help of a pulsed ruby laser:
a) hologram reconstruction with the analyzed area in the frame
b) result of the automatic evaluation (the numbers give the area values)

Experiments to apply this automatic scheme to acoustically produced cavitation bubble fields are under way.

This work was sponsored by the Deutsche Forschungsgemeinschaft.

I thank Dr. W. Lauterborn for many discussions and steady encouragement in pursuing this work.

6. References

1 Lauterborn, W., Ebeling, K.J., Proc. XII. Int. Congr. on High Speed Photography, Toronto 1976, SPIE Vol. 97, p. 96-103

2 Bader, F., Ph.D. Thesis, Göttingen 1973

3 Abele, L., Lange, C., Informatik Fachberichte, Vol. 17, Springer Berlin (1978), p. 329-333

4 Haussmann, G., Lauterborn, W., Informatik Fachberichte Vol. 17, Springer Berlin (1978), p. 275-280.

5 Haussmann, G., Ph.D. Thesis, Göttingen 1979

Determination of Bubble Sizes by Far Field Diffraction of Photographic Recordings

R. Butt

Drittes Physikalisches Institut, Universität Göttingen, Bürgerstr. 42-44
D-3400 Göttingen, Fed. Rep. of Germany

K. Hinsch

Fachbereich IV, Universität Oldenburg, Ammerländer Heerstraße 67-69
D-2900 Oldenburg, Fed. Rep. of Germany

Abstract

An analog optical method for size determination of small objects is presented. The objects, e.g. cavitation bubbles, are stored in a transparent photographic recording from which the far field diffraction pattern is evaluated. The proposed method is first tested with computer generated model bubbles and then successfully applied to real bubbles.

1. Introduction

Knowledge of the size distribution of acoustically produced cavitation bubbles is a good tool for comparing measured noise spectra with the bubble noise spectra predicted theoretically.

In the last years there have been many investigations taking photographic or even holographic records of cavitation bubbles. In acoustic cavitation, information about thousands of bubbles may be stored in a single hologram but it takes a great deal of effort to measure the size of each bubble. There are two different approaches towards this problem. HAUSSMANN [1] has scanned the real image of a hologram by a TV-camera which is controlled by a computer.

In the present study, rapid analog data processing by optical Fourier transformation is utilized. This minimizes considerably the amount of data for computation. The price, however, is a decrease in accuracy, for the bubbles can be ranked only into a limited number of sizes.

2. Basic Principles

The underlying method of counting and classifying small objects by their far field diffraction pattern was first developed by ANDERSON and BEISSNER [2] and will be briefly described here. The calculations are done in polar coordinates.

A lens may be regarded as an "optical analog computer": when a transparent pattern illuminated with collimated coherent light is placed in front of the lens the intensity of the diffraction pattern measured in its back focal plane is the squared two-dimensional Fourier transform (intensity spectrum) of the input pattern [3].

Let us assume identical circular objects having random spatial distribution on a transparency (such as a photograph of equal-sized bubbles). Then the expected value of the measured intensity is given by

$$I(\rho) = N \cdot G(\rho) \tag{1}$$

where ρ is the radius in the back focal plane (i.e. the spatial-frequency domain) and $G(\rho)$ is the intensity of the diffraction pattern (i.e. the squared Fourier transform) of a single object of the given class.

If there are objects of M different classes on the transparency (different sized bubbles, for example) which are all placed randomly, each class being characterized by its diffraction pattern $G_j(\rho)$, then all the expected intensity values add up.

$$I(\rho) = \sum_{j=1}^{M} N_j \cdot G_j(\rho) \tag{2}$$

We are interested in the number of objects N_j of each class. For this purpose the spatial frequency spectrum is sampled at a number of points, P, which is larger than M.

$$I(\rho_i) = \sum_{j=1}^{M} N_j \cdot G_j(\rho_i) \qquad \text{for } i = 1, \ldots P \geq M \tag{3}$$

We define

$$I_i = I(\rho_i) \quad \text{and} \quad G_{ij} = G_j(\rho_i) \tag{4}$$

so that we can express (3) in matrix notation.

$$I = G \cdot N \tag{5}$$

Minimizing the mean square error yields the inversion formula for the number of objects

$$N = (G^T \cdot G)^{-1} \cdot G^T \cdot I , \tag{6}$$

where G^T is the transposed matrix.

Of course, the measured intensity value at a single point in the frequency domain does not equal the corresponding expected value $I(\rho_i)$. The random character of the object positions results in coherent noise known as laser speckle. Therefore, smoothing of the measured values is necessary which is achieved by rotating the transparency resulting in a corresponding rotation of the spectrum and by averaging the intensity during rotation. This improves the measurement considerably.

3. Experiments with Model Bubbles

We first tested the proposed method with transparencies of known objects. The patterns were produced by a computer-driven plotter and consisted of circular black spots of five different sizes. These plots were photographed yielding small white disks of different sizes on a dark background.

These model bubbles have the advantage that the classes (sizes) present in the pattern are well determined. Furthermore, there is a simple method to determine the characteristic functions $G_j(\rho)$. By taking patterns of equal-sized spots which are plotted and photographed in just the same way, the characteristic spatial frequency spectra of the different classes are measured.

The investigations with model bubbles resulted in the important conclusion that the accuracy in determining the numbers N_j depends very much on the size differences between the classes. If the bubbles or spots are very similar in size their characteristic functions are very similar too. This leads to a quasi-singular matrix in the inversion foumula (6) which results in large errors.

We found that the differences between the bubble radii should be about 20%. Then the characteristic spectra of the classes are easy to distinguish and the error in determining the numbers lies between 5% and 10%.

4. Experiments with Real Bubbles

When photographs of real bubbles are to be analysed the evaluation is much more difficult. In this case the bubbles do not a priori belong to known classes of bubble sizes. We first have to choose classes which are suited for the given problem and which have clearly differentiated spectra.

From the experiments with model bubbles we know which classes should be suitable for the inversion process. The second problem is to determine the characteristic functions of the classes chosen. This determination has to be done very carefully for it has great influence on the accuracy of the evaluation.

Therefore, we looked for a method to generate bubbles of nearly equal size. When water of high gas content is poured into a glass container, small air bubbles are produced at the walls and grow slowly. By photographing them at different times patterns of certain bubble classes are recorded from which the corresponding spectra are measured. From many trials we chose those patterns which had marked dark and bright fringes in their diffraction patterns indicating a very narrow range of sizes.

Figure 1 shows some results from our experiments. The patterns are photographs of mineral water into which we introduced sugar to produce many bubbles. The big bubbles in the second pattern flatten as they rise. It is therefore difficult to define a class for them.

The measured spectra were inverted according to (6) giving the size distribution of the bubbles. The mean and the standard deviation of the results are plotted. In spite of the inherent inaccuracies in determining the bubble classes this analysis gives a good picture of the size distribution and the total number of bubbles in the given pattern.

For evaluating cavitation bubble images stored in holograms one should first produce a photograph. One problem is to avoid defocused bubbles. Therefore it would be necessary to have a large depth of focus. Otherwise, the introduction of additional classes (i.e. defocused bubbles) might be considered which has not yet been done.

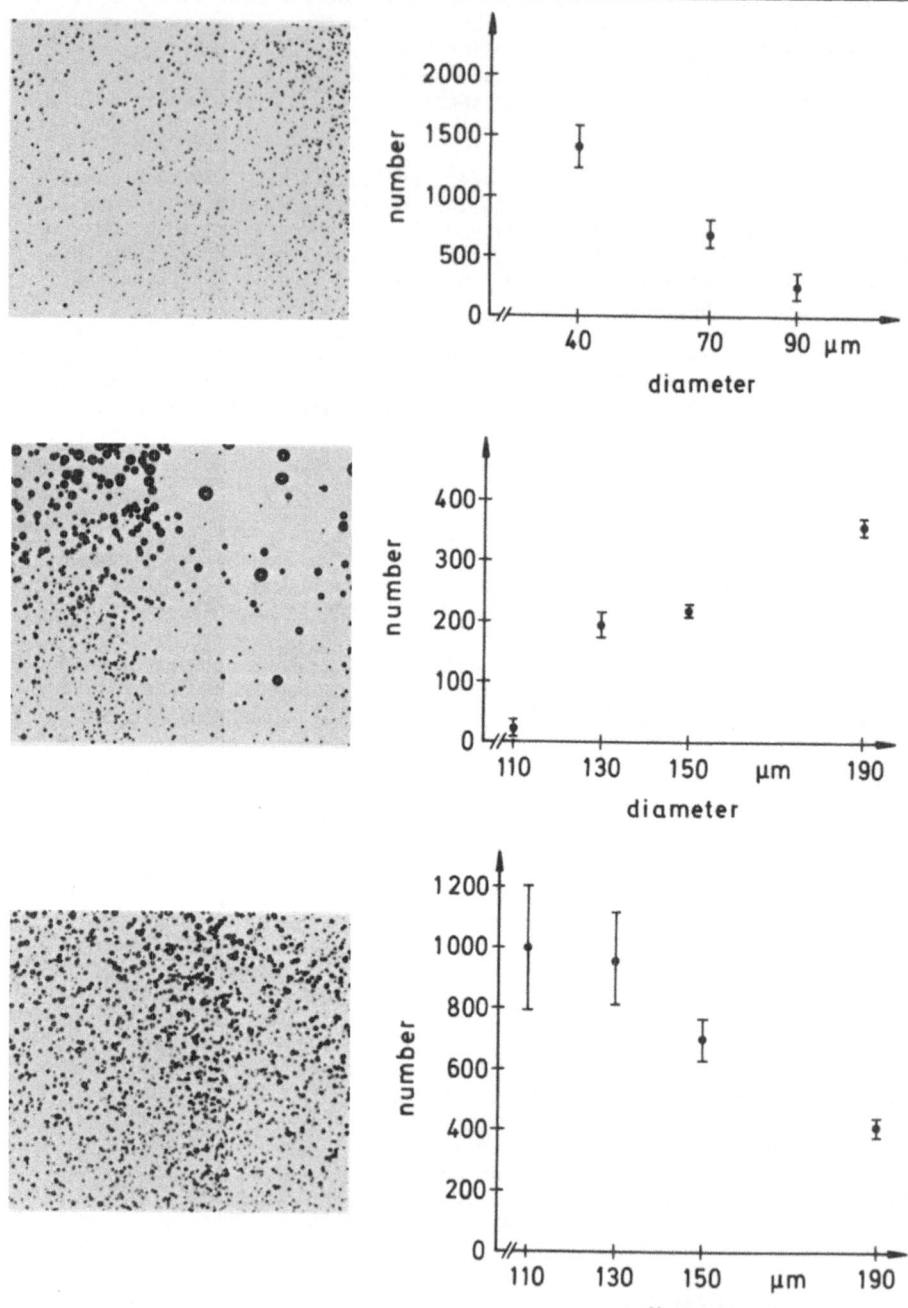

Fig.1 Bubble patterns and corresponding size distributions

5. Conclusion

It is shown that the determination of size distributions in a bubble field can be achieved by means of the analog optical Fourier transform. The analysis, of course, cannot be exact because the bubble sizes vary continuously over a range of radii, while the classification requires well-defined single valued classes.

Nevertheless, this method gives good qualitative results and takes full advantage of parallel data processing. Thousands of bubbles can be classified just as rapidly as can ten. It would take much more effort to count as many bubbles as are shown in Fig. 1 directly by means of a TV-camera and a computer, for example.

The accuracy of the evaluation increases with the care taken in defining the classes and in determining their characteristic frequency spectra. Thus, the capability of this method will be fully utilized when there is a large number of similar patterns to be evaluated.

References

1 G. Haussmann, Determination of Bubble Size Spectra by Digital Processing of Holograms, in this volume.
2 W.L. Anderson, R.E. Beissner, Counting and Classifying Small Objects by Far Field Scattering, Appl. Opt. 10, 1503 (1971).
3 J.W. Goodman, Introduction to Fourier Optics, New York 1968.

Complementing Discussion Contribution to the Papers of H. Medwin, P. Schippers, and A. Løvik

E.-A. Weitendorf

Hamburgische Schiffbau-Versuchsanstalt
D-2000 Hamburg, Fed. Rep. of Germany

Since several years extensive cavitation research with propellers has been carried out at the Hamburgische Schiffbau-Versuchsanstalt and the Institut für Schiffbau of the Hamburg University. The experimental investigations were performed in close cooperation with KELLER [1], using his Laser-scattered-light-(LSL)-technique. The results of this common model investigation [2,3] revealed the large influence of the free air content of the test water on the propeller cavitation and hereby excited pressure fluctuations. Having the intention to find additional laws of similitude for cavitation including the free air content, one has also to know the free air content for full scale conditions. Corresponding investigations were carried out on board of the container ship "SYDNEY EXPRESS" [4].

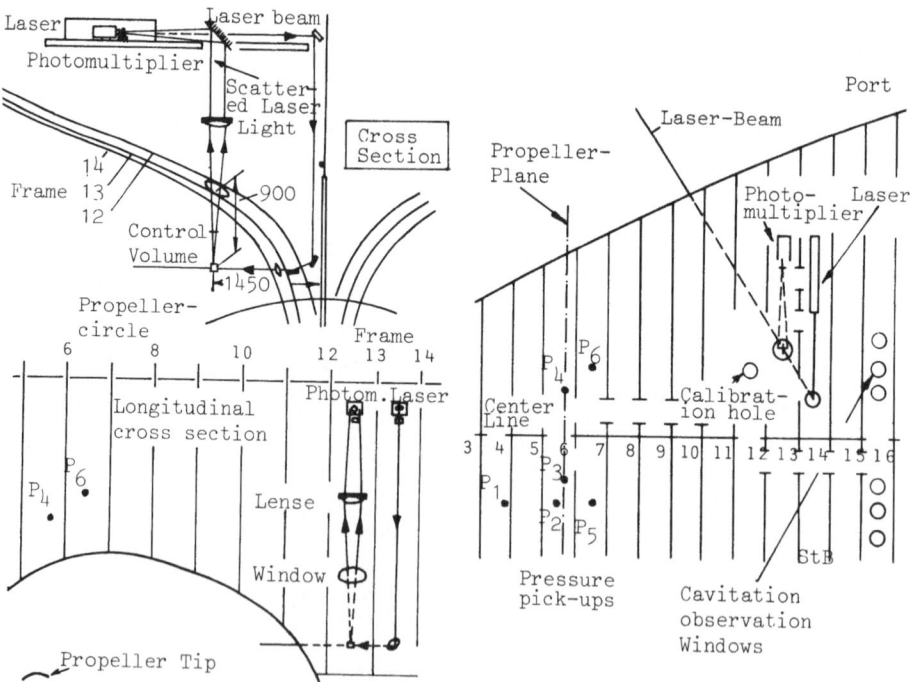

<u>Fig.1</u> Arrangements for the full scale tests on CTS "SYDNEY EXPRESS"

Fig. 1 shows the arrangement of the measuring instruments in the ship: The pressure pick ups P1 to P6 between the frames 3 to 7, the Laser-scattered-light unit between the frames 12 and 14 and the cavitation observation windows between the frames 15 and 16. The optically defined control-volume of 0.98 mm^3 (see the cross-section in Fig. 1) was located 1450 mm off the center line and 900 mm below frame 12.

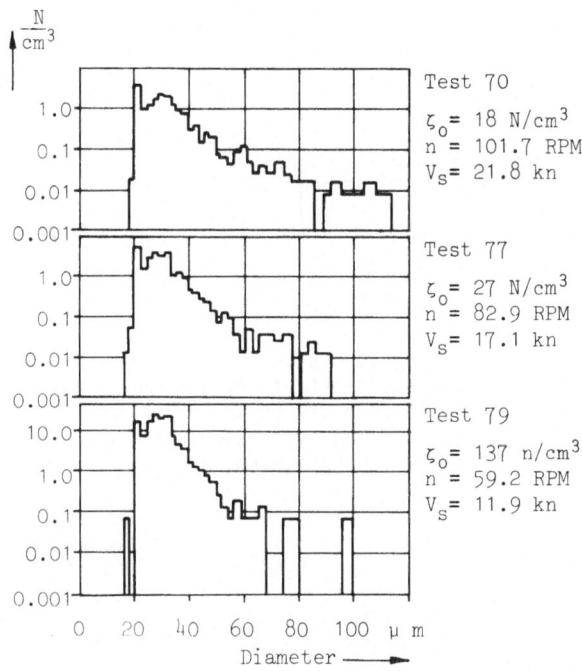

Test 70

ζ_0 = 18 N/cm^3
n = 101.7 RPM
V$_s$ = 21.8 kn

Test 77

ζ_0 = 27 N/cm^3
n = 82.9 RPM
V$_s$ = 17.1 kn

Test 79

ζ_0 = 137 n/cm^3
n = 59.2 RPM
V$_s$ = 11.9 kn

Fig.2 Nuclei distribution for different ship speed

Fig. 2 shows the distributions of nuclei diameters between 20 and 117 μm measured in this control-volume at ship speeds of V$_s$ = 21.8 kn, 17.1 kn and 11.9 kn. Thereby the total number N of measured nuclei per cm^3 increases from ζ_0 = 18 N/cm^3 via 27 N/cm^3 to ζ_0 = 137 N/cm^3. Furthermore, the relative maximum of the nuclei distribution between 30 and 40 μm of test 70 becomes an absolute maximum at test 79. This also occurred in the seaway (see [4]). The range of nuclei between 30 and 40 μm consists of bubbles. The range of nuclei with diameters of 20 μm and below consists, however, of suspended particles and bubbles. This can be seen from the results of oceanographic investigations shown in Fig. 3. This figure shows also the results of the "SYDNEY EXPRESS" measurements. Some of these suspended particles with diameters below 20 μm (upper diagram in Fig. 3) act as pore nuclei at the cavitation process, into which the liquid evaporates if the pressure is reduced.

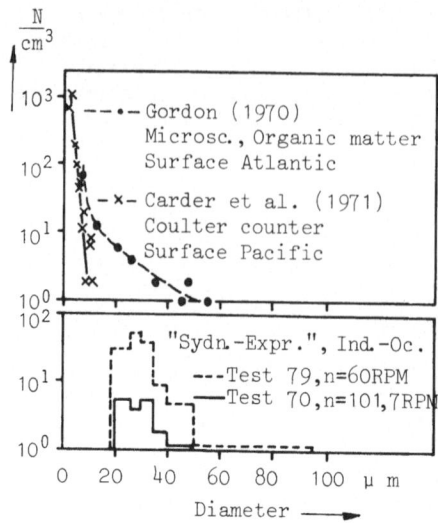

Fig.3 LSL-technique compared
with other investigations

The presumption of increasing bubble and particle density below diameters
of 20 μm was unsafe to a certain extent, during the evaluation of our
measurements. That the bubble density below nuclei diameter of 20 μm still
increases, was once more confirmed by (besides the oceanographic results
of Fig. 3) this CIUA-Congress in Göttingen, and here mainly by the paper
of MEDWIN.

Regarding Fig. 4 of LØVIK's paper it has to be mentioned that especially
the depressurized towing tank near Wageningen suffers from a too low free
air content for propeller cavitation investigations (see [5], p.301). To
overcome the difficulties of the depressurized towing tank special means
have to be tried. So the comparison in Fig. 4 of LØVIK's paper remains
improvable in the context of model propeller cavitation investigations.
Perhaps a better comparison would be reached by the figure on page 300 of
[5]. Concerning all three papers, i.e. MEDWIN, SCHIPPERS and LØVIK, it
should be noticed that these results do not contain nuclei diameters below
10 μm, which mainly take part in the cavitation process of full scale propellers. This question is discussed in [4] and [6].

Another result of our research [4,6], based on the full scale investigations, was that for equal propeller loadings substantial differences of
the cavitation extent can apparently never occur in different sea regions.
The reason is the always present large number of nuclei, which can be enlarged by the strong negative dynamic pressures on the full scale propeller.
This means that the free air content is of greater influence in model cavitation investigations than in full scale. Furtheron our investigations
[6] revealed that there are remarkable differences between model and full
scale cavitation phenomena and hereby caused hull pressure fluctuations.
The reason is the scale effect of the cavitation due to different absolute
dynamic pressures on the propellers in model and full scale.

References.

1 Keller, A.P., "Experimentelle und theoretische Untersuchungen zum Problem der modellmäßigen Behandlung von Strömungskavitation. (Experimental and Theoretical Investigations on the Problem of Cavitation in a Flow with Models)", Versuchsanstalt für Wasserbau der Technischen Universität München, Rep. 26/1973.

2 Keller, A.P., and Weitendorf, E.-A., "Der Einfluß des ungelösten Gasgehaltes auf die Kavitationserscheinungen an einem Propeller und auf die von ihm erregten Druckschwankungen, (Influence of Undissolved Air Content on Cavitation Phenomena at the Propeller Blades and on Induced Hull Pressure Amplitudes)", Institut für Schiffbau, Universität Hamburg, Rep. 321A, 1975.

3 Keller, A.P. and Weitendorf, E.-A., "Influence of Undissolved Air Content on Cavitation Phenomena at the Propeller Blades and on Induced Hull Pressure Amplitudes", IAHR Symposium on Two Phase Flow and Cavitation in Power Generation Systems, Grenoble 1976, pp. 65-76.

4 Weitendorf, E.-A. and Keller A.P., "A Determination of the Free Air Content and Velocity in Front of the "SYDNEY EXPRESS" - Propeller in Connection with Pressure Fluctuation Measurements", 12th Symposium on Naval Hydrodynamics, Washington D.C, June 1978

5 "15th International Towing Tank Conference", Proceedings, Report of Cavitation Committee, NSMB, The Hague, September 1978

6 Weitendorf, E.-A., "Conclusions from Full Scale and Model Investigations of the Free Air Content and of the Propeller Excited Hull Pressure Amplitudes due to Cavitation", ASME International Symposium on Cavitation Inception, New York, December 1979.

Part IV

Particle Detection

Acoustical Detection of Astrophysical Neutrinos in the Ocean

A. Parvulescu and G.D. Curtis

Hawaii Institute of Geophysics, University of Hawaii
Honolulu, HI 96822, USA

Abstract

The Deep Undersea Muon and Neutrino Detection (DUMAND) Project is briefly
described, starting with its inception circa 1974. Acoustical detection is
preferred over optical means for astrophysical observations, and extensive
hydrophone arrays in kilometer - cube sections of deep ocean can both detect the neutrinos and give their direction of arrival. Statistics for
expected events, and the unique problems associated with detecting an occasional acoustic impulse are discussed. Preliminary field and laboratory
tests are described. Current thinking as well as novel schemes for detector
geometries and signal processing are presented.

1. The Neutrino

1.1 Description and Detection

The neutrino is an elusive particle. Physicists now consider that the neutrino is perhaps the most common particle in the universe and yet its existence was predicted only about 20 years ago, determined experimentally just
a few years later and its characteristics subsequently measured with enormous difficulty in accelerator and reactor experiments. It is so elusive
because its mass is apparently zero, its charge is zero, and it has no other
quantum numbers; it possesses only energy. The cross section of the neutrino
is so small that the neutrino of average energy will travel through typical
rock for approximately three light years before it has a 50 % probability
of interaction, so that the experiments have generally required very large
target volumes or masses and extensive shielding from other interactions
of nature. Some of the experiments with natural neutrinos have been designed
to operate at great depths under the earth such as at the bottom of gold
mines in South Africa and salt mines in Montana, and have used hundreds or
thousands of tons of target material such as carbon tetrachloride. Yet, even
with an enormous shielding of three kilometers of rock and with those thousands of tons of target materials the number of neutrinos confirmed was
only two over several years of observation. We would like to see more neutrinos. In order to see them we must find a larger target that we can afford. We cannot buy carbon tetracloride in much larger quantities nor can
we contain it in larger volumes.

During the after-hours discussion at a Neutrino conference the suggestion
was made to use a free target material: water, specifically, seawater. And
to use free shielding material, specifically, to use the ocean itself. The

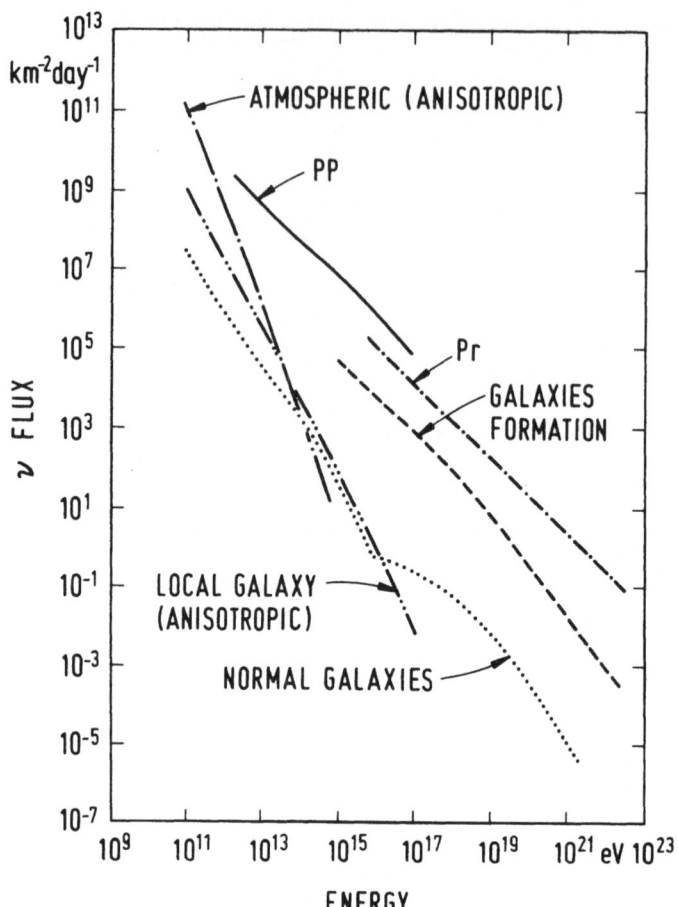

Fig.1 Source, flux and energy of neutrinos (LEARNED, 1978 [5])

idea is to try to get at the neutrino by shielding against natural cosmic rays and against natural radioactivity as cosmic rays of other than neutrino origin are usually absorbed rather rapidly (within meters) by water. It was also suggested that we devise a means of detecting the neutrino at the bottom of the sea, and within the ocean water used as a target. Preliminary discussion in the Conference indicated that the ultrahigh energy neutrino would generate a detectable amount of secondary radiation in water, with characteristics that would make identification possible.

The suggestions have been followed. The possibility of optical detection through Cerenkov radiation is being pursued actively by scientists and universities all over the world as part of the DUMAND (Deep Undersea Muon and Neutrino Detection) Project, which has developed as an organization of the scientists interested in these novel ideas. The proceedings of these conferences and workshops have been published [1,2,3,4]. The DUMAND Project scientists are investigating suitable photomultiplier tubes for Cerenkov detection at the bottom of 5-km oceans, and the computer processing of such data among other problems. The scheme can be summarized as follows: the optical method has potential to detect neutrinos of atmospheric origin, i.e.

high energy neutrinos in the range 10^{12} electronvolts (eV) to about 10^{15} eV. Neutrinos that are produced by ultra-high energy protons reaching the atmosphere from outer space (see Fig.1) and the nature of the optical project would require that approximately a cubic kilometer of seawater at the seabed be instrumented with 160.000 photomultiplier tubes arranged in a geometric pattern. By coincidence of the photon detection and the temporal and spatial alignment of the detector, such detections would then be used to interpret the signals as coming from suitable neutrinos. A large number of photomultiplier tubes is needed because light is absorbed quite rapidly in the water, the absorption distance being 20, 30 or 40 meters in various parts of the ocean.

1.2 Astrophysical Neutrinos

We are concerned here with the possibility that neutrinos of far greater energy than those of atmospheric origin exist in a natural state. We are concerned with neutrinos of energies of 10^{17} eV to 10^{22} eV. Those energies

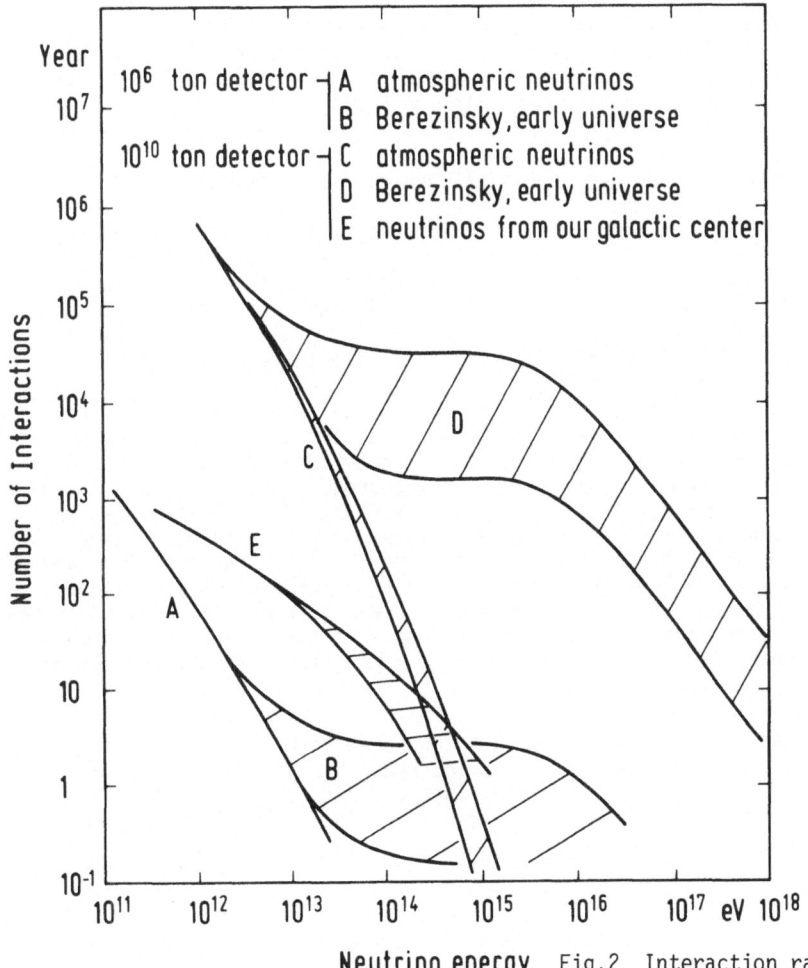

Neutrino energy Fig.2 Interaction rates in DUMAND

can be imparted to the neutrino by catastrophic astrophysical events; by stars collapsing into the Black Hole condition, by colliding galaxies, and possibly by phenomena at the center of our own galaxy. In addition, cosmic rays in the same order of energies (10^{17} to 10^{22} eV), ultrahigh energy protons and neutrons, populate the universe and can interact with the background $3^{\circ}K$ radiation left over from the "Big Bang." Such interactions between the $3^{\circ}K$ photons and the ultra-high energy charged (or neutral) particles will also generate neutrinos in the same energy range. Some of those neutrinos would traverse the earth; a very small percentage would interact with particles in the earth, the atmosphere, or the ocean. The expected rates at which those ultra-high energy neutrinos will interact are very low, so that to detect such neutrinos even a cubic kilometer of seawater (10^9 tons) is insufficient for any observation within a human lifetime. A rough estimate is that one interaction might occur per cubic kilometer every 10 to 100 years. An optical detector could only be instrumented for about 10^9 tons or about a cubic kilometer of seawater before the price becomes totally prohibitive. If we could instrument 10^{11} or 10^{12} tons of water we could hope that events of that extraordinary nature could be detected about once a day. Figures 1 and 2 show estimates of flux and event rates for energies of incoming neutrinos of various origins and detector size. The slope of some of these curves is very steep; note that at any one point the integral from threshold to infinity is almost entirely determined by the neutrinos close to the minimum energy.

2. Why We Need to Detect Astrophysical Neutrinos Acoustically

This conference is concerned with underwater acoustics and the substance of this paper is to acquaint you with acoustical effects of neutrinos. We will describe the recent experimental verifications of the theory and explain how we might construct a detector that would listen to the neutrino's sounds in 10^x tons of seawater and interpret them as *neutrinos*, with a definite direction of arrival in space and with a definite energy level.

First, let us speculate on what we might actually discover about the universe when we have a neutrino telescope - be it an acoustical or optical detector. This speculation is analogous to what happened when instead of just optical telescopes we built radio telescopes, ultra-violet and infrared telescopes and x-ray telescopes. The discovery of radio stars was *absolutely* revolutionary. The discovery of x-ray stars was revolutionary. The discovery of pulsars and quasars could not possibly be predicted from any extrapolation of knowledge at the time. The enormous flux of various particles and waves, whether x-rays or radio waves, is completely beyond extrapolation of optical knowledge and is analogous to the discovery of the Van Allen belt when rockets went up beyond the atmosphere. If anything, we were overloaded by far too much detection!

All the estimates of event rates that we have here are based on existing knowledge, and are very conservative. The neutrino, in terms of the effective wave lengths, is so much smaller than anything else we have ever observed that any of these predictions can only, in our opinion, be overconservative. Truly, we are going to open a "nu" window to the universe.

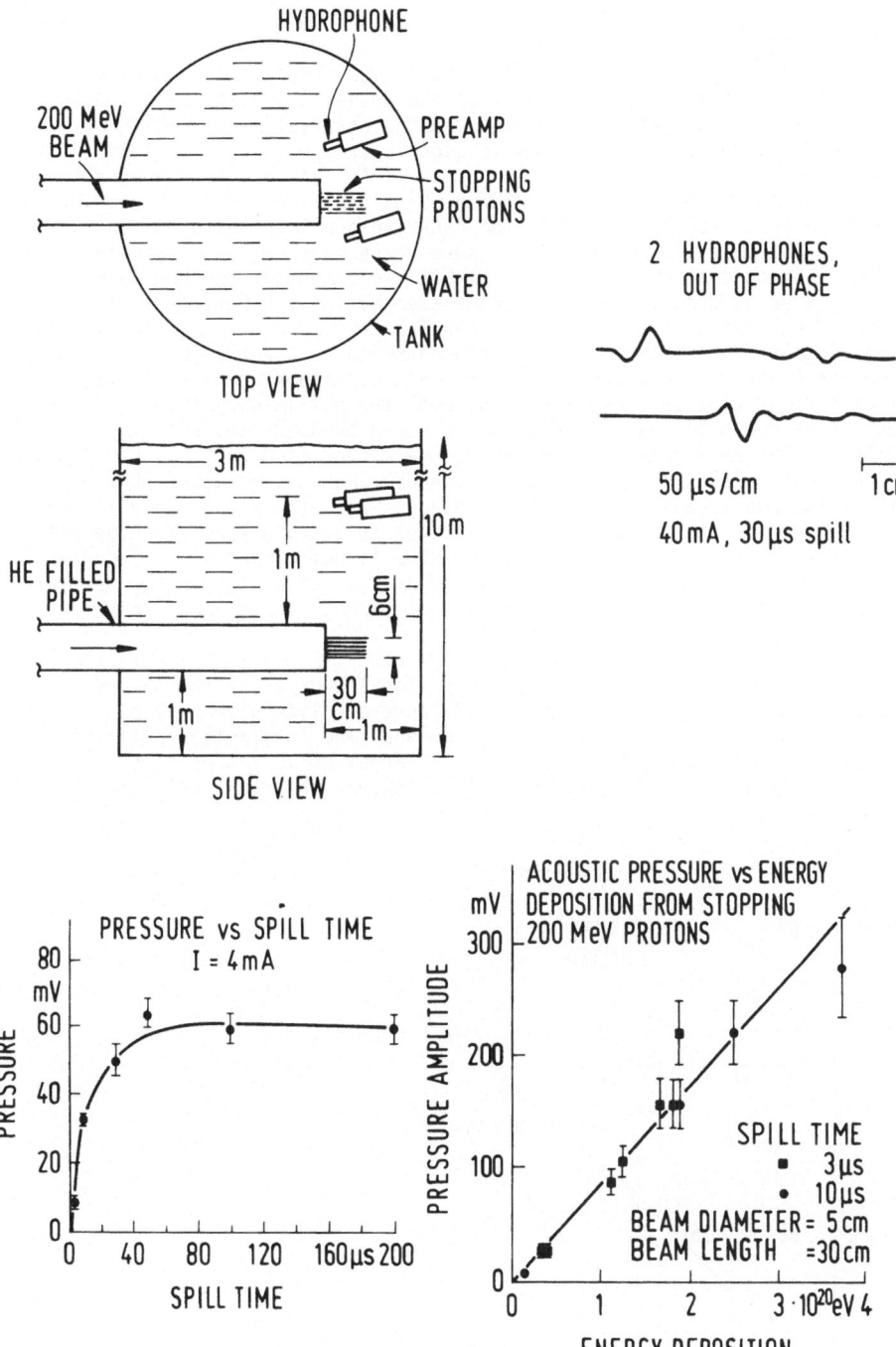

Fig.3 LINAC proton experiments (SULAK et al., 1978 [11])

2.1 Acoustic Detection

That sub-atomic particles, elementary particles, can produce sound when
interacting with a target was found by the Soviet scientist ASKARIAN in
1957 [6] and, as an independent experimental observation, by HOFSTADTER
and BERON, in the report of acoustic signals from particles colliding with
a solid [7]. The possibility that neutrino interaction with seawater could
produce sound was proposed by BOWEN in 1975. Contributions in the 1976
DUMAND Workshop [2] by BOWEN [8] and ASKARIAN and DOLGOSHEIN [9] gave preliminary estimates of the sound pressure to be expected by a nuclear interaction. The estimates were revised within two months of the DUMOND workshop.
Quickly organized experiments at Brookhaven National Laboratory determined
that indeed, showers of protons, either from the Linac or from the AGS (Alternating Gradient Synchrotron), produced sound in a water container as
shown in Fig.3. The measured wave forms and impulse spectra of these acoustical signals were in close agreement with the preliminary theory. Experiments were conducted at the Harvard Cyclotron by SULAK and his students
while further refinements of the theory were conducted both by the Russians
and by BOWEN, LERNER, W.V. JONES and BRADNER. We have no doubt that particles stopped by a liquid target eventually heat the liquid locally and
therefore create an expansion that radiates as a sound wave. We have not
been able to determine (in a tank) the directional radiation characteristics of the sound that would be produced by the ultra-high energy neutrino,
but we have theorized the sound to be that from a line source. These factors
will be detailed in a subsequent section.

Supposing that the neutrino acoustic signal behaves as the theory predicts - how can we actually detect it? There are literally billions of neutrinos going through each of us every second. The interaction rates of those
neutrinos are extremely *low*; even with a target of 10^{11} tons of ocean, interactions will still be few. They will, however, have a finite energy-de-

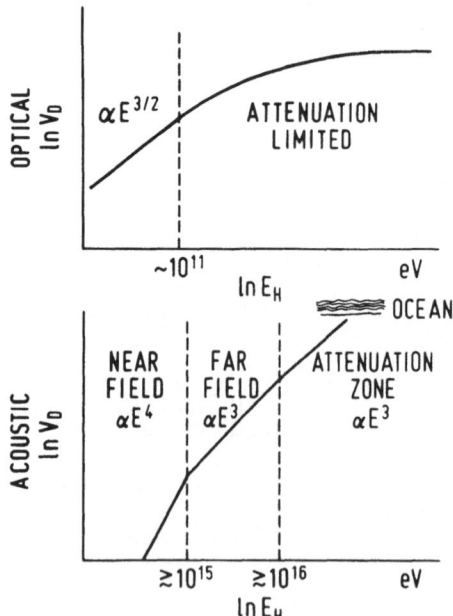

Fig.4 Detectable volume
(LEARNED, 1978, [5])

pendent rate. The interactions produced by high energies can be heard at great distances. What is the volume of water around each such interaction in which the sound is, say, the 20 dB above ambient usually considered necessary to detect it reasonably and unambiguously? Fig.4 indicates the relation of volume (of sea water) to neutrino energy for detection by acoustic means; a curve for optical detection is included for comparison.

The volume is not merely a question of distance but also of the acoustic radiation pattern. Because the portion of space into which the sound is radiated is very thin (as will be shown subsequently), there is a relatively small volume in which the sound from a neutrino event can be heard at significant levels. So, *detectable volume* is a concept which has been clearly a function of both signal and noise characteristics. We shall examine each of these in turn.

2.2 Acoustic Radiation from the Interactions

An acoustical effect may be produced by interaction of one high-energy particle with a medium, or by interaction between particles. The essence of the process is that nuclear particles interact and transfer their initial energy to other particles including photons; the energy of the subsequent products is distributed among other particles and the eventual result is that the energy is deposited in the medium as heat or radiated from it as photons. The estimates of 1976 have been further refined in a series of papers by ASKARIAN, DOLGOSHEIN, KALINOVSKY, and MARKOV [10]. Other estimates were obtained by W. Vernon JONES and are included in [4]. The estimates all state that in the interaction of one incident neutrino there will be a muon that in many cases carries about half of the total energy of the incident neutrino. The muon is very penetrating; it will shed its energy over a path length of the order of kilometers and thus will produce only slight ionization of any point along the kilometer, detectable only by means other than acoustic. Most of the other energy of the initial neutrino interaction will be transferred to hadronic particles and will be localized in a region generally cylindrical in shape, roughly 5 m long, and 5 cm or less in diameter (see Fig.5).

The theoretical results are presented in detail in the ASKARIAN and the W. Vernon JONES reports, but they are derived, inevitably, by Monte Carlo calculations starting from different assumptions. There are, therefore, differences in the resulting estimates, and uncertainties about the accuracy and validity of the hypotheses used. Nevertheless, the general features are unquestioned; that the neutrino energy is eventually deposited as heat in a long and thin volume with circular (nearly cylindrical) symmetry. The exact shape of this region will affect the wave form that can be detected acoustically in various directions away from the heated cylindrical region, but most of the acoustical energy will be contained in a very flat disk at right angles to the heated cylindrical region. This region has become known colloquially as the "hot rod." The regions of space away from this disk will exhibit substantially different waveforms, with greatly decreased acoustical pressures, so that initially we don't need to concern ourselves with detecting a signal away from the major disk of insonification.

Experiments carried out in the fall of 1976 at Brookhaven National Laboratories confirmed that when large numbers of particles (in this case protons) are stopped by water they indeed produce an acoustical wave with characteristics anticipated by theory. The waveform generated by protons

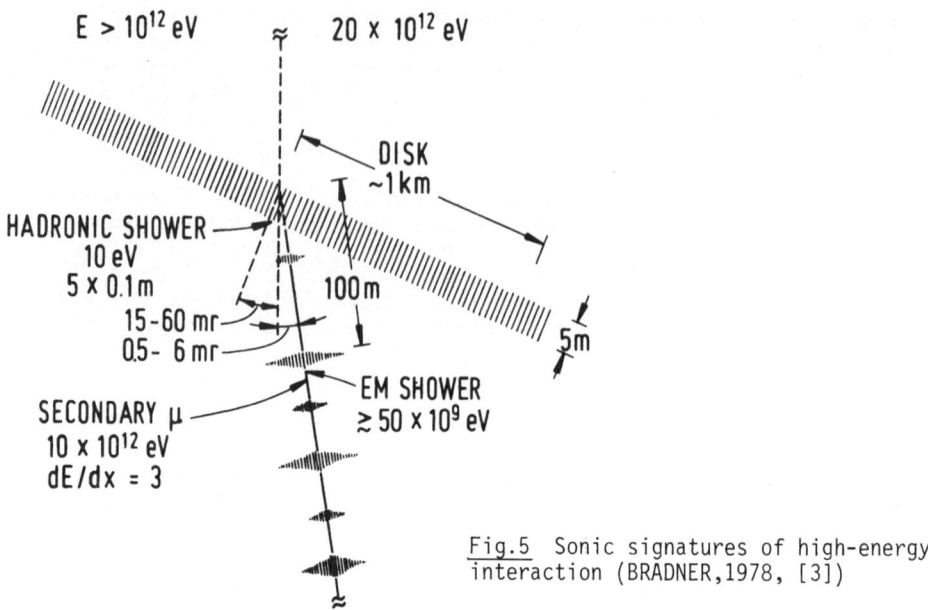

Fig.5 Sonic signatures of high-energy interaction (BRADNER,1978, [3])

heating a volume of 30 cm by 6 cm at the Linac (Fig.3) demonstrates that the duration of the pulse and the waveform details correspond to what seems to be the spatial extent of the proton beam in the water. And when the duration of the proton beam is increased in the tank, the waveform again corresponds to the theoretical prediction, namely it is the time derivate of the heat deposition function.

A similar experiment was then conducted at a higher incident energy [11]. The Linac experiment used 200 million electron volts or 2×10^8 eV; later we used the Alternating Gradient Syncrotron (AGS) at Brookhaven, using the so-called Fast Extraction Beam (FEB) output where the "spill" of protons could be shortened to as little as 3 μs, and within a very thin (3 mm diameter) area.

The geometrical configuration was different; the tank was much smaller and had much accessory equipment very close to the hydrophone. Fig.6 shows that the acoustic pulse length is about 3 μs corresponding to the pulse duration; the rest of the signal is reverberation in the tank. In Fig.7 the same acoustical signal is analyzed in overlapping decade bands to show something of the energy spectral content of the process. For comparison, we show in Fig.8 theoretical waveforms from LEARNED [12], as also predicted by ASKARIAN et al. [10] and BOWEN [8]. These are the theoretical waveforms broadside to the hot rod. At angles away from the normal, the waveform changes from its characteristic duration proportional to the diameter of the cylinder. The agreement between amplitudes in theory and experiment seems to be better than 3 dB, which at this stage of sophistication and government funding is quite satisfactory.

The waveform and the amplitude do depend on the detailed distribution of the energy in the heat deposition region. The neutrino will produce a normally distributed cross section in its "cylinder" with progressively changing widths, whereas the laboratory protons are stopped in different spatial

244

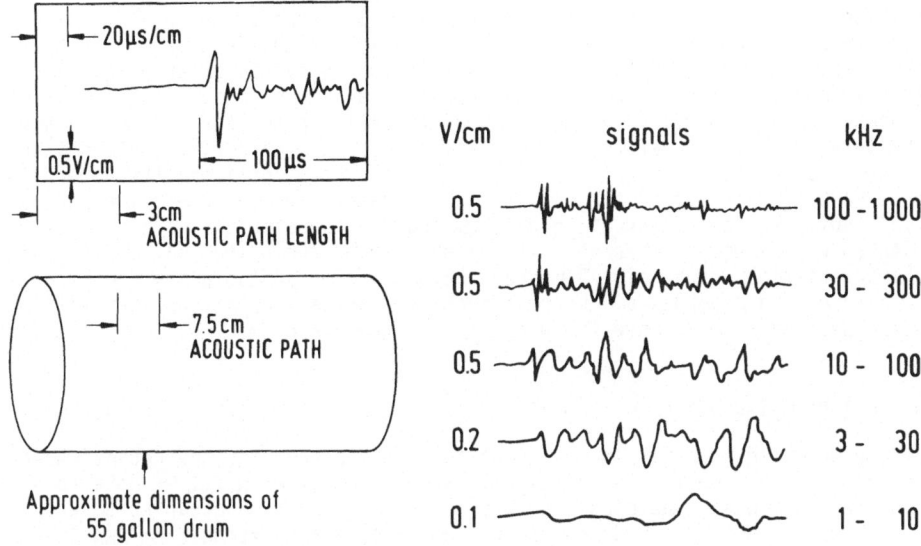

Fig.6 Brookhaven National Laboratory
Alternating Gradient Synchrotron
(Oct. 1976) (Sulak et al., 1978 [11])

Fig.7 Analysis of signals from
Fig.6

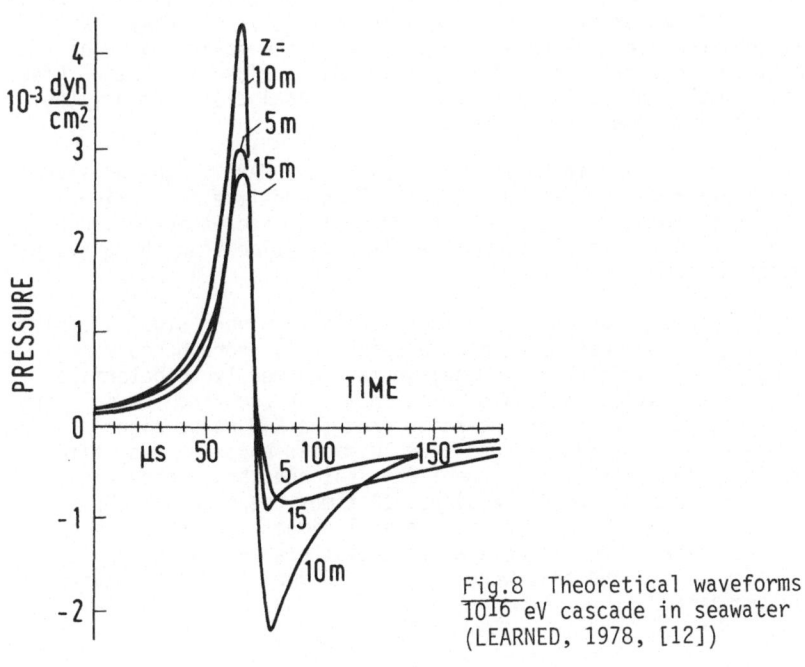

Fig.8 Theoretical waveforms
10^16 eV cascade in seawater
(LEARNED, 1978, [12])

fashion. This is because the protons are arriving through an iris in the cyclotron with a rather uniform distribution across the iris, which then drops rapidly away from the axis. These are 10^{11} protons rather than 10^5 or 10^6 byproduct particles from one neutrino. So we are considering only what happens when the particle energy becomes thermalized and forms a sound wave. The details of the waveform are surprisingly close considering the mismatch between assumptions. Further studies at Harvard have confirmed the dependence on heat capacity of the medium and its coefficient of thermal expansion [11]. But, the most interesting problem now for acousticians is not to design an experiment to measure these waveforms and their spectral and spatial characteristics accurately. Some have been designed and some are now being carried out in Brookhaven. The next problem - whose foundations are still being planned - is to move the project into the ocean.

3. Noise In the Ocean

Before we build an array in the ocean to detect a neutrino and decide where to locate it and what level of signals we can detect with it, we have to consider the noise. The classical knowledge about ambient ocean noise was summarized by WENZ in 1962 and brought up to date and extended by ARASE and ARASE [13] and BRADNER and HOWARD [14].

Our present knowledge of oceanic noise is essentially of a statistical nature. We know of RMS noise estimates obtained by various experiments, but the experiments were designed around certain assumptions regarding spatiality and sources of the noise.

Apart from the steady-state random thermal noise, there are no stationary ambient conditions, but there are two other important categories: the random non-stationary and the capricious non-stationary. The non-stationary "random" processes we usually think of refer to days of calm, days of heavy rain (which is a high frequency source at the surface of the ocean), days of average conditions, microseismic activity, etc. By capricious we mean those which are totally unpredictable and lacking statistical distribution; they cannot be tied in with the curves of Fig.9.

The noise we really have to be concerned with in DUMAND is the non-distributed, non-stationary, capricious signals which are like a neutrino cascade. And, we do not know enough about the origins and extent of such noises. As we learn more, we can plan more discrimination than the primary means discussed so far.

To that end a group of us conducted some very preliminary studies [15]. By using existing hydrophones mounted on a seabed (the hydrophones were there for a completely different purpose) we managed merely to determine that indeed brief acoustical pulses of high amplitude occurred far more frequently than expected from an extrapolation from the RMS noise signals, seen in Fig.10. Signals with extreme values like 10 sigmas happened many times a second, whereas statistically the probability that one should occur is approximately 10^{-11}. In other words, a signal 10 times the probable RMS error should happen once every 10 days and not several times a second as in our field experiments. Therefore, we determined that there were some totally independent sources of sound that were not just thermal or surface agitation of the ocean seawater.

246

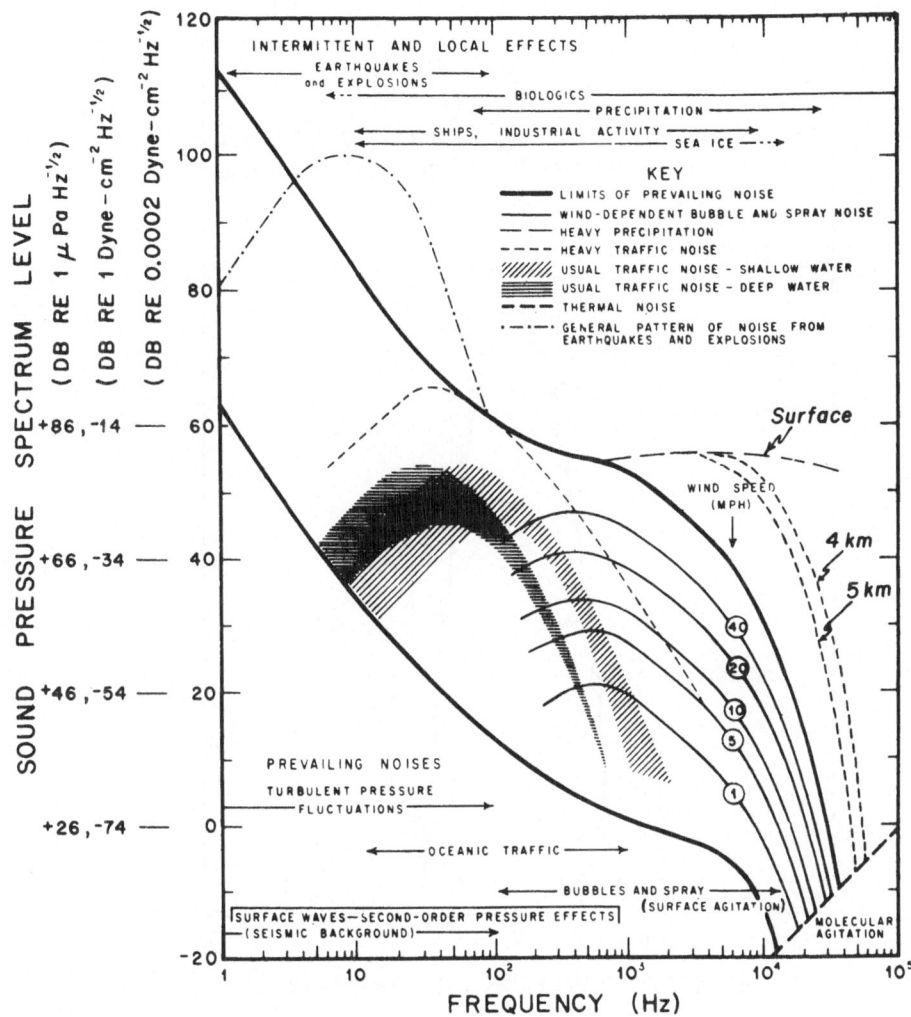

Fig.9 Noises in the ocean (BRADNER & HOWARD, 1978 [14])

One source we know something of, and which has been discussed in the summer workshops, is biological. Whales, and especially porpoises emit clicks which mimic the signal we expect from a neutrino. Both have been reported to produce broad-band levels exceeding 200 dB//1 μPa. Even if it is assumed that these animals remain near the surface ~ 4 KM above our detectors, the received levels have concerned those analyzing the problem. While we share their concern, we do not consider it as serious a problem as previously depicted.

The sounds emitted by whales are concentrated below 20 kHz and therefore the energy available in our band of interest is a small fraction of the reported (wide-band) levels at best. Porpoises, however, not only can produce

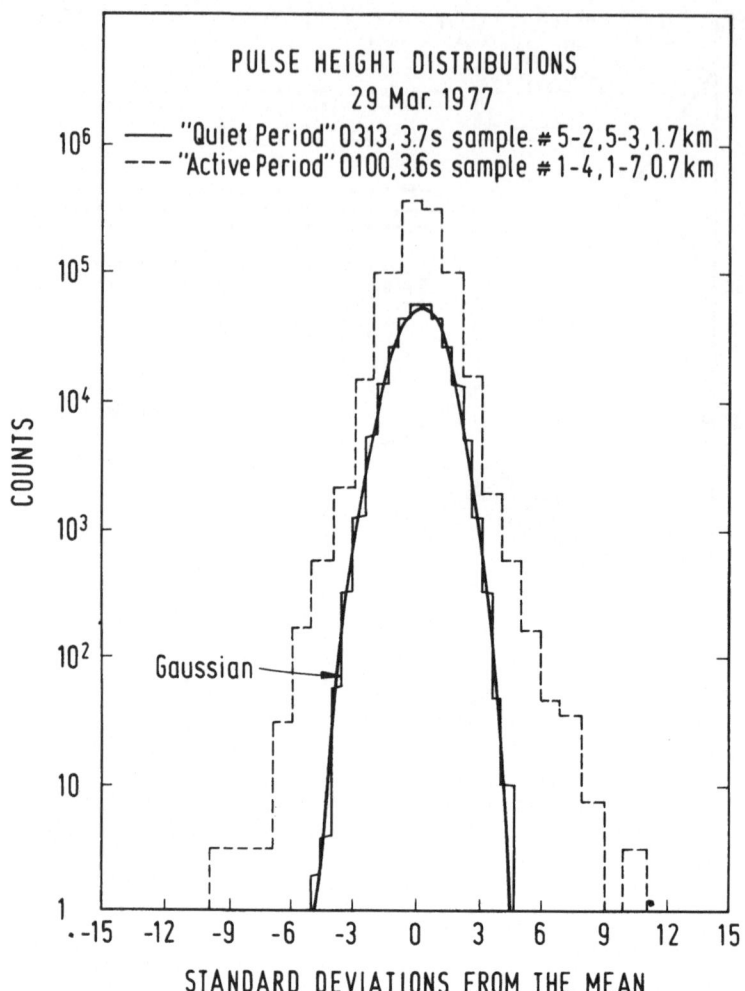

Fig.10 Noise samples from ocean near Kauai, Hawaii (STENGER, 1977, [15])

a click but they shape it for optimum amplitude (in relation to their ambient) in the band from 30 kHz to over 100 kHz as shown by AU [16].

Fortunately, these air-breathing mammals stay near the surface so we have the benefit of some 120 dB attenuation, as derived and used in Fig.9. We gain a few dB more because we are not listening wide-band. More importantly, the animal has excellent beamforming capability and normally does not usually emit signals more than a few degrees downward - and the ocean surface is not a clean reflector for his upward emissions.

But, if a worst-case signal of, say 80 dB//1 μPa does reach our array, can the system segregate it from a neutrino event? This is roughly the source level predicted for a 10^{16} event in [10]. First, most porpoise signals are actually click trains or whistles and would be clearly distinguishable although they might mask an event. An isolated click (which has not been reported to occur in the ocean) would appear as a point source in the far field and so we have another significant difference from the radiation from the cascade-generated disk. However, it would be of, or above, the order of the known ambient (thermal and surface) and would present a processing problem. As yet, we should acknowledge that a biological signal reaching the array - especially near its periphery - could be confused with a neutrino event. So could other noises about which we do not have even the "estimating knowledge" of the porpoise noise. We have no choice but to look carefully at capricious noise in relation to our very capricious, infrequent signal.

4. Instrumenting the Ocean

Now comes the question of the use of these acoustical signals at the bottom of an ocean where the depth is 5 km and the pressure 500 atmospheres (5×10^7 Pa). We must define the nature of the measurement to be made. We want to know that the neutrino existed; we want to know from what direction it arrived, and its initial energy. To begin with, that is all we need. We will assume that the neutrinos with which we are concerned are those at ultra-high energies, 10^{17} eV or more, which are rare, and therefore require us to have acoustical equipment that can detect their presence in a large volume of water. We are considering here something on the order of 10^2 to 10^4 cubic km, i.e. we are talking of a target mass of 10^{11} to 10^{13} tons of water. This now becomes a measurable fraction of the total water of all the oceans, which is of the order of 10^{18} tons. We want to detect the acoustic signal sufficiently well that we can estimate some of its parameters. That is, we want not merely to detect the signals, in the sense of signal detection theory, but also to perform an "estimation" of signal parameters so that we may deduce the energy and arrival direction of the neutrino cascade. The general problem has been reviewed by PARVULESCU [17].

In order to detect a signal we need a signal-to-noise ratio greater than zero at the output of some basic component of the system. No matter how good the processor is, detection theory has told us that there will be errors in the detection. The first kind of error is to miss the signal which actually existed. And the second error is the "false alarm" when there was no signal. The false alarm rate is of critical importance in phenomena of infrequent nature. Because the "cost" of evaluating or using false alarms can be extremely severe, detection theory - especially as applied in sonar - describes how to evaluate the relative rate of detection and of false alarms. To obtain detections with reasonable false alarm rates, we need a signal-to-noise ratio of about 20 dB. In trying to estimate the parameters of the signal in addition to detecting its presence, we require at least a similar signal-to-noise ratio because parameter estimation has a continuous distribution of errors. Calculations have been made by BLACKINTON and LEARNED, BRADNER, BOWEN, and STENGER [18,19,4,15], estimating the acoustic event rates that would be detectable by a single hydrophone. Perhaps the most descriptive illustrations are Fig.11 (LEARNED [12]) which shows the distances at which acoustic signals from various levels of neutrino energy can be observed, and Fig.12 (also from LEARNED), which provides an interpretation of volumes of ocean that can be observed by hydrophone arrays. Thus, we develop the concept of "detectable volume."

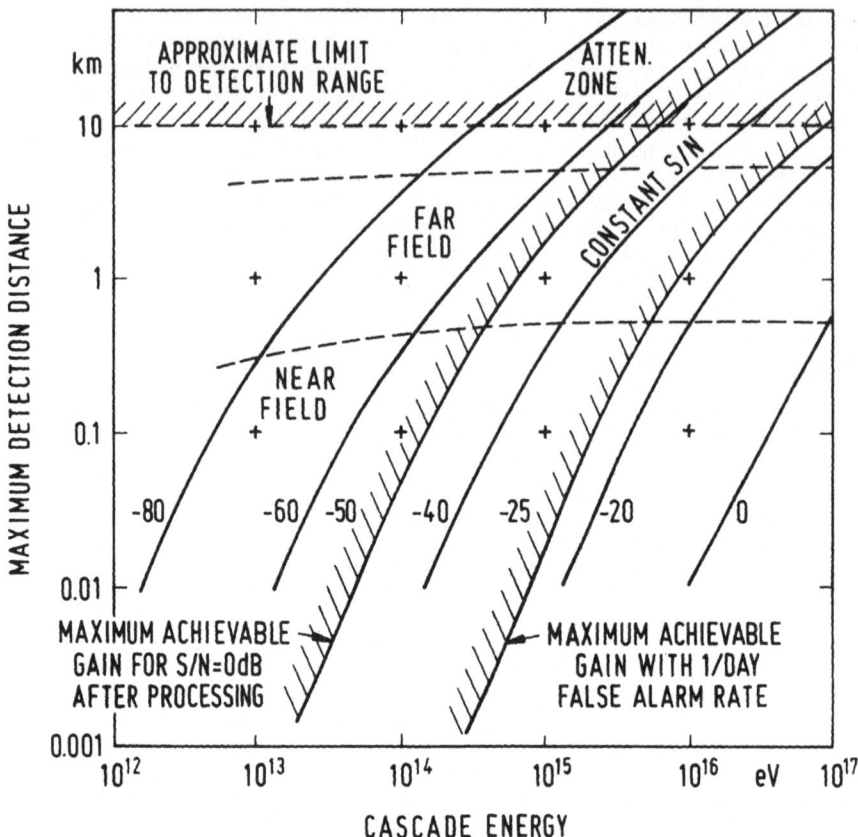

Fig.11 Maximum detection distance versus energy, for various signal-to-noise ratios (LEARNED, 1978 [12])

4.1 DUMAND Hydrophone Arrays

We require a coherent array for practical detection of the neutrino signal with a usable signal-to-noise ratio; we also need enough receiving points to insonify the detectable volume of seawater required to find an event in a reasonable time span. Of course, "reasonable time" may be a day, a year, the lifetime of the equipment in the ocean, or our lifetime, depending on your viewpoint. Realistically, we must orient our sensor system design to give us every advantage in detecting with high reliability the lowest level signals possible amidst the assorted noises previously discussed.

Let us examine the problem of this detection and see what properties of the signal we seek will help us to find it, and what it will take to accomplish this. One property that has been mentioned is the spatially unique disk radiation from the hot rod. There are very peculiar properties which are quite different from those of ordinary array theory. Array theory is usually addressed at a point source at infinity, at receiving a plane wave at a delta function in angle. Here, we don't have a far-field source even when we are at a great distance, in a region where it approaches a plane wave; we are looking at reception at many places in an expanding wave front.

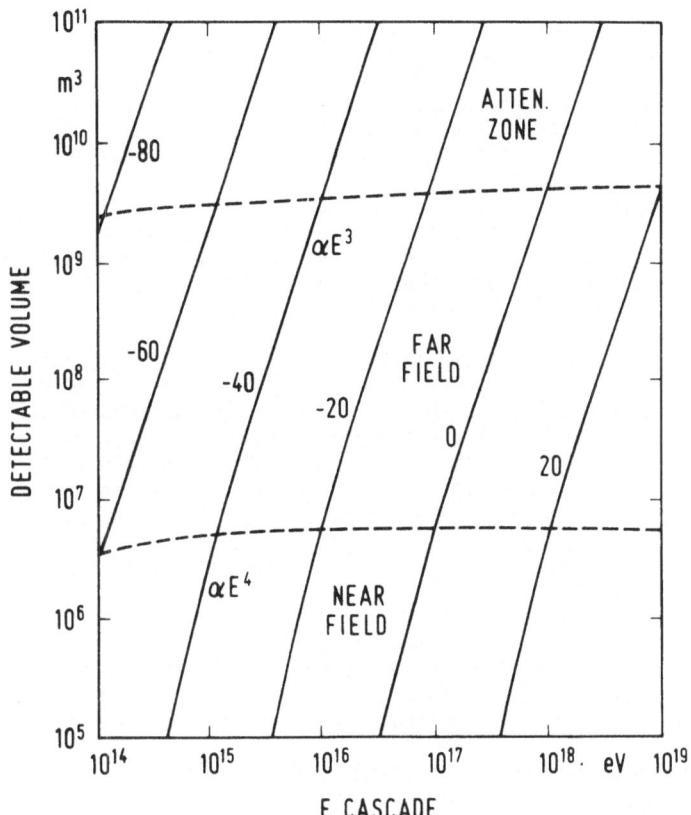

Fig.12 Detectable volume versus energy (LEARNED, 1978 [12])

We may need receivers that are spatially very distant, in order to triangu-
late with a very large, sparsely occupied receiving array. We have practical
reasons for a few, densely occupied receiving clusters with large gaps in
between; these include array gain and deployment factors.

The array can take several forms and we should not feel constrained at
this time to assume that it will necessarily be cubical or plane, or even
that it will be made of individual (discrete) hydrophones. Perhaps a sheet
of polymer laid on the ocean floor will be the most feasible small array!
First we will look at some of the factors involved in the sensing and detection (processing) systems without too much regard to the physical details involved. But, for some very cogent reasons we feel that the DUMAND
acoustical array will be made up of a number of hydrophone clusters.

4.1.1 Sub-Arrays

Let us first discuss clusters, or sub-arrays. If we were to depend on single
hydrophones to detect the presence of the acoustic field from the cascade,
our only recourse would be to put the hydrophones extremely close together.
But then we would have an enormous processing task, weeding out false alarms

and obtaining directional information. So we must - in order to get a +20 dB S/N according to Fig.11 - put the hydrophones 10 m apart for 10^{17} eV events. If we want to cover 100 cubic km or 10^{11} tons of ocean, and we put the hydrophones every 10 m, we need 10^8 hydrophones - obviously too many. Yet if we space them too far apart, they won't hear the same event, and if we place them closer we will not fill the volume. Thus, we soon realized that we needed to cluster the hydrophones to obtain adequate coherent gain (i.e. the signal is coherent while the noise is not, in the insonified area) from each cluster, and to distribute the clusters in the volume in some (hopefully) optimum fashion. The basic theory, trade-offs, and possible spacings and ranges were presented by STEWART [20], by BLACKINTON and LEARNED [18] and by BRADNER and LEARNED [21].

The reader should note the distinction between the *apparent* volume passively isonified by a cluster (or any array) and the *detectable volume* in which the actively insonified disk and the receiving array volume coincide. BLACKINTON and LEARNED [18] stated it a reasonable assumption to consider this detectable volume, for a given cluster, to be equal to the volume of a cascade-induced acoustic disk detectable by that cluster. This assumption leads to the geometrical conclusion that approximately 1 % of the ocean viewed by, say, a cubical array is actually its detectable volume. This is so because the disk will be of any orientation and location, due to the random nature of arrival of the neutrinos.

Thus, the overall "array" will be treated in terms of clusters of various configurations, deployed in three dimensional space near the bottom of the ocean. In some of the referenced literature the clusters are considered as a single hydrophone but it is recognized that they will generally be an array. An additional advantage of clustering, of course, is that it minimizes cabling and concentrates signal conditioning electronics.

4.1.2 Volumetric Arrays (Three-Dimensional Clusters)

Consider now a cluster about 5 m by 5 m by 5 m. This is about as large as practical (deployment problems aside) since the thickness of the sound disk is of about that order until the radius exceeds a kilometer and then it starts diverging. For a wavelength of 5 cm and half-wavelength spacing, we get 200 hydrophones cubed or 8×10^6 hydrophones in the cluster. This is good news in one sense: we get almost 69 dB (coherent) gain with such a cluster. Theoretically, from Learned's curves (Fig.12) each such cluster would cover the aforementioned 10^{11} m^3 of seawater, with a 20 dB signal-to-noise for 10^{16} eV cascades. The detection range might extend up to 10 km for a 10^{16} eV cascade; realistically an array gain (per cluster) of 50 dB is considered a practical limit that sets a range limit more like 3 km and a volume of $\sim 2 \times 10^8$ m^3 for the 10^{16} event. This is still a good result, but there are a lot of hydrophones to be processed! Alternatives include fewer hydrophones per cluster (since we seem to be exceeding feasible array gain) and clusters of different form. Obviously, there are many, many trade-offs to consider in this problem of instrumenting the ocean.

4.1.3 Planar Arrays (Two-Dimensional Clusters)

If we look at a 5 meter two-dimensional cluster, a planar array rather than a volumetric one, the number of hydrophones is 200^2 or 4×10^4 and the maximum coherent gain is 46 dB which means that we are looking at a volume of $< 10^8$ tons on the LEARNED- curves (Fig.12) for a cascade of 10^{16} eV and 20 dB signal-to-noise at the output. What is the number of such clusters for the whole array? Can we space them about 3 kilometers with that gain where we can receive up to -26 dB signal-to-noise? If we look within one kilometer we can see very weak signals and there are many more of these, so we don't need to insonify so much of the ocean. When we are in the near field of the source the signal will not be so coherent, and you will not have as much gain, but the signal will be much stronger so you don't mind too much.

The change of slope in Fig.4 is due to the fact that in the far field, instead of cylindrical propagation in just one dimension, you get spherical spreading. What also happens is that for a cluster in the near field, say at 50 m the phase distortion is sufficient so that the signals do not give you the same delay at the many hydrophones, so only a small portion of the array is coherent. But as you are nearer, you have a stronger signal; e.g., if you go from 300 meters to 100 meters (Fig.11) you gain almost 10 dB, so you could have only a tenth of the area.

A planar cluster has been considered favorable in two particular respects: Its beamforming capability is inherently simple in a bi-or uni-directional manner (e.g., at the bottom looking upward) and it seems easier to fabricate and deploy than a three-dimensional cluster. However, any additional consideration of factors such as these must be done in context of the entire array of which these clusters are elements.

4.1.4 Linear Arrays (One-Dimensional Clusters)

With a one-dimensional cluster of course we anticipate that the gain decreases. Assume the number of hydrophones is 200, again on the grounds of statistical independence of the outputs, which provides a gain of 23 dB. The number of beams is also 200. That makes the processing of the string of hydrophones much less complex. But the beams are now conical, so we must compute the intersection of three such beams to verify detection and determine direction if we look at it as a multiple array. And, of course, we have to process with suitable delays to "form beams".

For this lower gain (and "savings" in number of hydrophones), we must work at a kilometer and go to 10^{17} eV as a threshold. Or we can look for even higher energies at greater ranges - but such events are extremely rare so we must occupy more of the ocean (more clusters) before we can expect to hear one.

Obviously, the vertical cluster is the easiest to arrange and deploy along a cable in the mode discussed in the recent DUMAND workshops [4]. If we arrange these clusters vertically, they will cover a radius of 1 km for 10^{17} eV, per Fig.12. And if we space the vertical arrays 1 km apart, the maximum distance between a cascade and the three nearest arrays will be 700 meters, so this might be a good choice for a fully capable "simple" array design if we make it very extensive or wait a while to detect a neutrino.

But let us consider other factors. Since these clusters are actually centered (by our processing scheme) every 5 meters on a 1 km vertical string, simply detection of the *epoch*, in conjunction with the vertical locations insonified, is sufficient to determine cascade position and orientation. This approach may lead to simplification of the processing and verification of an event. Again, the utility of the array is dependent not only on the processing of the cluster to achieve the gain shown, but of the entire string and array to achieve the actual event data desired in one fashion or another.

Another unique approach to arranging and processing a linear array involves spacing the hydrophones in a non-uniform distribution in two lines one-half wavelength apart as shown in Fig.13. Here a waveform from line A is cross-correlated with that from line B as shown. This forms a spatial chirp or so-called MESS (Matched Equivalent Space Signal) processing. The signal-to-noise improvement is proportional to the number of hydrophones intercepted by the signal. Both the physical arrangement and the processing appear to be very compatible with a linear cluster.

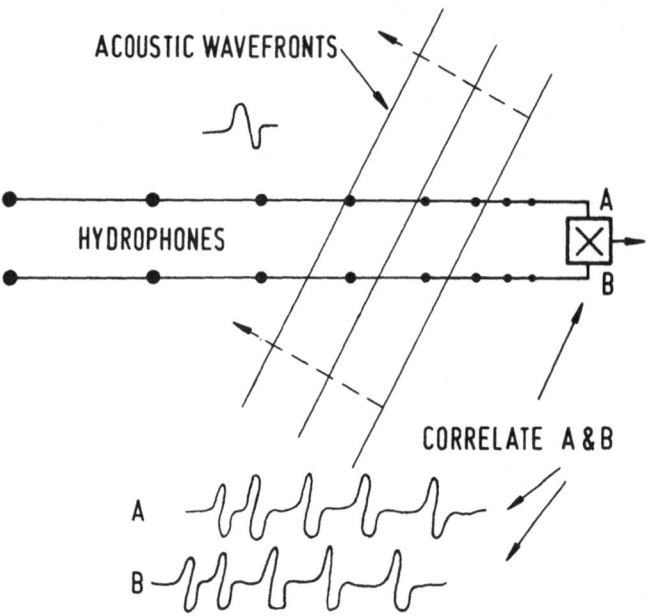

Fig.13   Spatial chirp (MESS) acoustic signal processing

4.1.4.1 Long-Line DUMAND

A special case of the arrangement of vertical line arrays which may offer advantages in deployment and processing is a parallel line system as shown in Fig.14. Here, with 1 km high strings set at 1 km spacing (two lines 866 m apart) the detection volume criteria discussed previously are maintained and the cable zig-zagging between strings can provide processing of coincident events in real time. The detectable volume will be 3 km^3 per km of length. We leave it to you to decide how many kilometers we can afford; there should be between one galaxy formation event per KM per year, up to (upper limit estimate) 10^2 total astrophysical events per KM per year.

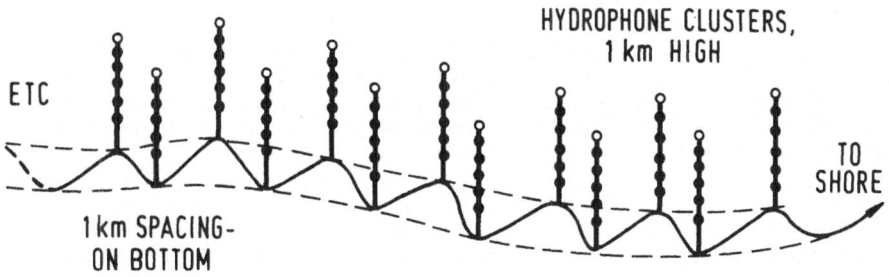

HYDROPHONE CLUSTERS,
1 km HIGH

ETC

TO
SHORE

1 km SPACING-
ON BOTTOM

Fig.14 Long-Line DUMAND

5. Processing

We expect to perform as much pre-processing as possible - without unduly
restricting the utility of this research device - because the volume of data
inputs demands that "noise" (all unwanted signals) be constantly reduced.
This pre-processing includes filtering, starting if possible with the hydrophone itself to emphasize the roughly 30 to 60 kHz band of interest, pre-
whitening across this octave, preamplification, threshold setting, and some
sort of DIMUS processing. An adaptive system is useful to reduce false
alarms and processing volume or both. Beamforming and wave front measurement are necessary for spatial information, and we need to preserve the
absolute level in processing if we are to estimate the energy. This "cascade
processing" will eventually result in a reasonable volume of significant
data transmitted to a shore station for further off-line processing. At
several points, we must make careful compromises between forwarding more
data than we can afford, and throwing out information we need.

With the proper information from each cluster, we can further process
potential events from clusters that receive them, as part of the overall
array. This array (just as in the case of the cluster) may take several
physical forms, as discussed previously. And of course, the manner of processing or reconstruction of the event is closely tied in with this form,
including its spacing or density and cluster form and sensitivity. Clearly,
processing, complexity, and estimating ability are closely interrelated.

For example, to determine the direction of an incident neutrino we must
determine the orientation of the plane in which the sound is radiating.
The normal to the acoustic disk is the neutrino direction. To localize the
acoustic disk we can intersect pencil beams from volumetric or planar
arrays and identify the neutrino direction by directional acoustical information. It is also conceivable that if three or more acoustical components
receive the signal the mere positional information of three points coplanar
in the acoustical disk also determines the spatial position of the disk and
therefore the neutrino direction. If we could obtain coherent gain against
noise without the added processing involved in estimating the direction of
the acoustic pencil beam and limit ourselves to noting time and position of
arrival of the acoustic signal, we can determine the neutrino direction
equally reliably by knowing the position of each sub-array that performs a
detection and the relative time of arrival of the signal at those positions
[22].

Lining up such signals - which may be done post-facto - eventually represents off-line processing. We expect to continuously divide the analysis between such tasks and on-line processing, which is limited by the amount of useless information we must handle along with the signal. So we will try to put it on-line after eliminating the false alarms and deciding a probable detection has occurred.

To summarize, we will pre-process in the clusters as much as possible, up to the extent of screening for a probable detection. In this era we have minimal electronic problems in doing so. We will assemble the resultant data with similar data from the entire array and process it further as appropriate to the array configuration. We then record the data from the entire array (since we have now reduced them to a manageable number), present them, and analyze them off-line as desired.

6. Hardware

The components of the hardware are within the state of the art. We know how to do all the things that have to be done; they have never been assembled in large-scale systems. Long ocean bottom cables have been reliable for almost a century; hydrophones have been laid on the sea bottom and in use for decades. The electronics for large-scale processing have been built and used for years. Their size and power drain is a fraction of that used a decade ago. The assembly, installation, and operation of a large system of this nature is more a matter of cost than feasibility. The huge number of hydrophones required might seem forbidding until we realize that when we specify hydrophones we really mean printed electrodes on a continuous layer of sensitive polymer or ceramic substrate; segments of the arrays can have the sensor components directly parallel.

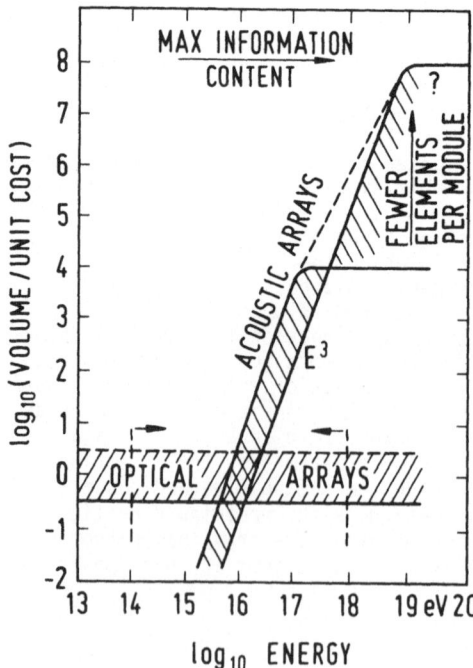

Fig.15 Relative cost of optical and acoustic arrays, on a sensitive volume per unit cost basis

The processing that is required as we proceed to eliminate "junk" information systematically and correlate the possible event signals - it will not be simply parallel or sequential processing - is compatible with microprocessor and related digital chip technology. We must not overlook of course the interconnection problem and that is a very good reason for limiting the volume of information that must be passed on at each stage. Connectors, in-array cabling, (including fiber optics data links) and deployment of the clusters and strings have been examined by the ocean engineering people of the DUMAND workshops. Two prospective sites near the Hawaiian islands of Maui and Hawaii have received preliminary evaluation, but extensive and detailed examination remains to be done. These aspects present a challenge, but one which can be handled.

Fig.15 indicates the expected volume-per-cost relative to energy levels detected. (Cost of waiting for a very rare, high eV event has not yet been factored in!) It can be assumed that the detection rate as well as the cost of the acoustic-related hardware is proportional to the detectable volume and thus a fair notion of the hardware cost per event can be found from these curves.

7. Summation and Conclusions

Neutrinos should produce sound. Sound generated by the nuclear particles will be detectable at sufficiently close distances with sufficient sophistication of equipment and can be distinguished from capricious noise or ordinary random noise if the processing equipment is sufficiently sophisticated. Although the design of possible systems to do the job has already been started and is only in its infancy at present, the problem has already created enough curiosity so that it will continue.

We predict that sooner or later the neutrinos from astrophysical sources at ultra-high energies, 10^{17} eV and higher, will be detected, and that in view of the enormous volume of target required and enormous thickness of shielding required the only practical detector and shielding material can be water. The masses of target required will be of the order of 10^{12} tons and therefore only an acoustic detector will be able to observe those neutrinos. We predict that we shall see the system, certainly not next year, but probably within the next 20 years. In the meantime, the investigations will be carried out at a low level of funding and complexity but a high level of intellectual sophistication. We anticipate that more and more acousticians will think about these factors and make valuable contributions to simplify and improve today's sketchy designs.

8. Acknowledgements

We thank Grant Blackinton for his constructive comments and the Hawaii Institute of Geophysics Staff for their assistance and support. This is HIG contribution no. 978.

9. References

1 P. Kotzer, ed., Proceedings of the 1975 Summer Study on Cosmic Ray Neutrino Interactions in the Ocean Depths and on Associated Oceanographic Physics and Marine Engineering (DUMAND), Western Washington State College, Bellingham, Washington, USA (SS 75).

2 A. Roberts, ed., Proceedings of the 1976 DUMAND Summer Workshop, University of Hawaii, Honolulu; Office of Publications, Fermi Nation Accelerator Laboratory, Batavia, Illinois USA (SS 76).
3 H. Bradner, ed., Proceedings of the La Jolla Workshop on Acoustic Detection of Neutrinos, July 1977, Scripps Institution of Oceanography, La Jolla, California, USA (SS 77).
4 A. Roberts, ed., Proceedings of the 1978 DUMAND Workshop (in three volumes), Vol. 1: Array Studies, Scripps Institution of Oceanography, La Jolla, California, USA (SS 78).
5 J. G. Learned, "Project Dumand and the Trade-Offs Between Acoustic and Optical Detection", University of California, Irvine, Physics Technical Report 78-37, June 1978.
6 G. A. Askarian, Atomnaya Energiya $\underline{3}$, 152 (1957).
7 B. L. Beron and R. Hofstadter, Phys. Rev. Letters $\underline{23}$, 184 (1969).
8 T. Bowen, "Sonic Particle Detection", in [2] (SS 76).
9 B. A. Dolgoshein, "Acoustic Detection from $E_0 > 10^{16}$ Particle Showers", Appendix A to a report compiled by L. R. Sulak in [2] (SS 76). (c.f. G. A. Askarian and B. A. Dolgoshein, Preprint No. 160, Lebedev Institute of the USSR Academy of Sciences, Moscow, 1976).
10 G. A. Askarian, B. A. Dolgoshein, A. N. Kalinovsky, and N. V. Markov, "Acoustic Detection of High Energy Particle Showers in Water", Nuclear Inst. and Methods, North Holland Publishing Co. (in press).
11 M. Levi, L. Sulak, et al.; T. Bowen, et al.; H. Bradner; A. Parvulescu; W. V. Jones; and J. Learned, "Experimental Studies of the Acoustic Signature of Proton Beams Traversing Fluid Media", IEEE Trans. Nuclear Sci. $\underline{25}$, 325 (1978).
12 J. G. Learned, "Acoustic Radiation by Charged Atomic Particles in Liquids, an Analysis", UCI Technical Report 77-44 (Phys. Rev., in press, 1979).
13 E. M. Arase and T. Arase, in *Acoustic and Vibration Progress*, 1974, R. W. B. Stephens and H. G. Leventhal, ed.
14 Hugh Bradner and R. S. Howard, "Attenuation of Surface-Generated Noise Received Deep in the Ocean", J. Acoust. Soc. Am. $\underline{64}$, 322 (1978).
15 V. J. Stenger, "DUMAND Acoustic Event Spectrum and Rate with Energy-Dependent Detection Volume", in [3] (SS 77).
16 Whitlow W. L. Au; R. W. Floyd, and J. E. Hann, "Propagation of Atlantic Bottlenose Dolphin Echolocation Signals", J. Acoust. Soc. Am. $\underline{64}$, 411 (1978).
17 A. Parvulescu, "Underwater Acoustical Astrophysics", J. Acoust. Soc. Am. $\underline{61}$, 580 (1977).
18 Grant Blackinton and J. G. Learned, "Considerations of the Rate of Events Seen by One (Possibly Clumped) Acoustical Detection Module" in [3] (SS 77).
19 Hugh Bradner, "Acoustic Detection and Counting Rates for Cascades", in [3] (SS 77).
20 James L. Stewart, "DUMAND Calculations", in [3] (SS 77).
21 Hugh Bradner and J. G. Learned, "Acoustic Detection in DUMAND", in [4], (SS 78).
22 A. Parvulescu, "The Signals and Noises of DUMAND, and a Proposed Space-Time Processing Scheme", J. Acoust. Soc. Am. $\underline{64}$, S106 (1978).

Part V

Inhomogeneities in Ocean Acoustics

Inhomogeneities in Underwater Acoustics

L. Bjørnø

The Acoustics Laboratory, Technical University of Denmark, Building 352
DK-2800 Lyngby, Denmark

Introduction

A realistic picture of underwater sound transmission between a
source and a receiver is strongly influenced by the acoustic
properties of the communication channel formed by the sea. Beside the strong influences on sound propagation arising from
the inhomogeneous channel boundaries, the sea surface and bottom,
also the volume inhomogeneities of the channel contribute considerably, and frequently decisively, to its time and space varying transmission properties.

These volume inhomogeneities, the influence of which may lead
to strong variations in signal amplitude and phase, occur in a
broad variety of space and time scales, of physical origin -
naturally occurring or man made - and they show individual features of deterministic or random character, which one by one or
together contribute to the overall acoustic properties of the
ocean.

The aim of this paper is to give a general introduction to
volume inhomogeneities, briefly defined as variations in space
and time of sound velocity and of density, discussing their origin, their physical characteristics and their consequences to
underwater sound propagation, with the hope through this presen-
tation to be able to bridge between the two main topics of this
underwater acoustics conference, the cavitation and bubble dynamics and ocean inhomogeneities in a broad sense.

1. Thermodynamic Properties of Seawater

For acoustic wave propagation in the ocean the most obvious parameter of concern is the *isentropic velocity of sound*, which
can be measured directly with high precision or which can be
calculated from measurements of other thermodynamic parameters.
For oceanographic research the *density* is the most interesting
property, since density differences contribute fundamentally to
drive the motions in the oceans. Other and interrelated properties are *temperature* and *salinity*, which coupled with the local
pressure determine the local density. Sensitivity of sound velocity on these parameters is roughly: $4.6 \ \mathrm{m \cdot s^{-1} \cdot c^{-1}}$,
$1.3 \ \mathrm{m \cdot s^{-1} \%^{-1}}$ and $0.18 \ \mathrm{m \cdot s^{-1} \cdot bar^{-1}}$.

Heat conduction and salt diffusion attempt to level out the variations in temperature and salinity, but only small and different scales are covered by these processes. During a 24 hour period heat conduction at $20^{\circ}C$. will cover a length of 11 cm (thermal diffusivity $\alpha \simeq 1.43 \cdot 10^{-3}$ cm$^2 \cdot$s^{-1}) and salt diffusion will show a length scale of 1 cm (diffusion coefficient $D \simeq 1.3 \cdot 10^{-5}$ cm^{2}s^{-1}).

2. The Local Ocean Climate

The largest scale variations in the ocean, the general circulation, showing seasonal fluctuations in the upper structure, are only limited by the size of the basin. These variations are determined by balances in heat input and efflux, wind stress forcing, and non-equilibrium with connecting basins. Sources of great water masses are fixed largely by air-sea interactions where heat and mass fluxes across the air-sea interface at certain times of the year result in the formation of large quantities of water having a rather repeatable temperature and salinity characteristics and thus determining the water density through its temperature and salinity. Gravity then acts on the density differences, and motions due to gravity or wind stress are influenced by basin shape, bottom topography, and rotation of the earth.

The long-term mean vertical structure shows a depth dependence with horizontal scales defined by the general circulation in the ocean basin and with seasonal variation in the upper few hundred meter while in the deeper parts of the ocean, when viewed in appropriate parameter space, the properties have not varied measurably through this century of accurately recorded measurements, for instance of the T-S (temperature-salinity) relation, which, for this part of the ocean, allows a one-parameter (temperature) determination of velocity of sound and density.

A characteristic example on the nonequilibrium with connecting basins is the Mediterranean outflow in the North Atlantic. Deep Mediterranean water is warmer and saltier than North Atlantic deep water, because evaporation exceeds precipitation and runoff in the nearly enclosed Mediterranean. The sea level is preserved by inflow of North Atlantic surface water, while a considerable amount of Mediterranean water flows outwards over the sill in the Strait of Gibraltar and is mixed with the North Atlantic water as it flows down the continental slope of the Iberian Peninsula and out into the North Atlantic at a depth determined by its own density. This mixing process leads to a varying density and sound velocity in the deep North Atlantic sea with distance from the Strait of Gibraltar.

3. Mesoscale Variations

Ocean fluctuations on spatial scales of 50 to 500 km generally are classified as being mesoscale variations, since this is between the large scale general circulation and the smaller scale represented by internal waves etc. The time scale represented by this variability usually is of the order of many days to se-

veral months. Characteristic features of this variability are
fronts and eddies, which in analogy with atmospheric conditions
have given the name, the ocean weather system, to the mesoscale
motions. Generally, these features can be tracked for weeks or
even months at a time, and they may therefore be considered to
be deterministic perturbations from the large scale structure.

An oceanographic front is a zone separating two regions having
different thermodynamic quantities, which may lead to density
gradients as for instance found in the Gulf Stream north wall,
where the density front gradients are balanced by motion of the
sea. Fronts are quite variable on the large scale also, as their
positions can fluctuate several hundred kilometers. Boundary
currents and their associated fronts appear to be instable in
the sense that perturbations from the mean location in the form
of travelling waves can amplify with time. The front can bend
around to form a meander, which can extend into either the cold
or the warm side of the front, where it can be pinched off,
forming an eddy. This eddy retains characteristics of the water
on the other side of the front in its interior and eddies appear
to retain their coherent form with a lifetime of months to several years before dissipating. In the open ocean, eddies move at
speeds of one to eight kilometers a day and they usually have
a ring-like structure with a diameter of 50 to 250 km. The dynamics of the eddies appropriately appear to be associated with
Rossby waves, though their features are more Soliton-like than
wave-like. In general, Rossby waves seem to form the basic structure of mesoscale fluctuations. Mesoscale eddies cannot always
be considered to have small amplitudes, and large ones do not
appear to disperse as a combination of linear Rossby waves would.
Nevertheless, mesoscale eddies may still satisfy the dispersion
relations for linear Rossby waves as indicated by fitting to
observations [1], [17]. A comprehensive review of Rossby waves
may be found in [2]. Mesoscale eddies may cause large temperature profile changes with changes in the position of the thermocline. A large amplitude eddy may raise isotherms several
hundred meters, and the raised parcels of water retain their
salinity characteristics as long as the eddy was not recently
shed from a front separating distinct acoustically different
water masses. Thus, while the T-S combination is fixed, the temperature and salinity change significantly at a given depth. Due
to density structure differences in the mesoscale features remarkable deflections of the surface may occur amounting to 20 cm
or more in large mesoscale features and to a meter across strong
boundary currents [3].

The influence of eddies on acoustic transmission is considerable. When an eddy crosses a transmission path, the intensity
of sound received may change as much as a factor of 10 [13].
Moreover, the existence of current and current shear causes
changes in the travel time and pulse shape for signals travelling
up current or down current. In general there is a considerable
theoretical and experimental support for the idea that acoustic
phase or travel time are directly related to dynamic events taking place in the ocean. However, detailed comparison of theory
and experiments are still lacking.

4. Internal Waves

There are several types of internal wave motions in the ocean. The internal tide, being a long-wave motion which appears to be significant at least in the vicinity of continental shelves, is generated by flows induced by the surface tide over the bottom topography. The length of these waves is of an order of magnitude of 150 km in deep water and they possess an amplitude of several meters in the open sea and often somewhat more on the continental shelf.

Random internal wave motions on a shorter scale are more prevailing. A horizontal scale of 100 m to 10 km and a vertical scale of 1 to 100 m with time scales from tens of minutes to many hours characterize internal waves. The associated horizontal currents are typically of an order of magnitude of 0.05 m·s^{-1}, and with an extremely complex vertical structure. These waves exist only because of increasing density with increasing depth of the ocean and because of rotation of the earth, as both of these effects tend to restore a water parcel to an equilibrium position once it has been moved from that position.

The degree of density stratification of a fluid with average potential density $\overline{\rho}(Z)$ as a function of depth Z is quantified by:

$$N^2 = \left(\frac{-g}{\rho_0} \frac{d\overline{\rho}}{dZ} - \frac{g^2}{a^2} \right) \simeq \frac{-g}{\rho_0} \frac{d\overline{\rho}}{dZ} \tag{1}$$

where $N(Z)$ is the Brunt-Väisälä (or buoyant) frequency, g is the acceleration due to gravity and ρ_0 is a reference density. $N(Z)$ is the frequency with which a vertically displaced fluid element would be expected to oscillate because of restoring buoyancy forces. If the displacement is not vertical, the restoring force is less and the frequency of oscillation is reduced. These physical ideas are intrinsic to the dynamic equations leading to the dispersion relation [5] for plane waves:

$$\omega^2 = (N^2 k^2 + f^2 m^2)/(k^2 + m^2) \tag{2}$$

which connect the angular frequency ω with the horizontal wavenumber k and the vertical wavenumber m, respectively. $f \doteq 2\Omega\sin$ (latitude) is the Coriolis frequency (or inertial frequency) associated with the earth's angular velocity: $\Omega = 2\pi/(24$ hours). The permissible range of frequencies is $f \leq \omega \leq N$. At frequencies slightly greater than f the particle motion is nearly horizontal and circular, while at higher frequencies the particle motion becomes increasingly inclined to the horizontal in an ellipse, tending to an up-and-down motion at $\omega = N$.

From (2) the group velocity for internal waves can be determined by:

$$(\partial\omega/\partial k, \partial\omega/\partial m) = km\omega^{-1}(N^2 - f^2)(k^2 + m^2)^{-2}(m, -k) , \tag{3}$$

which is orthogonal to the vector wavenumber with components k and m.

264

The Brunt-Väisälä frequency $N(Z)$ is a function of depth in the ocean, and it has a maximum at a thermocline. An internal wave of frequency ω propagating downwards into a region in which the local N is less than ω, will experience internal reflection which combined with reflection at the sea surface, or at some other level, where N falls below ω, means that a discrete set of modes for internal waves is possible, with the frequency for each given as a function of the horizontal wavenumber and the mode number. Although little is known about the distribution of internal wave energy among various modes, it might be anticipated that when the thermocline is fairly sharp, the lowest mode will dominate. Higher modes appear to be more difficult to excite, and moreover, they involve higher rates of shear and they show increased likelyhood of local instabilities and degradation into turbulence.

Internal waves can, in principle, be generated at the ocean surface, for instance by travelling pressure fields in the atmosphere or by variable wind stress, at the ocean bottom, for instance by surface tides or by quasi-steady currents advecting a stratified ocean over its bottom topography, or they may be generated in the interior of the ocean by processes not yet fully investigated.

Interactions between internal waves can lead to a nonlinear transfer of energy among themselves, to higher wavenumbers [6], but the energy transfer rates are very dependent upon the spectral shape, and hence, it is not possible to identify a definite, quantifiable cascade of energy to the more dissipative higher wavenumbers [7]. Also the formation of strongly nonlinear internal waves of a permanent form, the so-called *Solitons* are known theoretically and have been verified through measurements performed at sea.

The most likely way for internal wave dissipation is through the occasional *Kelvin-Helmholtz* shear instability due to the random superposition of different internal waves bringing the local *Richardson number* $(N^2/(\partial u/\partial Z)^2)$ - expressing the relative importance of buoyancy and shear effects - down below its critical value of 0.25.

Internal waves influence the structure of the ocean in several ways. They contribute to the mixing of cold, dense water formed in polar regions when this water sinks and spreads over the bottom of the ocean basin. They influence the intermittent uplift of phytoplankton into the sunlit layer of the upper ocean, and thus through increased photosynthesis lead to an enhanced productivity and improved conditions for life in the ocean [8]. The effect of internal waves on acoustic wave propation has recently been studied to some extent. The acoustic signal received at one point in the ocean interior from a source at another point can vary considerably in amplitude, travel time and direction from which it arrives. Estimates of sound velocity fluctuations due to internal waves have led most past work to be more concerned with the temperature fluctuations than with the fluid velocity fluctuations themselves. However, most of the energy in this fluid velocity fluctuation frequency band is near the inertial frequency where the fluid motions are essentially horizontal. The vertical shear due to these waves is of the order of 1 $cm \cdot s^{-1} m^{-1}$ in the thermocline, but it is coherent over several km in the horizontal, and so it is expected that the influence is most significant on nearly horizontal ray paths [9].

5. The Ocean Fine Structure

If we go down only slightly in spatial scale, the region of fine structure occupies vertical lengths of one to tens of meters and horizontal lengths of tens to several hundreds of meters. A characteristic feature of the presence of fine structure is the existence of a typical layer-structure - a stratification - in the vertical density or temperature profile. These layers are regions of relatively uniform density, separated by sheets of high density gradients. The thickness of the layers may vary strongly and especially large thicknesses - 25 to 40 meter - are for instance found in the base of the Mediterranean outflow.

Ocean fine structure and internal waves show an overlap in spatial scales and internal waves are the most likely source for the establishment of the fine structure. Large amplitude internal waves having a small vertical wavelength can distort the mean density profile into a step-like profile, which can propagate along with the wave pocket. Once the layer-structure is present it provides selective sites for enhanced shear during passing of waves, thus contributing to the generation of Kelvin-Helmholtz instabilities on the sheets between the layers, which may cause enhanced mixing [3], [11]. A characteristic mixing phenomenon in the layer structure is the formation of *salt fingers*.

If the scale of fluctuations goes down to some cm in the vertical and some meters in the horizontal scale a parameter range called the *microstructure* is reached [10], [12]. This range is difficult to observe, but the mechanisms for generation and maintenance of the microstructure apparently are similar to those for the fine structure.

A step-like structure may also be formed by isolated, horizontally flattened, turbulent patches in statically stable regions below the main thermocline. In each patch, the temperature and salinity are approximately uniform, but may differ from one to another [5]. Small scale turbulence is a significant source of signal fluctuations of higher frequencies, above 5 kHz, but also signal refraction and scattering from thermal microstructures contribute essentially to high-frequency signal fluctuations.

Single- and multipath fluctuations are caused by various scales of inhomogeneities. Calculations of signal strength and diffraction parameters for real ocean conditions have been presented in [14], where it is predicted that a single steep propagation path splits into many micro-multipaths at ranges greater than 300 km at all frequencies above 50 Hz.

6. Near-Surface Mixed Layers

Near the surface, there exists a layer of well mixed water at
most times of the year and locations in the ocean. When the wind
blows across the surface of the water, a tangential surface stress
is developed both directly from the interfacial stress, and in-
directly by the rate of momentum loss from the surface waves
for instance by wave breaking. Below the surface, a turbulent
mixed layer develops. If the underlying region is statically
stable or neutral ($N^2 \geq 0$), the interface between the turbulent
and the non-turbulent regions will become very sharp, and it
will remain so as the turbulence erodes the non-turbulent region by entrainment in an energy balance in which a flux of turbulent energy works against the gravity at the bottom of the
mixed layer. The temperature and the salinity in the mixed layer
are virtually uniform as a result of turbulent diffusion, and an
increasing contrast in acoustical qualities between the mixed
layer and the water immediately below may occur by continued
turbulent erosion, finally leading to the formation of a thermocline. Also the thermal energy balance across the ocean surface will strongly influence the mixed layer. The incoming radiation from the sun penetrates into an appreciable depth, but
the bulk of the energy is absorbed within the first few meters
of the sea. At the interface itself, heat is lost by infrared
reradiation and evaporation. Heat is either lost or gained by
molecular exchange with the air in contact with the surface, or
by turbulent transfer further away from the interface. These
processes are influenced not only by the physical nature of the
interface, but also by motions as waves and turbulence in this
vicinity, thus forming a very complicated and not yet fully understood interaction scheme.

Turbulence as a dissipation mechanism in ocean flows may also
be found near the ocean bottom, where it is influenced by the
local topography [20]. Again the response of the ocean to the
development of ocean currents as well as the near-surface large-
scale wind and thermal disturbances is dependent on the transfer
of matter, momentum and energy by irregular smaller scale motions of one or another kind.

The mixed layer, usually between ten and several hundred meters deep, has the ability to behave as an acoustic duct, the
mixed layer sound channel. The sound velocity increases weakly
with depth below the surface because of the pressure influence,
and then it decreases sharply in the thermocline.

Besides the temporal and large-scale geographic variability
in thickness and temperature, there is a considerable variability on smaller spatial scales due to forcing. Not only storms,
but fronts and eddies in the mesoscale can cause temperature
changes of several degrees Celcius and depth changes of an order of magnitude of 2. Moreover, internal waves propagating in
the thermocline may cause fluctuations of several meters in
mixed-layer depth. Another near-surface phenomenon of interest
to underwater acoustics is the so-called *afternoon effect*. The
solar heating of the few meters of the top of the sea leads to
a diurnal temperature variation showing the formation of a sharp

diurnal thermocline between the top of the sea and the remainder
of the mixed layer. This diurnal layer grows warmer and deeper
in the afternoon - several degrees Celcius warmer than the deeper
mixed layer and depths of up to 20 m may be reached depending
on the daily heat input - and during the night the heat is radiated from the surface to the relatively cold sky resulting in
a thermal convection which quickly destroys the diurnal layer.

7. Gas Bubbles

Beside the thermodynamically and kinematically influenced inhomogeneities,treated in the previous sections,volume inhomogeneities with other physical origin are present in the ocean. These
inhomogeneities can be either naturally occurring like gas
bubbles of various radii, marine animals, sand or dust particles,
etc., or they may be man-made like wakes, various kinds of pollution, etc. These inhomogeneities may influence the amplitude
and phase of underwater acoustic signals for instance by inter-
cepting and reradiating portions of the acoustic energy incident upon them, thus creating scattering and reverberation on a
par with thermodynamically and kinematically influenced inhomogeneities.

Free gas bubbles appear in various forms in the sea. They
occur immediately below the sea surface, where they are produced
by breaking of waves, by precipitation, by cosmic rays or by de-
caying of organic matter. They may also be vented gas of zooplankton or fish, the product of photosynthesis or produced in
wakes from ships. Free gas bubbles are quite small, since the
larger bubbles tend to rise fast to the surface. They form only
a very small percentage, by volume, of the sea in which they
occur, but nevertheless, because gases show a markedly different
density and compressibility than seawater, and because of the
resonance characteristic of the bubbles, suspended gas content
in seawater has a very strong effect upon underwater sound propagation.

When struck by an acoustic wave a bubble will scatter and
absorp acoustic energy in an amount governed by the scattering
and absorption cross sections of the bubble, which together
give the bubble extinction cross section, characterizing the
portion of the incident acoustic energy being intercepted
by the bubble. The cross sections depend upon the frequency of
the acoustic wave, the size of the bubble and the physical and
thermodynamic properties of the gas and its surrounding seawater, and at resonance the scattering and absorption cross sections of a typical bubble in the sea are of the order of magnitude of 1000 times its geometrical cross section.

Beside absorption and scattering suspended gas bubbles show
a strong dispersion effect. As reported in [15] a sound velocity
variation from around 700 $m \cdot s^{-1}$ to over 2000 $m \cdot s^{-1}$ takes place
in a cloud of bubbles of a uniform size around the resonance
frequency.

In general bubbles are not of uniform size and by covering
a broad spectrum of diameters,clouds of gas bubbles show reso-

nance frequencies over a very broad acoustic frequency spectrum and thus yielding a very pronounced influence on underwater sound propagation.

Also bubbles trapped in marine life for instance in gas-bubble-carrying plankton or the gas-filled swim bladder of fish contribute to volume inhomogeneities in the ocean leading to volume scattering, absorption and dispersion [16].

The existence of characteristic *deep scattering layers* formed by marine life have been found. The scattering in a deep scattering layer at frequencies in excess of 20 kHz is likely to be due to zooplankton or to smaller marine animals, while the dominant scatterers below 10 kHz are various fish types that possess a swim bladder. The amount of a trapped gas bubble that becomes resonant at a certain frequency is depending upon the type, the size and the depth of the fish. Due to the tissue around the bubble the resonance effect of trapped bubbles is far less than that of free gas bubbles.

The deep scattering layer, which frequently shows a multi-layered structure, generally exhibits a diurnal migration in depth, being at a greater depth during the day than during the night, and showing a rapid depth change near sunrise and sunset. This diurnal depth migration frequently exceeds 100 m and tends probably to be just sufficient to keep light intensity constant at the depth of the layer for maintaining the photosynthesis and life conditions.

8. Other Naturally Occurring or Man-Made Inhomogeneities

Marine animals without trapped gas bubbles may, due to their deviation from their environment in compressibility and density, also form inhomogeneities and thus lead to sound scattering and absorption [19]. This attenuation is much weaker than observed by trapped gas bubbles, and seems to be proportional to the frequency squared and may therefore be described as an increase in the shear and particular the bulk viscosity relative to clean water.

Suspensions of nonbiological materials like dust and sand particles also lead to a relative increase in the shear and bulk viscosities of the seawater and only negligible dispersion will be observed.

Ship wakes will due to their contents of bubbles, for instance later stages of cavitation bubbles, form inhomogeneities which, in particular due to turbulent diffusion, decay in time and will be spread in horizontal direction, particularly, when they are formed in a stratified region [18]. Not the wakes arising from the increasing traffic across the oceans should be a source of strong worry, but the inhomogeneities created by the strongly increasing man-made pollution will probably in the future attract more attention from oceanographers and acousticians.

Big solid items, wrecks, used cars, waste from industry cover the seabed, in particular the continental shelves, around the world and especially the seas around the industrialized nations.

Chemical pollution caused by industrial waste water containing heavy metals, salts, organic materials, etc. kills the life in all scales in the sea - that life which had survived ruthless fishing - , it changes the acoustic absorption and dispersion conditions - creation of new relaxation frequencies - and it creates new inhomogeneities in the ocean.

Moreover, oil contaminations in a large scale seem to be more frequently occurring and the gradual contamination in the atmosphere, the break down of the ozone layer will lead to climatic changes, increased melting away of the ice from the Poles and thus to a change in the oceanic climate and in meso· and smaller scale inhomogeneities in the ocean.

These dark perspectives really ought to give food for thoughts.

Conclusion

The experience of recent years has tought us that sound is greatly influenced by the conditions of the sea through which it propagates. Many relations between oceanic physical properties and acoustic transmission observation have been found, but like a Pandora's box, attempts to solve one problem disclose the existence of other problems. The field of inhomogeneities in underwater acoustics must be expected to be just as stimulating and inspiring a future field of research as it has been during the past, a field where research victories and defeats will be met also in the future in attempts to disclose the acoustic secrets of the sea.

References

1. McWilliams, J.C. and Flierl, G.R.: Optimal quasi-geostrophic wave analysis of MODE array data. Deep-Sea Res. *23*, 285-300, (1976).
2. Dickinson, R.E.: Rossby waves - Long-period oscillations of oceans and atmospheres. Ann. Rev. Fluid Mech. *10*, 159-195, (1978).
3. Dugan, J.P.: Oceanography in underwater acoustics. In: *Ocean Acoustics*, J.A. DeSanto (Ed.), Springer-Verlag, 1979.
4. Garrett, C. and Munk, W.: Internal waves in the ocean. Ann. Rev. Fluid Mech. *11*, 339-369 (1979).
5. Phillips, O.M.: *The dynamics of the upper ocean*. Cambridge Univ. Press. 1969.
6. Olbers, D.J.: Non-linear energy transfer and the energy balance of the internal wave field in the deep ocean. J. Fluid Mech., *74*, 375-379 (1976).
7. McComas, C.H. and Bretherton, F.P.: Resonant interactions of oceanic internal waves. J. Geophys. Res. *82*, 1397-1412 (1972).
8. Kamykowski, D.: Possible interactions between phytoplankton and semidiurnal internal tides. J. Mar. Res. *32*, 67-89 (1974).

9. Sanford, T.B.: Observations of strong current shears in the deep ocean and some implications on sound rays. J. Acoust. Soc. Amer., *57*, 1118-1121 (1974).

10. Mellberg, L.E, and Johannesen, O.M.: Layered oceanic micro-structure - its effect on sound propagation. J. Acoust. Soc. Amer., *53*, 571-580 (1973).

11. Ewart, T.E.: Acoustic fluctuations in the open ocean - A measurement using a fixed refracted path. J. Acoust. Soc. Amer., *60*, 46-59 (1976).

12. Berman, A. and Guthrie, A.N.: On the medium from the point of view of underwater acoustics. J. Acoust. Soc. Amer., *51*, 994-1009 (1972).

13. Porter, R.P.: Acoustic probing of space-time scales in the ocean. In: *Ocean Acoustics*, J.A. DeSanto (Ed.), Springer-Verlag, 1979.

14. Flatte, S.M., Dashen, R., Munk, W.H. and Zachariasen, F.: *Sound transmission through a fluctuating ocean*. Stanford Res. Inst. JSR-76-39 (May 1977).

15. Fox, F.E., Curley, S.R. and Larson, G.S.: Phase velocity and absorption measurements in water containing bubbles. J. Acoust. Soc. Amer., *27*, 534-542 (1955).

16. Weston, D.E.: Sound propagation in the presence of bladder fish. In: *Underwater Acoustics*. V.M. Albers (Ed.), Plenum Press, N.Y. 1967.

17. Maxworthy, T. and Browand, F.K.: Experiments in rotating and statified flows. Oceanographic applications. Ann. Rev. Fluid Mech. *7*, 273-305 (1975).

18. Lin, J.-T. and Pao, Y.-H.: Wakes in stratified fluids. Ann. Rev. Fluid Mech. *11*, 317-338 (1979).

19. Haslett, W.G.: The fine structure of sonar echoes from underwater targets, such as fish. Proceedings Ultrasonics International 1979, IPC Science and Technology Press Ltd., Guilford, 1979.

20. Inman, D.L., Nordstrom, C.E. and Flick, R.E.: Currents in submarine canyons: An Air-Sea-Land interaction. Ann. Rev. Fluid Mech. *8*, 275-310 (1976).

Sound Propagation in an Inhomogeneous Ocean

R.H. Mellen

Naval Underwater Systems Center
New London, CT 06320, USA

1. Introduction

The historical development of underwater sound technology reflects a growing awareness of the inhomogeneous nature of the medium. The earliest known example is a 1919 paper by LICHTE [1] the title of which can be translated as "On the Influence of Horizontal Temperature Stratification in Sea Water on the Range of Underwater Acoustic Signals.[1]" In this paper, the author presents experimental data obtained by German ships prior to World War I and shows downward refraction by negative thermal gradients to be responsible for the short ranges attained under summer conditions. Similar diurnal behavior observed by U.S. scientists almost two decades later became known as the Afternoon Effect (Fig.1). LICHTE also shows that, in deep water, the pressure effect overcomes an adverse temperature gradient.

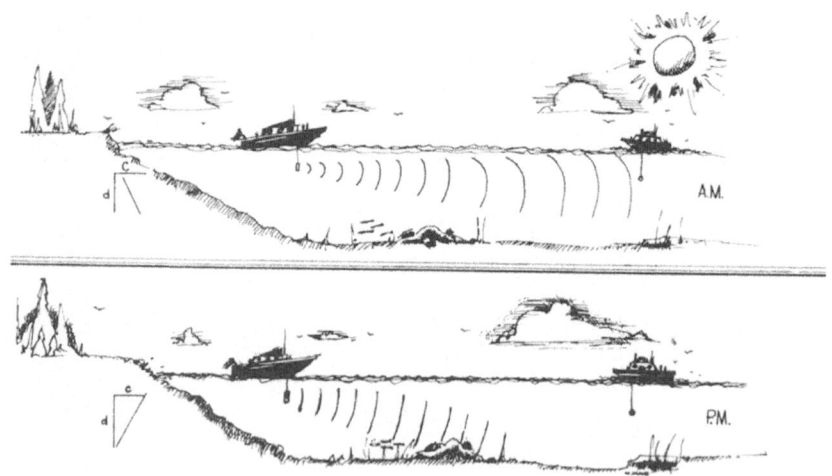

Fig.1 "Afternoon Effect" showing change from upward to downward refraction caused by solar heating

---

1 Über den Einfluss horizontaler Temperaturschichtung des Seewassers auf die Reichweite von Unterwasserschallsignalen.

The deep sound channel (DSC) was first investigated by EWING and WORZEL [2] who showed that the combined effects of downward refraction in the thermocline and upward refraction in barocline leads to refractive trapping of sound waves. Very long ranges are possible under these conditions because of cylindrical rather than spherical spreading and because boundary losses are avoided. A system called SOFAR (Sound Fixing and Ranging), making use of the DSC propagation mode, was employed in the Pacific Ocean during World War II for the purpose of locating downed aircraft. The subsequent growth of interest in long range sound propagation resulted in a rather intensive investigation of the DSC which was directed toward understanding the attenuation mechanisms.

One of the first results was the discovery of the low frequency attenuation anomaly by THORP [3] indicating another relaxation near 1 kHz (Fig.2).

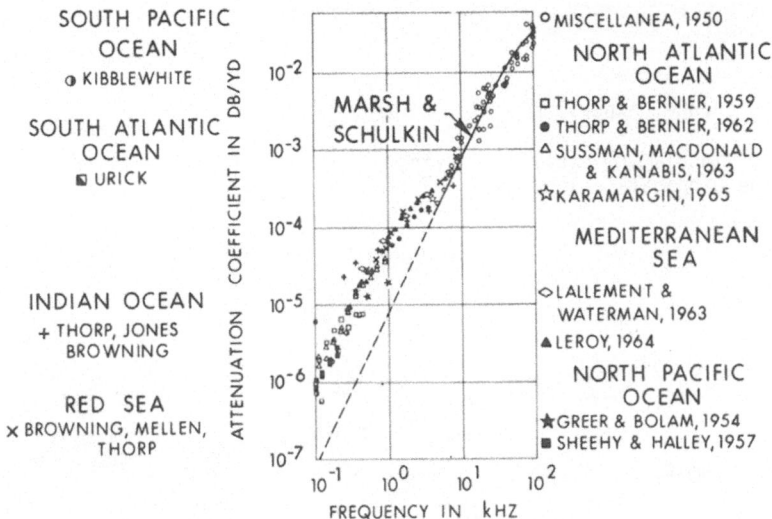

Fig.2 Low frequency attenuation anomaly. Dashed line is $MgSO_4$ absorption

A similar high frequency anomaly observed in ocean measurements had earlier been shown by WILSON and LEONARD [4] to be caused by magnesium sulfate and the relaxation mechanism was intensively investigated here at Göttingen by EIGEN and TAMM [5]. The low frequency anomaly was shown by YEAGER et al. [6] to arise from a relaxation associated with the boric acid-borate equilibrium. Because this reaction depends on pH chemical inhomogeneities are also important. For example, in the North Pacific where the pH is approximately 7.7 on the DSC axis compared to 8.0 in the North Atlantic, we have found the low frequency absorption to be only half as large [7].

At very low frequencies, DSC measurements at extremely long range, show evidence of additional excess attenuation that is clearly not chemical in origin. The most likely mechanism in this case appears to be scattering by random temperature inhomogeneities within the sound channel with subsequent absorption of the scattered energy by the ocean bottom [8].

2. Sound Channel Loss Measurements

A typical experiment involves a stationary listening ship and a moving source ship (or aircraft) from which charges are dropped to explode near the channel axis (Fig.3). Signals are received via the multipath arrivals shown by the rays.

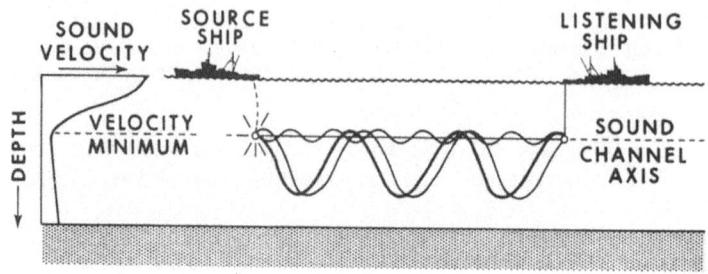

Fig.3 Sound channel experiment

The received signals are usually analyzed by means of bandpass filters to determine the propagation loss as a function of frequency and range (Fig.4).

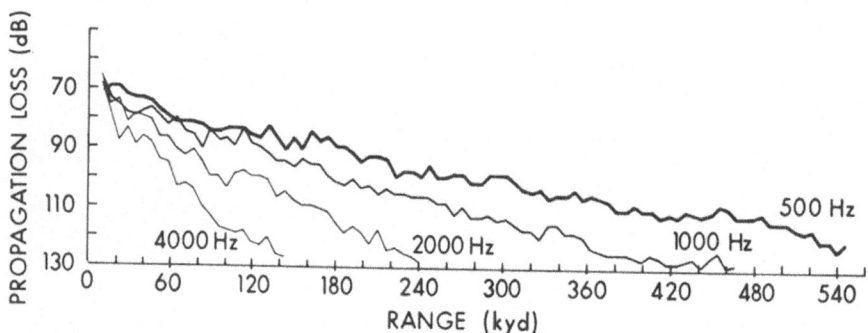

Fig.4 Propagation loss vs. range

In most cases a simple cylindrical model serves to describe the geometrical spreading loss over the range of interest. This loss is subtracted from the propagation loss and the attenuation coefficient determined by linear range regression. The attenuation coefficient therefore represents the sum of all losses that are proportional to range i.e. absorption, diffraction and scattering (Fig.5). In sea water, absorption is dominated by chemical relaxations and is only now beginning to be well understood. Diffraction loss is the result of low frequency channel cut-off when wavelengths become too long for efficient trapping and bottom losses begin to affect propagation. In the deep sound channel, diffraction loss is difficult to measure because the cut-off frequency is only a few Hertz and sources capable of generating such low acoustic frequencies are rare. However in surface ducts the cutoff frequency can be quite high. A surface duct is a mixed layer of 10-100 m thickness in which trapping involves upward refraction and surface reflection. If the surface is not very rough, the only attenuation mechanisms are absorption

274

SOUND ATTENUATION IN THE OCEAN

Fig.5 Deep-sound channel attenuation mechanisms

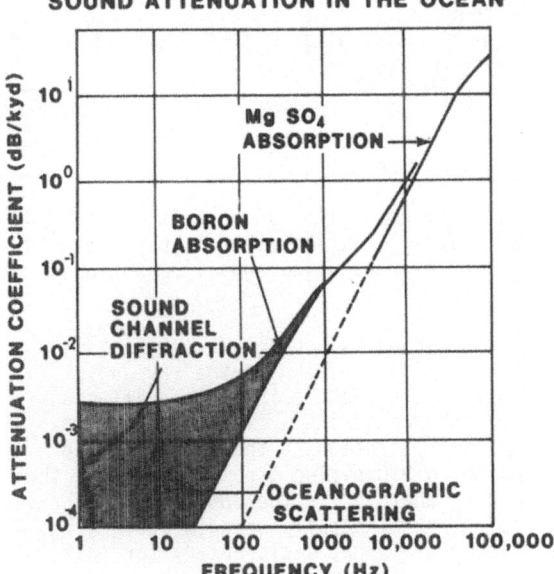

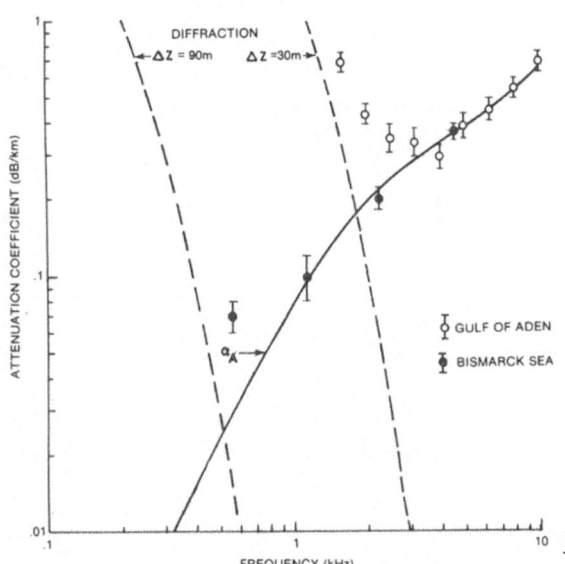

Fig.6 Surface duct attenuation

and diffraction. We find good agreement between experiment and theory which indicates that both the deterministic diffraction model and the absorption model are satisfactory (Fig.6) [9].

Long range measurements in the DSC, show evidence of scattering loss only below a few hundred Hz. However, in shallow water where sound channels are weaker the losses are greater and therefore measureable at higher frequencies. By subtracting out known values of absorption we find the extra loss to be generally independent of frequency. It is therefore apparent that the scatter mechanism involves inhomogeneities that are large compared to the acoustic wavelength and therefore ray theory should apply.

3. Ray Diffusion in Sound Channels

The theory of ray diffusion is discussed by CHERNOV [10].

One approach to the sound channel problem is the random-walk ray method such as MOCASSIN program of SCHNEIDER [11,12] . This method is particularly useful in calculating effects of smearing near caustics and illumination in shadow zones.

A perhaps simpler approach to sound channel attenuation is by solution of the ray diffusion equation $\partial\omega/\partial r = D\,\partial^2\omega/\partial\theta^2$ where ω is the ray angle probability density, r is range and θ is the ray angle [8]. The diffusion constant is $D \approx \mu^2/a_0$ where μ^2 is the variance of index of refraction and a_0 is a characteristic correlation length. Application to the uniform sound channel with lossy boundaries reduces to an eigenvalue problem and gives the attenuation coefficient α = const. x D/θ_0^2 where θ_0 is the bottom cut-off ray angle. The isotropic (turbulence) model contains only a single scale size. However, as MUNK has pointed out, this model is not satisfactory for scattering by highly stratified internal waves which appears to be the dominant mechanism in the DSC.

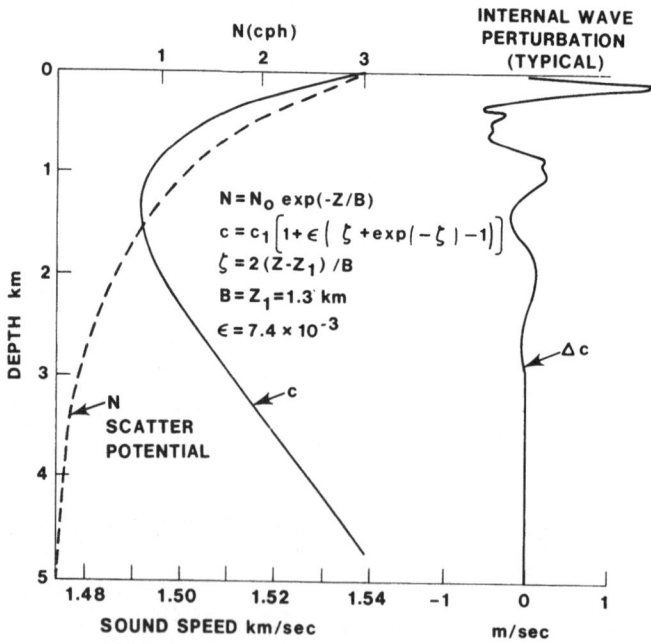

SOUND CHANNEL PARAMETERS (MUNK)

Fig.7 Garrett-Munk internal wave model

4. Internal Waves

MUNK and ZACHARIASON [13] have applied a random internal wave model of GARRETT and MUNK [14] to the problem of acoustic fluctuations. The internal wave field arises from a conditional stability of the thermocline which is characterized by a depth-dependent buoyancy frequency N (Fig.7). When some dynamic process causes disturbance, the isotherms are subject to displacements that are small and rapid near the surface and large and slow at great depths. The temperature gradient decays rapidly with depth; hence, the sound speed perturbations encountered by acoustic waves also decay with depth. The scatterers also are highly stratified and have much larger horizontal than vertical scales. Extension of the ray diffusion model gives the modified diffusion constant $D \approx \mu^2 b_0/a_0^2$ where b_0 and a_0 are the horizontal and vertical correlation lengths respectively (Fig.8) [15]. All terms are depth-dependent;

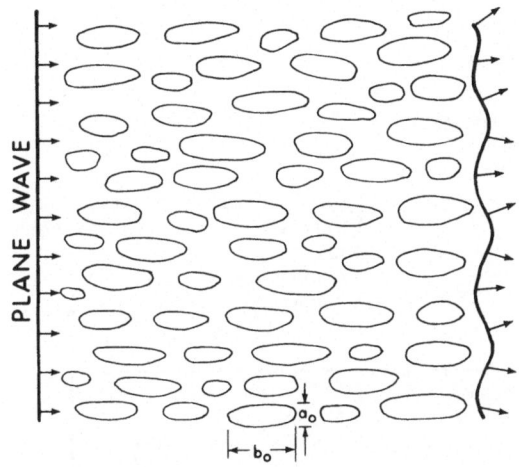

Fig.8 Scattering model showing isocorrelation contours. Values on the axis are $a_0 \approx 0.2$ km, $b_0 \approx 4$ km and $\mu^2 \approx 10^{-8}$

hence, to find the effective diffusion "constant" one integrates over a ray cycle and calculates an average value. The internal wave model has no adjustable constants other than a latitude correction. If infinite bottom loss is assumed the final result is a single value of attenuation for a given latitude (Fig.9). Comparison with experimental values show that the calculated value is quite small but still consistent with the lowest experimental values that have been measured. We see other data, however, that are an order of magnitude larger. KIBBLEWHITE et al. [16] find a regional dependence of loss that tends to correlate with water masses. A similar dependence was found in the KIWI experiment by BROWNING et al. [17] showing high attenuation in the central region of the South Pacific Ocean which correlates with Antarctic Polar Water (Fig.10). Such variability is not part of the internal wave model.

Other problems are also evident. When finite bottom loss is taken into account, the discrepancies in Fig.9 become even greater [18]. The model also predicts infinite values of diffusion constant at the equator where the horizontal scale size increases without limit. Because of ray curvature the calculated effective diffusion constants are finite but still very large. No such latitude dependence has been observed in the acoustic experiments. Finally, the internal wave model applies only to deep ocean conditions; hence, we have no means of scaling for shallow water where the scatter losses are greater and therefore easier to measure.

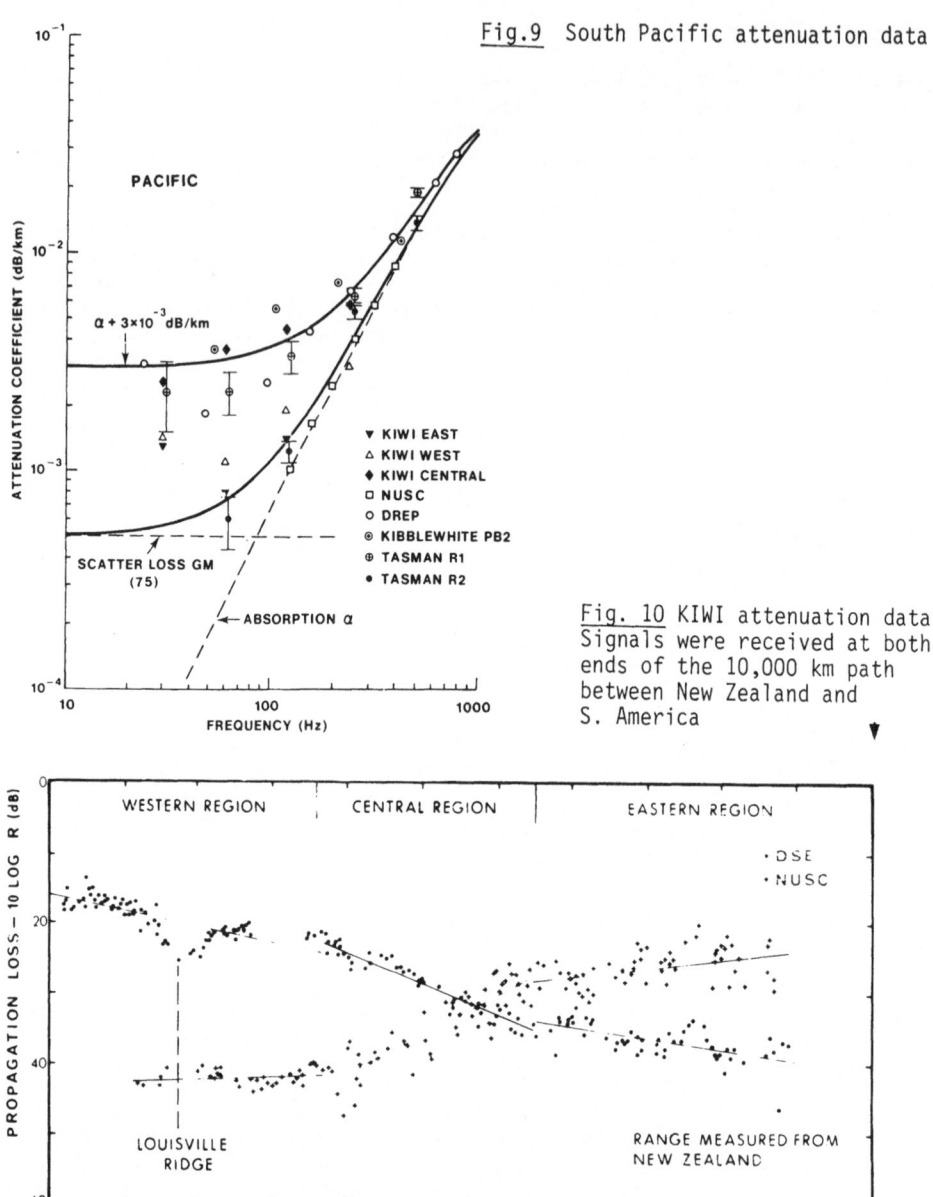

Fig.9 South Pacific attenuation data

Fig. 10 KIWI attenuation data. Signals were received at both ends of the 10,000 km path between New Zealand and S. America

The effects of particle velocity associated with the internal wave motion have so far been neglected. WORCESTER [19] made simultaneous measurements along reciprocal paths and found poor correlations between received signals. To account for such non-reciprocal behavior an effective Mach number U/C_0 must be added to the RMS refractive index μ. Preliminary results, however, appear to show that internal wave motion still does not appear to account

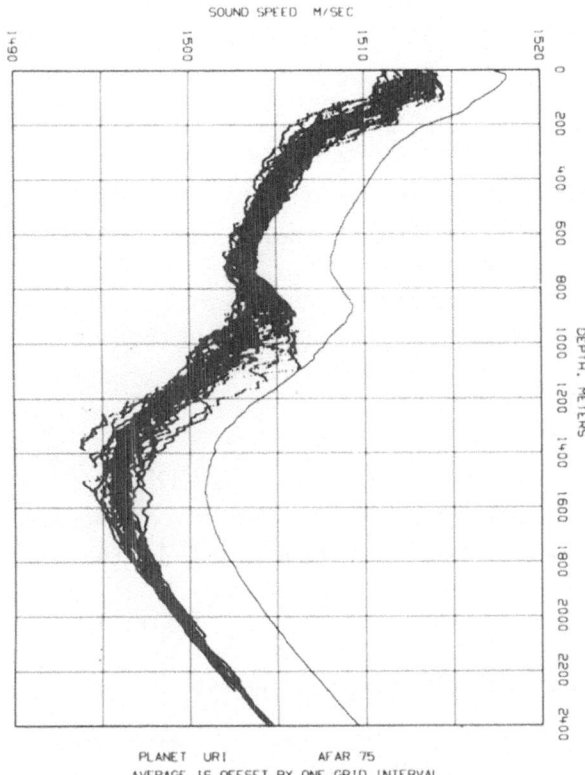

SOUND SPEED M/SEC

PLANET URI   AFAR 75
AVERAGE IS OFFSET BY ONE GRID INTERVAL

Fig.11 Azores Fixed Acoustic Range sound speed profile data taken at one location over a period of time

for the discrepancies, either in the case of signal amplitude fluctuation spectra or sound channel attenuation. Other oceanologic phenomena may therefore be involved. For example, sound speed profile measurements at the AZORES Fixed Acoustic Range show much greater variance near the channel axis than predicted by theory (Fig.11). Such an increase, caused by an intrusion of Mediterranean water in this case, might account for higher losses observed in the DSC.

5. Conclusion

We presently face the problem that our attempts to connect the oceanologic and acoustic phenomena are not yet completely satisfactory. If the difficulty is purely acoustical, then better accounting of the acoustic field perturbation may account for the discrepancies. On the other hand it may turn out that phenomena other than internal waves may be involved. In any case acoustic oceanography is coming into its own and should indeed prove useful in future studies and monitoring of ocean processes [20].

6. References

[1]Lichte, H., Phys. Zeit. 17, 385 (1919)

[2]Ewing, M. and J. L. Worzel, Geol. Soc. Am. Mem. 27 (1948)

[3]Thorp, W. H., J. Acoust. Soc. Am. 38, 648 (1965)

[4] Wilson, O. B. and R. W. Leonard, J. Acoust. Soc. Am 26, 223 (1964)

[5] Eigen, M. and K. Tamm, Z. Elektrochem., Ber. Bunsenges. physik. Chem. 66, 93 (1962)

[6] Yeager, E., F. H. Fisher, J. Miceli and R. Bressel, J. Acoust. Soc. Am. 53, 1705 (1973)

[7] Mellen, R. H. and D. G. Browning, J. Acoust. Soc. Am. 61, 704 (1977)

[8] Mellen, R. H., D. G. Browning, and J. M. Ross, J. Acoust. Soc. Am. 56, 80 (1974)

[9] Mellen, R. H. and D. G. Browning, J. Acoust. Soc. Am. 65, 1624 (1978)

[10] Chernov, L. A., Wave Propagation in a Random Medium, Ch. 2 (McGraw-Hill, New York. 1960)

[11] Schneider, H. G., Acustica 35, 18 (1976)

[12] Schneider, H. G., J. Acoust. Soc. Am. 62, 871 (1977)

[13] Munk, W. H. and F. Zachariasen, J. Acoust. Soc. Am. 59, 818 (1976)

[14] Garrett, C. and W. H. Munk, J. Geophy. Res. 80, 291 (1975)

[15] Mellen, R. H., D. G. Browning and L. Goodman, J. Acoust. Soc. Am. 60, 1053 (1976)

[16] Kibblewhite, N. C. Bedford and S. K. Mitchell, J. Acoust. Soc. Am. 61, 1169 (1977)

[17] Browning, D. G., R. W. Bannister and R. N. Denham, J. Acoust. Soc. Am. (submitted)

[18] Mellen, R. H. and H. G. Schneider, J. Acoust. Soc. Am. 62, 1038 (1977)

[19] Worcester, P. F., J. Acoust. Soc. Am. 62. 895 (1977)

[20] Munk, W. H. and G. O. Williams, Nature, 267, 774 (1977)

Acoustic Fluctuations in the Ocean

Y.J.F. Desaubies

Department of Ocean Engineering, Woods Hole Oceanographic Institution
Woods Hole, MA 02543, USA

1. Introduction

The aim of this survey is to review some of the recent - within the last
decade, say - experimental and theoretical investigations of acoustic
fluctuations in the ocean. There is no attempt to be exhaustive but rather
to emphasize those contributions in which theory and experiment corroborate
each other. Moreover the discussion will be limited to fluctuations in
acoustic signals propagating through the ocean interior where the scattering is essentially in the forward direction. Interactions with the surface
and bottom are not discussed.

The fluctuations to be considered stem from the changes in index of
refraction induced in the ocean by a variety of processes, ranging in time
scales from minutes to weeks, in space from centimeters to hundreds of
kilometers. However, as will be discussed below in more detail, the physics
and statistics of the smallest scale processes is poorly known and they are
expected to scatter only sound of relatively high frequency (several kHz);
because of chemical absorption such signals do not propagate very far.
Some work is presently in progress on the scattering by ocean finestructure
but is not yet to the point of being reviewed.

At the other end of the spectrum large scale disturbances are often
observed in the ocean, known as mesoscale eddies. They are the oceanic
analogue of the atmospheric storms. These features are sufficiently large
and vary slowly enough that they can in principle be tracked, mapped and
described deterministically. Sound transmission through eddies poses no
conceptual problems.

Internal waves, on the other hand, are ubiquitous, have periods from
15 m to 15 hrs, wavelengths from kilometers to hectometers; they can be
described satisfactorily only as a stochastic process. They constantly
perturb the sound velocity field by displacing and distorting the isotachs.
Thus they have a significant effect on sound propagation and have been the
object of several experimental and theoretical investigations. These will
be reviewed in some detail here.

As these notes were being written, an excellent book, edited by FLATTE
[1], was published in which the scattering of sound by internal waves is
treated comprehensively; there is also a remarkable chapter summarizing
the various ocean processes relevant to underwater acoustics. Several of
the topics touched upon here will be found treated in great detail in that
monograph.

2. The Ocean Medium

The characteristics of acoustic propagation through the ocean are
determined by the spatial and temporal distributions of the sound speed
and, to a lesser extent, by the ocean velocity field. The sound speed is
a function of temperature, salinity, and pressure which are the quantities
commonly studied in physical oceanography. One of the outstanding features
of these fields is their variability. Significant fluctuations are always
observed, as a function of depth, range, and time. These fluctuations
occur over an extremely wide range of temporal and spatial scales: from
minutes to years, from centimeters to megameters (an example is given by
WUNSCH [2]). This circumstance poses some difficulty in the meaningful
definition of a mean state; nevertheless one may assume that in some sense
such mean profiles exist, even though they may vary slowly.

The sound velocity field, for instance, has a strong depth dependence
often with a minimum at some intermediate depth, called the sound channel,
which allows long range propagation. The mean sound profile also changes
according to the large scale ocean features such as the major current
systems (e.g. Gulf Stream, Kuroshio), the presence of various water masses
(e.g. the Mediterranean outflow, the 18° water in the Sargasso Sea [3,4],
and the boundaries between them, i.e. the large scale oceanic fronts).
Thus the ocean is anisotropic and inhomogeneous.

The processes just referred to vary sufficiently slowly in space and
time that they can in principle be mapped and considered to be known
deterministically. For a given transmission experiment the acoustic field
can be predicted on the basis of ray theory, as is discussed in standard
textbooks [5,6]. The fluctuating part of the sound velocity field can
be discussed in terms of its depth variability (measurements in the vertical
are among the most common in oceanography). The microstructure refers to
the smallest scales resolved (of the order of centimeters to decimeters),
it involves turbulent motions and dissipation by molecular processes; the
finestructure is an ill defined transition regime with scales ranging from
1 m to perhaps 20-50 m. The structures observed in that wavelength band
are assumed to be due to irreversible mixing and/or to the reversible
straining of the mean field by small scale internal waves [7]. The
internal wave field, with vertical wavelengths between a few meters to the
depth of the ocean, continuously distorts the mean field so that any given
measured profiles shows fluctuations on corresponding scales. The topics
of micro and finestructure and internal waves have been recently reviewed
in a number of papers where an extensive bibliography can be found [1,8,9,10]

These various small scale regimes of ocean variability extend also in the horizontal and temporal dimensions. The appropriate space-time description of the fine and microstructure is poorly known, except for the fact that because of the strong stratification the horizontal extent is an order of magnitude larger than the vertical, and that these structures appear to be highly intermittent (in time) and patchy (in space). Because of this either a deterministic or a statistical representation is difficult to obtain. The situation with the internal wave field is quite different - and more encouraging.

Internal waves obey a system of hydrodynamical equations from which a dispersion relation and a set of relations between the pressure, density and the velocity components can be established. These equations cannot be solved however (except in highly idealized models) because the forcing function is poorly known. During the last decade a large number of observations of the internal wave induced temperature and velocity fluctuations have been made from fixed (moored), and vertically or horizontally moving sensors. These measurements have been consolidated by GARRETT and MUNK [11, 12] into an empirical model that describes remarkably well the mean statistical properties of the internal wave field. The product of the model is a set of expressions for the power spectral densities of the three velocity components in a three-dimensional wavenumber space $\underline{k}=(k_1, K_2, k_3)$. The equations of motion allow one to relate these spectra to the spectra of the fields of displacement, temperature, etc., the linear internal wave dispersion relation $\omega = \omega(\underline{k})$ allows the transformation of the wavenumber spectra in $\underline{k}$ space into frequency-wavenumber spectra (ω, k_1, k_2), (ω, k_2, k_3), etc... The model gives expressions also for the cross-spectra of different variables observed at different points and/or different instants.

The interest of the model for acoustic calculations is that it provides the required statistical representation of the stochastic internal wave field. In a loose sense it is the equivalent of the KOLMOGOROV spectral representation of three-dimensional turbulence for electromagnetic wave propagation studies in the atmosphere. The major differences are that the internal-wave-induced fluctuations are strongly anisotropic, depth-dependent, and with time dependence constrained by the dispersion relation. From the spectra given by the model of GARRETT and MUNK [12] (often referred to as the GM75 model) the correlation and structure functions of the fluctuating medium can be calculated.

In the theory of propagation through a random medium, the correlation function of the fluctuations (or the structure function, or the spectrum) is the only descriptor of the medium needed. Thus there are two reasons why internal waves have recently received considerable attention in the literature: firstly because they are the ocean process whose scales are most likely to match the acoustic wavelength and hence to produce scattering; secondly because they are the only process (besides surface waves) for which an appropriate statistical representation is available, i.e. the GM model.

To be sure other ocean structures influence sound transmission. The
fine and microstructure already discussed are likely to affect the highest
frequencies (above 5 kHz say); this constitutes a difficult problem since
no satisfactory statistical description of these fields exists. The
larger scale processes such as tides, mesoscale eddies (quasigeostrophic
perturbations with length and time scales of the order of 200 km and 3
weeks), major current systems, etc... do not scatter the sound but modulate
the rays and give rise to slow travel time variations. This is conceptually
a deterministic problem.

The remainder of this discussion will focus on the effects of internal
waves.

3. Theoretical Developments

The theory of wave propagation through random media finds applications in
very diverse fields such as electromagnetic wave propagation in the
atmosphere and through plasmas, sound transmission in the ocean, and in a
number of ocean wave problems [13] . The earlier literature on the subject,
mostly in the context of electromagnetic waves in an isotropic turbulent
atmosphere, has been reviewed in two monographs by CHERNOV [14] and
TATARSKI [15] .

Since the medium through which the waves propagate is random the
governing equations are stochastic. They cannot be solved deterministically: solutions can be obtained only for the expected value of the field,
the various correlation functions, and the higher order moments. The
medium itself is represented by the correlation function of its index of
refraction. If the fluctuating part of the index is denoted by $n(x,t)$
the correlation is

$$\langle n(x,t)\, n(x+\xi, t+\tau)\rangle \equiv \varepsilon^2(z) \, \langle \mu(x,t)\mu(x+\xi,t+\tau)\rangle$$
$$\equiv \varepsilon^2(z) \, \rho(\xi,\tau;z)$$

where $\langle \mu \rangle = 0$, $\langle \mu^2 \rangle = 1$ and the argument z, which represents the
vertical coordinate, in ρ emphasizes that all properties in the ocean are
strongly depth dependent; $\varepsilon^2(z) = \langle n^2 \rangle$ is the mean square index
fluctuation, typically of order 10^{-8} of 10^{-9} in the ocean. Equivalently,
the structure function $D(\xi,\tau;z) = 2\varepsilon^2 (1-\rho)$ or the spectrum

$$\phi_\mu(k,\omega;z) = \int d\xi\, d\tau\, \rho(\xi,\tau;z)\, exp\left[i(k\cdot\xi - \omega\tau)\right]$$

are used in some formalisms.

Correlation lengths can be introduced as qualitative measures of the
important scales of the medium. For instance a vertical correlation length

284

could be defined in a number of ways, either from an expansion of ρ for small vertical lag ξ_3

$$\rho(\xi_1, \xi_2, \xi_3) \simeq \rho(\xi_1, \xi_2, 0)\left[1 - (\xi_3/\ell_v)^p + \cdots\right]$$

or

$$\rho(0, 0, \xi_3) \simeq 1 - (\xi_3/\ell_v)^q + \cdots$$

or from an integral scale: $\ell_v = \int_0^\infty \rho(\xi)\,d\xi_3 \,/\, \rho(\xi_1, \xi_2, 0)$

It is clear that the scales so defined are not necessarily equal and in fact could differ by an order of magnitude. Indeed a medium with a band limited spectrum - between k_ℓ and k_u, say - cannot be characterized by a single scale. With this <u>caveat</u> in mind, "typical" correlation lengths for internal waves are found to be of order 300 m in the vertical and 3 km in the horizontal, both being depth dependent.

The acoustic pressure field is represented as a modulated plane wave, where $p(x,t)$ is a complex random function. In the presence of the sound channel and for other wave forms (beam, propagation from a point source) some geometric factors must be introduced, but a fluctuating complex wave function can still be defined. A number of related real functions are often introduced by writing $P = p(x,t) \exp[i(q \cdot x - \sigma t)]$

$$p = A\,e^{iS} = p_r + i\,p_i \qquad \ln p = \ln A + iS = \chi + i\beta$$

The theory can only predict statistical moments of the field p, such as the mean field $\langle p \rangle$, the mutual coherence function

$\Gamma_{12} = \langle p(x_1)\, p^*(x_2) \rangle$, the fourth moment $m_4 = \langle p(x_1)\, p^*(x_2)\, p(x_3)\, p^*(x_4) \rangle$
etc... These moments can be obtained by two distinct procedures: either by formally solving the wave equation for p and then forming the required moments, or by deriving from the start a differential equation satisfied moment, and then solving it. The former method is used in the usual perturbation schemes such as geometric acoustics (first BORN approximations), the RYTOV approximation [14] and in the more general and recent method of path integrals [16] ; the latter approach leads to the method of moment equations [17] which is related to the radiation transport Eq. [18] .

3.1 The Weak Scattering Regime

For high enough frequencies and ranges that are not too long - (these statements will be refined soon) - simple approximate solutions to the wave equation can be obtained in the form of series expansion in the small parameter. The simplest approximation is known as geometric (or ray) acoustics where it is assumed that as the wave propagates through the random medium it is simply refracted, with no significant diffraction. The two parameters that are most easily calculated are the amplitude A and

phase S: phase fluctuations are the result of the integrated effect along
the path of the small fluctuations in index of refraction; amplitude
fluctuations are of higher order and are related to the random changes in
the ray tube cross sections.

The phase fluctuation is given simply by

$$S = q \int_{R} n(\underline{x}) \, ds$$

where the integral is along the ray. This is a linear relation between S
and n so that if n is normally distributed, so is S. Hence $<S> = 0$
and

$$<S^2> = q^2 \iint_{R} <n(\underline{x}) \, n(\underline{x}')> ds \, ds' \simeq q^2 \epsilon^2 \ell_H R$$

where R is the range and ℓ_H an horizontal correlation length (the definition of ℓ_H must be modified when the ray curvature is significant
[19]). This mean square phase is an important parameter for all
scattering regimes [17]. In the ray approximation it can also be shown
that the amplitude variance $< A^2 >$ increases proportionally
to the square of the range.

The ray approximation is limited to situations where diffraction
effects can be neglected, a condition which can be expressed [14,15,1]

$$R \ll q \ell_v^2 \qquad \text{or} \qquad \Lambda \equiv R / q \ell_v^2 \ll 1$$

(again, ℓ_v must be appropriately defined to account for ray curvature
[19]). The wave parameter Λ has been introduced by CHERNOV [14] and
generalized by DASHEN [16]; it is essentially the square of the ratio of
the radius of the first FRESNEL zone to the scale of the inhomogeneities.
When $\Lambda \gg 1$ (which corresponds to FRAUNHOFER diffraction) the ray
approach is no longer valid and a more general expansion must be used.
The RYTOV method [13,14] solves the equation for $\ln p = \chi + i S$ whereby
the conditions of validity are less restrictive than in the particular
case of geometrical acoustics. The books by CHERNOV [14] and TATARSKI
[15] discuss this approach in great detail in the context of isotropic
homogeneous fluctuations. Specific applications to the scattering of sound
by internal waves have been given by MUNK and ZACHARIASEN [19],
DESAUBIES [20], and USCINSKI [21].

In the first of these references [19] the solution for lnp is expressed
in integral form. The moments of interest for comparison with data ($<\chi^2>$,
$< S^2 >$ for instance) are then calculated with the aid of
stationary phase approximations, in which specific account is taken of the
complications introduced by the geometry of the rays. Equivalent results
are obtained by DESAUBIES [20], but they are limited to configurations
where the ray curvature is small, i.e. to shorter ranges. The approach is
different too: the parabolic approximation to the wave equation is made

286

[22] : the resulting equation is Fourier transformed whereby it becomes an ordinary differential equation with the range as variable. The solution is thus expressed in the spectral domain and emphasizes the contributions of the various scales of the medium to the scattered field.

Any of the calculations made in the first BORN or RYTOV approximation predict that the energy in the scattered field increases continuously with range. This is clearly unphysical: the energy in the incoming, coherent wave being finite there is a limit to how much energy can be fed into the scattered field. When all the energy has been scattered the regime is described as saturated, and the approximations just discussed are no longer valid. Thus a condition for the validity of the RYTOV approximation is that

$$\sigma_I^2 \equiv \left[\langle I^2 \rangle - \langle I \rangle^2 \right] / \langle I \rangle^2 \quad \ll 1$$

where $I = p p^*$ and σ_I is known as the scintillation index.

In summary, the conditions of validity of the weak scattering theories can be shown to be [1] :

- for the ray acoustics $\quad \Lambda \ll 1$ and $\Lambda \langle S^2 \rangle \ll 1$

- for the RYTOV approximation: $\quad \langle S^2 \rangle \ll 1$

3.2 The Strong Scattering Regime

The mathematical treatment of the weak scattering case is physically equivalent in the ray-particle analogy to single scattering: the wave propagates undisturbed until it encounters a "scatterer" and then continues without further modification by the medium. Of course, in reality the medium continuously modifies the propagating wave; even though the amount of scattering over a correlation length is very small (this is measured by $q^2 \varepsilon^2 \ell_H^2 \ll 1$), for propagation over long ranges the cumulative effect of these small focusings and defocusings can become significant.

The general literature on the subject has been reviewed by PROKHOROV et al [23] and FANTE [24] ; useful discussions can also be found in References [1, 21] . In applications to the case of internal waves in the ocean the mathematics is considerably simplified by the use of the parabolic approximation which reduces the elliptic Helmholtz equation to the parabolic type. This is particularly advantageous for numerical computations. The parabolic method has been pioneered in underwater acoustics by TAPPERT [22] , its validity requires that the scattering be only at small angles in the forward direction, i.e. $q \ell_v \ll 1$.

DASHEN [16] solves the parabolic equation in terms of FEYNMAN'S path integral which is an infinite dimensional integral. In calculating the various moments of the field p of interest it is found, by stationary phase arguments, that most contributions to the integral come only from finite regions in space which are defined by the perturbed ray equation. These so-called micro-paths are due to the random structures in the medium and are not to be confused with the deterministic multipaths related to the sound channel. Thus, in the saturated regime the received signal can be interpreted as resulting from the superposition of several random micropaths; the statistics of the acoustic field does no longer depend on the detailed properties of the medium, but rather reflects this random summation. As a result of the central limit theorem the real and imaginary parts of $p = p_R + \hat{\imath} \, p_i$ are jointly Gaussian, therefore the intensity $I = p_R^2 + p_i^2$ is Rayleigh distributed. However in a medium with a spectrum of scales there is a "partially saturated" regime where there are micro-multipaths, but they are within a correlation length: they do not fluctuate independently but are correlated. This occurs, for internal waves, when $\Lambda \langle S^2 \rangle \gg 1$ but $\Lambda \langle S^2 \rangle^{1/2} \ll 1$.

A different, but equivalent, physical interpretation is given by YURA [25] . He points out that as the wave propagates through the medium it looses transverse coherence since the different parts of the phase surface encounter independent scattering events. Thus, the coherent wave front becomes a set of uncorrelated wavelets which are the equivalent of the multipaths.

Rather than solving the parabolic equation and then forming the moments of the field, one can derive differential equations for the moments and then solve. The latter is the approach taken by USCINSKI [17,21] where further references are given (see also [23] and BERAN [26]). The analysis of BERAN [26] includes the effects of a sound channel, that of USCINSKI [21] treats the anisotropic case.

Those models establish that

i) the mean field

$$\langle p \rangle = e^{-\frac{1}{2} \langle S^2 \rangle}$$

where the mean square phase is calculated in the geometric approximation. Since in the saturated regime $\langle S^2 \rangle \gg 1$ the mean field is vanishingly small. The quantity $\langle S^2 \rangle / R \simeq q^2 \varepsilon^2 \ell_H$ is called the attenuation constant.

ii) $\quad \Gamma_{12} = \langle p\,(\underline{\Delta}_1)\, p^*(\underline{\Delta}_2) \rangle = e^{-\frac{1}{2} D\,(\underline{\Delta}_1 - \underline{\Delta}_2)}$

where $\underline{\Delta}_i$ is the space, time vector $\underline{x}_i$, t and D the phase structure function of geometric acoustics:

$$D\,(\underline{\Delta}_1 - \underline{\Delta}_2) \equiv \langle (S\,(\underline{\Delta}_1) - S\,(\underline{\Delta}_2))^2 \rangle .$$

The moment Γ_{12} is the transverse coherence of the field and is directly

related to the visibility defined in optics $[27,28]$. The power in the
field is given by $\Gamma_{11} = \langle p\ p^* \rangle = 1$ by conservation of energy.

iii) the fourth order moment is required to calculate the statistics
of the intensity. In the fully saturated regime the signal is the result
of the interference of several random micromultipaths so that the
statistics are Gaussian. Then the fourth order moment can be expressed
in terms of the second order ones. An exception arises in the partially
saturated regime which occurs in the scattering by internal saves. Since
the micropaths are correlated the central limit theorem does not apply;
this case is discussed in $[1]$.

Finally we mention the use of the radiation transport equation $[18]$
This equation is essentially a FOURIER transform of the equation for the
mutual coherence function; it expresses the changes along a ray of the
directional distribution of intensity. The changes are due to scattering
into and out of the beam and are expressed by a scattering kernel, which
in turn is directly related to the power spectral density of the index of
refraction. Since this is precisely the quantity given by internal wave
models and since the method is easily implemented numerically it renders
the technique particularly attractive, as pointed out by WILSON and
TAPPERT $[18]$.

4. Experimental Results

Because of the occurrence of the sound channel in the ocean, the
acoustic wave usually travels along several paths between source and receiver. For ranges of hundreds of kilometers as many as fifteen paths are
possible. Many experiments have been conducted under such conditions, using
CW signals. The received signal is then the sum of the contributions from
all the paths and it is impossible to separate the individual paths. Such
experiments have been reported by DEFERRARI and LEUNG $[29,30]$, PORTER,
SPINDEL and JAFFEE $[31,32]$, CLARK and KRONENGOLD $[33]$. In general the
signals are recorded as a function of time and processed to obtain time
series of phase and amplitude and corresponding spectra.

DYER $[34]$, DYER and SHEPARD $[35]$, and DYSON et al $[36]$ have
derived various statistics of the interfering paths, in generally good
agreement with the data. DYER $[34]$ finds that the multipath effect
dominates over the scattering, the distribution of p is normal, of p^2
exponential and of $|p|$ Rayleigh. DYER and SHEPARD $[35]$ consider the
joint probability density of amplitude and phase, and the zero crossings
statistics. They show that the amplitude fluctuates faster than the phase,
and that the ratio of phase period to amplitude period is Rayleigh distributed, both results in agreement with observations. DYSON, MUNK and
ZETLER $[36]$ treat more particularly the time correlations and spectra of
the signal. If internal waves are the fluctuating ocean process, only one
parameter determines the statistics of the acoustic field; it is the mean

square rate of phase fluctuation along a single path: $\langle \dot{S}^2 \rangle$ which they denote by ν^2. They calculate the parameter ν on the basis of [19] and obtain good predictions of the observed variance of intensity, phase, and the spectra of the Cartesian components of the pressure field. The amplitude statistics is dominated by the fade outs which occur when the multipaths interfere destructively.

In all of those models it is assumed that the many paths are statistically equivalent and uncorrelated, an hypothesis that is not necessarily always verified. In particular some paths might be more energetic than others. This might account for some of the discrepancies between theory and experiment. The main result remains nonetheless that in the presence of multipaths the acoustic field is determined more by the random superposition of independent signals than by the details of the ocean structure.

Therefore if one is to relate the properties of the ocean fluctuations to those of the acoustic field through a scattering theory - and thus understand the physics of the interaction - one must either consider experiments where individual paths are resolved or consider more elaborate statistical descriptions of the received signals.

A simple experimental configuration was obtained by EWART [37] who transmitted short pulses over a single wholly refracted path in the deep ocean. The receivers and transmitter were fixed, the range 18 km, the frequencies 4 and 8 kHz. Time series of log amplitude χ and phase S were obtained for a period of two weeks. The parameters Λ and $\langle S^2 \rangle$ defining the scattering regime can be estimated as in [1] ; they indicate that the simple geometric approximation should be valid. Calculations based on that assumption have been given by MUNK and ZACHARIASEN [19] DESAUBIES [20,38] and USCINSKI [21] . They obtain good predictions of the phase data but cannot account for the amplitude frequency spectra, which are predicted to decay as ω^{-3} and are observed to decrease as ω^{-1} with significant high frequency content. This anomalous behavior of the amplitude data has been attributed to advection of the internal waves by currents [1] , "sporadic" multipaths [19] , the effects of ocean finestructure [20]. Another interpretation yet is proposed by USCINSKI [21] who introduces parameters slightly different from those of DASHEN [16] . Consequently he concludes that the experiment of EWART [37] is in a strong focusing region at the parametric boundary between the geometric acoustics and saturated regimes (in his model USCINSKI does not describe the partially saturated regime). He then obtains a good agreement with the measured amplitude spectrum.

A set of measurements covering the partially saturated regime is discussed in [1] . The experiment took place in the Azores on a fixed range of 35 km with frequencies ranging from 400 to 5000 Hz; there was one fully refracted path. Comparison with the theory given by FLATTE et al [1] and DASHEN [16] deals with the various time lagged covariances of the pressure and intensity. The results confirm that the scattering by internal waves is dominant and that the regime is partially saturated, i.e. dominated by correlated multipaths. Similar conclusions were reached, albeit less definitely, by WORCESTER [39,40] who conducted a two way transmission experiment between two drifting hydrophones. Two deterministic rays were present, one predicted to be in the geometric regime, the other partially saturated. He also calculated a number of time correlations, but the statistical significance was low because of the short duration of the time series (8 hr); moreover the ships' drift introduced unknown effects in the data. Nevertheless he was able to conclude that the fluctuations were not inconsistent with the theory based on scattering by internal waves [1]. He also pointed out that the measurement of the difference in travel time is directly related to the velocity field of the water.

5. Conclusion

It is by now firmly established that internal waves play a major role in scattering sound in the ocean. Different scattering regimes can occur depending on the relative values of two parameters. One, $\langle S^2 \rangle$, is the mean square phase fluctuation along a single path, it is directly proportional to the index of refraction variance ε^2 and is thus a measure of the strength of the scatterers; the other, Λ, measures the size of the inhomogeneities relative to the size of the Fresnel Zone: it indicates the importance of diffractive effects. Ways of estimating these parameters are given in [1], they turn out to be highly sensitive to the ray geometry and to the characteristics of the internal wave field. The agreement between theoretical predictions and observed statistics is generally very good, with few exceptions. The effects of other ocean processes such as fine and microstructure, currents, intrusions, or eddies remain to be investigated in detail.

6. References

1. Flatte, S. M., editor, Sound Transmission Through a Fluctuating Ocean, Cambridge University Press, 299 pp(1979)
2. Wunsch, C., The spectrum from two years to two minutes of temperature fluctuations in the main thermocline at Bermuda, Deep Sea Research, 19, 577-593 (1972)
3. Worthington, L. V., The 18° water in the Sargasso Sea, Deep Sea Research, 5, 297-305 (1959)
4. Worthington, L. V., On the North Atlantic Circulation, Johns Hopkins Oceanogr. Stud. 6, 110 pp, The Johns Hopkins University Press, Baltimore, MD (1976)
5. Tolstoy, I. and Clay, C. S., Ocean Acoustics: Theory and Experiment in Underwater Sound, McGraw-Hill, New York, 293 pp (1966)

6. Clay, C. S. and Medwin, H., Acoustical Oceanography: Principles and Applications, John Wiley and Sons, New York, 544 pp (1977)

7. Desaubies, Y.J.F. and Gregg, M. C., Reversible and Irreversible Finestructure, J.of Physical Oceanography, in press (1979)

8. Gregg, M. C. and Briscoe, M. G., Internal waves, finestructure, micro-structure and mixing in the ocean, Reviews of Geophysics and Space Physics (to be published) (1979)

9. Garrett, C. H. R. and Munk, W. H., Internal Waves in the Ocean, Annual Reviews of Fluid Mechanics, 11, 339-369 (1979)

10. Garrett, C. J. R. Mixing in the Ocean Interior, Dynamics of Atmospheres and Oceans, in press (1979)

11. Garrett, C. and Munk, W. H., Space-time scales of internal waves Geophys. Fl. Dyn., 3, 225-64 (1972)

12. Garrett, C. and Munk, W. H., Space-time scales of internal waves: a progress report, J. Geophys. Res., 80, 291-7 (1975)

13. Mysak, L. A., Wave propagation in random media, with oceanic applica-tions, Rev. of Geophys. and Space Phys., 16, 233-261 (1978)

14. Chernov, L. A., Wave propagation in a random medium, McGraw-Hill, New York, 168 pp (1960)

15. Tatarskii, V. I., The effects of the turbulent atmosphere on wave propagation, Israel Program for Scientific Translation, Jerusalem. Available through N.T.I.S., Springfield, VA (1971)

16. Dashen, R., Path Integrals for Waves in Random Media, J. Math. Phys. 20 (5), 894-920 (1979)

17. Uscinski, B. J., The Elements of Wave Propagation in Random Media, McGraw-Hill, Inc., 153 pp (1977)

18. Wilson, H. L. and Tappert, F. D., Acoustic Propagation in Random Oceans Using the Radiation Transport Equation, J. Acoust. Soc. Am., in press(1979)

19. Munk, W. H. and Zachariasen, F., Sound propagation through a fluc-tuating stratified ocean - theory and observation, J. Acoust. Soc. Am. 59, 819-38 (1976)

20. Desaubies, Y. J. F., On the Scattering of Sound by Internal Waves in the Ocean, J. Acoust. Soc. Am., 64(5) 1460-69 (1978)

21. Uscinski, B. J., Acoustic Propagation Through Internal Waves, J. Acoust. Soc. Am., in press (1979)

22. Keller, J. B. and Papadakis, J. S., ed, Wave Propagation and Under-water Acoustics, Lecture notes in Physics V.70, Springer-Verlag (1977)

23. Prokhorov, A. M., Bunkin, F. V., Gochelashvily, K. S. and Shishov, V. I., Laser Irradiance Propagation in Turbulent Media, Proc. IEEE 63, 790-811 (1975)

24. Fante, R. L., Electromagnetic Beam Propagation in Turbulent Media, Proc. IEEE 63, 1669-1692 (1975)

25. Yura, H. T., Physical Model for Strong Optical-Amplitude Fluctuations in a Turbulent Medium, J. Opt. Soc. Am., 64, 59-67 (1974)

26. Beran, M. J., Coherence Equations Governing Propagation Through Random Media, Radio Science 10, 15-21 (1975)

27. Ratcliffe, J. A., Coherence, Correlation, and Visibility, Reps. Prog. Phys. 19, 188 (1956)

28. Beran, M. and Parrent, G., Theory of partial coherence, Prentice-Hall, Englewood Cliffs, NJ (1964)

29. DeFerrari, H. and Leung, R., Spectrum of Phase Fluctuations Caused by Multipath Interference, J. Acoust. Soc. Am., 58, 604-607 (1975)

30. DeFerrari, H. A., Effects of horizontally varying internal wavefields on multipath interference for propagation through the deep sound channel, J. Acoust. Soc. Am., 56, 40-46 (1974)

31. Porter, R. P., Spindel, R. C. and Jaffee, R. J., Acoustic-internal wave interaction at long ranges in the ocean, J. Acoust. Soc. Am., 56, 5, 1426-1436 (1974)

32. Porter, R. P. and Spindel, R. C., Low frequency acoustic fluctuations and internal gravity waves in the ocean, J. Acoust. Soc. Am., 61, 943-958 (1977)

33. Clark, J. G. and Kronengold, M., Long-Period Fluctuations of CW Signals in Deep and Shallow Water, J. Acoust. Soc. Am., 56, 1071-1083 (1974)

34. Dyer, I., Statistics of Sound Propagation in the Ocean, J. Acoust. Soc. Am., 48, 337-345 (1970)

35. Dyer, I. and Shepard, G. W., Amplitude and phase fluctuation periods for long range propagation in the ocean, JASA 61, 937-942 (1977)

36. Dyson, F., Munk, W. H., Zetler, B., Interpretation of multipath scintillations Eleuthera to Bermuda in terms of internal waves and tides, JASA 59, 5, 1121-1133 (1976)

37. Ewart, T. E., Acoustic fluctuations in the open paean, A measurement using a fixed refracted path, JASA, 60, (1): 46-59 (1976)

38. Desaubies, Y. J. F., Acoustic phase fluctuations induced by internal waves in the ocean, J. Acoust. Soc. Am., 60, 785-800 (1976)

39. Worcester, P. F., Reciprocal acoustic transmission in a midocean environment, J. Acoust. Soc. Am., 62, 895-905 (1977)

40. Worcester, P. F., Reciprocal Acoustic Transmission in a Midocean Environment: Fluctuations, J. Acoust. Soc. Am., in press (1979)

Mesoscale Inhomogeneities and Turbulence in Ocean Acoustics

P. Scully-Power

Naval Underwater Systems Center
New London, CT 06320, USA

Abstract

Consideration of the nonlinear potential vorticity equation for quasi-geostrophic motion of a continuously stratified ocean on a β-plane shows the dynamics to be controlled by two specific horizontal wavenumbers; the first depends on β and marks the transition from turbulence to waves while the second depends on the modal Rossby deformation radius and marks the transition from uncoupled to vertically coupled motion. An examination of the dynamical processes involved, namely the cascade of energy to small wavenumber, the concomitant cascade of enstrophy to large wavenumber and the baroclinic instability-induced cascade to large wavenumber, shows that the most persistent structures which evolve are near-circular eddies having a vertical structure similar to a first baroclinic mode Rossby wave, a horizontal length scale equal to the corresponding internal deformation radius, and for which the linear and nonlinear effects are of equal magnitude. These eddies also have a unique non-dimensional number associated with them. The effects of these inhomogeneities on the propagation of an acoustic field are thus predicted in a fully deterministic manner.

1. Introduction.

The world's oceans can be conceptualized in the broadest sense as a field of fluctuating fluid flow in which many length and velocity scales coexist. In its simplest form, a particular motion of the fluid can be characterized by a characteristic velocity U and a characteristic length L of the flow pattern, while the fluid itself is characterized by its kinematic viscosity ν and the velocity of sound c in the fluid. The two dimensionless parameters that can therefore be used to describe the field are the Reynolds number $R_e \equiv UL/\nu$ and the Mach number $M_a \equiv U/c$. Since, however, for the world's oceans $\nu \sim 10^{-2}$ cm ^{2}s^{-1} and $c \sim 1500$ ms^{-1}, the basic field is one in which $R_e \gg 1$ and $M_a \ll 1$. In this simple sense therefore, the ocean can be regarded as being potentially fully turbulent.

That this is not simple three-dimensional homogeneous turbulence however can be easily seen in one of the few long time series of data available from the deep ocean [1]. The spectra show a kinetic energy gap nearly a decade wide at all four depths of measurement. If this were three-dimensional homogeneous turbulence, the gap would have been filled in long ago by the local cascade of energy in the classical sense [2]. This inability of the ocean to dissipate energy at large Reynolds numbers strongly suggests therefore that the governing dynamics is that of two-dimensional turbulence since that class of turbulence has the distinctive feature of being unable to dissipate energy via a cascade to small scales, a property that was realized as early as 1917 [3].

As far as underwater acoustics is concerned, however, a description of the fluid flow field in terms of two-dimensional rather than three-dimensional turbulence would be of little help in predicting the associated stochastic acoustic field if it were not for the insights afforded by the recent revolution in two-dimensional turbulence studies [4]. The overthrow of the deeply ingrained notion of chaotic motion in turbulence by one of structured turbulence, in which the fine-scale turbulence is superimposed on well-defined larger scale two-dimensional coherent vortex structures (at least in turbulent shear flows), offers the distinct possibility of a largely deterministic description of the acoustic field in the ocean if only some analogous description of the flow field was found to apply there.

Fortunately at the same time that the revolution in turbulence was taking place (the early 1970's), another quiet revolution was underway in ocean dynamics studies. The evidence is now overwhelming that the dynamics of the ocean is intimately controlled by the presence of eddies in the flow field, the most energetic of which have mesoscale dimensions ($\sim$100 km) [5]. Since the eddies have a well defined structure [6], there now exists the possibility of relating ocean turbulence to the most energetic mesoscale inhomogeneities and thence predicting, in a fully deterministic manner, the major effects on ocean acoustics caused by these inhomogeneities.

2. Turbulence and Waves

Since the essential ingredient in turbulent flow is the vortex interactions, the pertinent equation governing ocean flow on a rotating earth is the potential vorticity equation. Its derivation is by no means trivial [7], but, taking the curl of the full momentum equation under the usual quasi-geostrophic approximation (earth's rotation force balanced by the pressure gradient, and time scales greater than a day) leads to the following formulation for free flow of a uniform density, incompressible fluid in two dimensions on a β-plane:

$$Dq/Dt = 0 \tag{1}$$

where the potential vorticity q is given by

$$q = \nabla^2\psi + \beta y . \tag{2}$$

Here $D/Dt = \partial/\partial t + \mathbf{u}\cdot\nabla$ is the Lagrangian acceleration, ψ is the stream function, $\mathbf{u} = (-\partial\psi/\partial y, \partial\psi/\partial x)$ is the horizontal velocity vector and $\beta = \partial f/\partial y$ (assumed constant) is the meridional component of the Coriolis parameter $f = 2\,\Omega\,\sin\theta$, where Ω is the magnitude of the planetary rotation and θ is the latitude.

Expansion of (1) leads to

$$\nabla^2\psi_t + J(\psi, \nabla^2\psi) + \beta\psi_x = 0 \tag{3}$$

where J is the Jacobian, and the subscripts refer to partial differentiation.

Since (3) is nonlinear, and scale analysis shows the nonlinearity to be significant for some values of the ocean's typical velocity field, there have been two major approaches to elucidating the dynamics governed by (3): pure two-dimensional turbulence and pure linear waves. Each provides some insights into the overall motion, but it requires the combination of both to resolve the true dynamics.

2.1 Pure Turbulence

With $\beta = 0$ in (3), the dynamics are that of free, two-dimensional turbulence, the theory of which has been extensively studied. There exists two cascades, one for energy and the other for enstrophy ($\frac{1}{2}$ squared vorticity). The energy cascade is to small wavenumber [8]; this 'red' cascade is in the opposite direction to the classical three-dimensional turbulence cascade and implies that turbulent eddies will continue to grow in size. The enstrophy cascade is to large wavenumber [9], which implies the distortion of vorticity contours in physical space into thin filaments. During these cascades, both energy and enstrophy are conserved. Scale analysis of the pure turbulence equation indicates a time scale of the turbulent processes, which leads to an equivalent turbulence dispersion relation $\omega \propto Uk$ (where ω, k are the frequency, wavenumber respecitvely). Hence, the migration of energy to small k is also a migration to small ω, a point that will be referred to later.

2.2 Pure Waves

With $J = 0$ in (3), the potential vorticity balance is linear, and the dynamics are that of free, two-dimensional waves called Rossby waves [10]. These are long waves with a dispersion relation $\omega \propto \beta k^{-1}$.

2.3 Turbulence - Wave Transition

The fact that the energetic eddies, though significantly nonlinear, fail to grow via the turbulent energy cascade to the size of their physical domain (the ocean basin) can only be explained by the consideration of turbulent and wave effects taken together. Thus, the full equation (3) must be investigated as a single entity. Scale analysis of (3) shows that the ratio ε of the nonlinear to linear advective terms can be expressed as

$$\varepsilon = 2k^2 U/\beta \tag{4}$$

and hence for $\varepsilon \ll 1$ only the linear terms are important, resulting in Rossby wave dynamics, whereas for $\varepsilon \gg 1$ only the nonlinear terms are important, resulting in turbulent dynamics. The dividing line between the turbulent and wavelike regimes therefore occurs at $\varepsilon = 1$ which, as has been pointed out in [11], is equivalent to the Rossby waves attaining unit steepness.

Eq. (4), with $\varepsilon = 1$, therefore prescribes a unique wavenumber $k = k_\beta$ where

$$k_\beta = (\beta/2U)^{1/2} \tag{5}$$

so that for $k > k_\beta$ ($\varepsilon > 1$) turbulence dominates, but for $k < k_\beta$ ($\varepsilon < 1$) waves dominate.

This provides a mechanism for countering the 'red' energy cascade, since turbulent eddies, as they expand in size, make themselves more and more vulnerable to the restoring effects of β, until at $k \sim k_\beta$, the nonlinear effects and wave dispersion effects balance and the turbulent 'red' cascade is halted by wave radiation. Furthermore, since the energy and enstrophy cascades are interrelated, the enstrophy cascade also halts at $k \sim k_\beta$.

This combination of both turbulence and waves in the full dynamical picture also carries within it the concept of anisotrophy. The turbulent energy migration to both small wavenumbers and small frequencies is not compatible in an isotropic medium with the dispersion relation for waves, which associates small frequencies with large wavenumbers. Hence the fluid field must adjust towards anisotrophy, in which north-south wavenumbers (and hence by the geostrophic relation, east-west currents) are favored. Thus the eddies are flattened by the β-effect, in the sense that their streamlines become ellipses aligned in the zonal (east-west) direction.

3. Stratification

Up to this point, the dynamics has only involved the motion of a homogeneous two-dimensional fluid. However, the most outstanding feature of the real ocean is the fact that it is horizontally statified. This therefore raises the question as to how the foregoing analysis must be modified to account for this feature. Since vertical shear is now possible, what is its effect on the overall dynamics? Furthermore, does the turbulence-wave transition still occur at wavenumber $k = k_\beta$?

The simplest possible model which allows for these effects is one consisting of two layers of homogeneous fluid having differing densities. Although this might seem to oversimplify the situation, it does bring out the significant dynamical effects, as was so well demonstrated in [12].

For a two-layer ocean under the usual Boussinesq approximation (where the fluid density variations are neglected in their effect on inertia, but not on buoyancy), (1) is still valid with (2) modified to the form

$$q = \nabla^2 \psi_1 - a_1^2 (\psi_1 - \psi_2) + \beta y \tag{6}$$

for the upper layer, with an equivalent form for the lower layer, where the subscripts 1, 2 refer to the upper and lower layers, respectively, and $a_1^2 = f^2/g'H_1$, g' being the reduced gravitational constant $g\Delta\varrho/\varrho$ for the layers of density difference $\Delta\varrho$, and H_1 is the depth of the upper layer. a_1^{-1} defines a length scale μ which is the well known Rossby deformation radius. It separates strong vertical coupling $k \ll a_1$ from weak vertical coupling $k \gg a_1$ in which the interface is essentially rigid. Thus the introduction of stratification produces a second specific horizontal wavenumber $k = k_a$ where

$$k_a = f/(g'H_1)^{1/2} = \mu^{-1} \tag{7}$$

so that for $k < k_a$ coupling of the upper layer to the lower layer dominates, but for $k > k_a$ the two layers behave as two uncoupled layers of two-dimensional fluid.

In addition, substitution of (6) into (1), expanding, and noting that $(\mathbf{u} \cdot \nabla)\psi \equiv 0$, yields

$$\partial/\partial t\, [\nabla^2\psi_1 - a_1^2(\psi_1 - \psi_2)] + J(\psi_1, \nabla^2\psi_1) + \beta\partial\psi_1/\partial x = 0 \tag{8}$$

which is the two-layer equivalent to (3).

The interesting point in comparing (8) to (3) however, is that the advective terms (both linear and non-linear) remain unchanged. Hence, the measure of nonlinearity ε is still given by (4) and thus the turbulence-wave transition still occurs when $k = k_\beta$.

In the more realistic situation of a continuously stratified ocean, (6) must be generalized to the form

$$q = \nabla^2\psi + \partial/\partial z[(f^2/N^2)\partial\psi/\partial z] + \beta y \tag{9}$$

where N is the Brunt-Vaisala (buoyancy) frequency $N = [-(g/\varrho)\partial\varrho/\partial z]^{\frac{1}{2}}$.

Now, in a linearized ocean $(J = 0)$, any distribution of the horizontal velocity $\mathbf{u} = (u, v)$ with respect to depth can be expressed as the sum of orthogonal normal modes. Each mode has its own characteristic distribution with respect to depth, proportional (say) to $c_n(z)$, $n = 0, 1, 2, \ldots$. The barotropic mode $n = 0$ has $c_n(z) = 1$, i.e., the horizontal velocity is independent of depth, which corresponds to a homogeneous, two dimensional ocean, and (9) reverts back to (2). The baroclinic modes $n = 1, 2, 3, \ldots$ are depth dependent and the eigenfunctions are determined by a Sturm-Liouville formulation [13].

Each baroclinic mode then satisfies the characteristic equation [14]

$$\partial/\partial z[(f^2/N^2)\, \partial\psi/\partial z] + a_n^2\, \psi = 0 \tag{10}$$

where the inverse of the eigenvalue a_n is the baroclinic Rossby deformation radius μ_n for that particular mode.

The potential vorticity for each baroclinic normal mode then takes the form (by combining (10) with (9))

$$q = \nabla^2\psi - a_n^2\, \psi + \beta y \tag{11}$$

which, when substituted into (1) gives

$$(\nabla^2 - a_n^2)\psi_t + J(\psi, \nabla^2\psi) + \beta\psi_x = 0. \tag{12}$$

Comparison of (12) with (3) shows, for reasoning similar to that given for the two-layer case, that the turbulence-wave transition for each baroclinic normal mode occurs when $k = k_\beta$ (independent of the mode number), although the modes cannot be simply superposed owing to the nonlinear nature of (12).

298

Thus, for a continuously stratified (real) ocean, there are two horizontal wavenumbers which control the dynamics: $k_\beta = (\beta/2U)^{1/2}$, which separates waves $(k < k_\beta)$ from turbulence $(k > k_\beta)$, and $k_{a_n} = \mu_n^{-1}$ which separates vertically coupled motion $(k < k_{a_n})$ from uncoupled motion $(k > k_{a_n})$ for each of the baroclinic modes. However, since the eigenvalues a_n of (10) are such that $\mu_1 > \mu_2 > \mu_3 \ldots$, the value of k_{a_n} increases with increasing mode number, and so higher order modes could be vertically coupled whereas lower order modes having the same horizontal wavenumber could be uncoupled. Hence, all modes are vertically coupled and the coupling process is complete only for $k < k_a$ where

$$k_a = k_{a_1} = \mu_1^{-1} \tag{13}$$

4. The Equilibrium End-State

In a stratified ocean, the existence of the two specific wavenumbers k_β , k_a implies that the energy-preserving cascade of turbulence to small wavenumber can follow two distinct paths; either the cascade is halted by the restraining effects of β before the wave number has become small enough to be influenced significantly by stratification, or stratification effects dominate before the termination of the cascade. Furthermore, since k_a is fixed (through (10)) for an ocean of given stratification N, the relative magnitude of k_β to k_a is determined solely by the energy level; the cross-over point $k_\beta = k_a$ occuring when $U = U_c$, which from (5) and (13) is given by

$$U_c = \tfrac{1}{2}\beta\mu_1^2 . \tag{14}$$

For small energy turbulence $U < U_c$, $k_\beta > k_a$ and the 'red' cascade encounters the β-effect at $k = k_\beta$ well before stratification becomes important. At that point the cascade stops and the energy is radiated away as a series of uncoupled Rossby waves.

The more interesting case occurs for higher energy turbulence $U > U_c$, for then $k_a > k_\beta$ and the 'red' cascade is dominated by stratification (at $k = k_a$) well before it reaches its limiting size $k = k_\beta$. When k reaches k_a , the fluid is fully coupled vertically, vortex stretching occurs, and the motion becomes barotropic throughout the fluid column. After converting to barotrophy, however, there is no further impediment to the cascade which then acts out its thrust to smaller wavenumbers until impeded by the β-effect at $k = k_\beta$, at which point the cascade stops and the energy is radiated away as a barotropic Rossby wave.

The only other dynamical situation possible is that where the initial field consists of large scale $(k < k_a)$ baroclinic energy. Here one would expect the energy to propagate simply as baroclinic Rossby waves. However, it turns out that these waves are unstable, and another process takes hold, that of baroclinic instability, which results in another cascade in the direction opposite to the turbulent energy cascade. This cascade causes the large scale baroclinic Rossby waves to break up into eddies of wavenumber $k \sim k_a$. At this point the dynamics is the same as for higher energy baroclinic turbulence; the motion converts from baroclinic to barotropic, expands to wavenumber $k = k_\beta$ and then radiates as a barotropic Rossby wave. This inverse cascade, together with the small and large energy turbulent cascades, have been demonstrated recently

[5] in an elegant series of computer models, which show the progressing cascades in a series of graphical visualizations of the evolving energy and enstrophy fields.

5. Unique Eddies

The form of the critical velocity in (14) is highly suggestive of the possibility of some unique underlying dynamics. In order to elucidate this, consider the dispersion relation for a baroclinic Rossby wave [15], here generalized to the case of a continuously stratified ocean

$$\omega = - \beta\ell/(\ell^2 + m^2 + a_n^2) \tag{15}$$

where ℓ, m are the wavenumbers in the x and y directions and a_n is the inverse of the Rossby deformation radius for the particular mode.

Now, the wavenumbers ℓ and m can be written in the form

$$(\ell, m) = (k \cos \alpha, k \sin \alpha) \tag{16}$$

where α is the orientation of the wavenumber vector, and so the zonal velocity ω/ℓ of the wave pattern is

$$\omega/\ell = - \beta/(k^2 + a_n^2) \tag{17}$$

which is independent of α. Hence the whole pattern moves westward with a velocity given by (17), irrespective of the direction of the wavenumber vector.

The dispersion relation (15) can, however, also be written in the form

$$(\ell + \beta/2\omega)^2 + m^2 = (\beta/2\omega)^2 - a_n^2 . \tag{18}$$

Since the left hand side of (18) is positive definite, there exists a high frequency cut-off $\omega_{c,n}$ determined by equating the right hand side to zero, and so

$$\omega_{c,n} = \tfrac{1}{2}\beta\mu_n \tag{19}$$

where μ_n is the Rossby deformation radius.

Substitution of (19) back into (18) then yields, for $\omega = \omega_{c,n}$,

$$k = \mu_n^{-1} . \tag{20}$$

Furthermore, since $\mu_1 > \mu_2 > \mu_3 \ldots$, the highest frequency possible for any baroclinic Rossby wave is ω_c, where

$$\omega_c = \omega_{c,1} = \tfrac{1}{2}\beta\mu_1 \tag{21}$$

which is the high frequency cut-off of the first mode and occurs when the wavenumber is the inverse of the deformation radius.

The dynamics pertinent to the case $\omega = \omega_c$ are especially significant here, for the westward drift in this case becomes (from (17))

$$\omega/\ell = -\tfrac{1}{2}\,\beta\mu_1^2 \tag{22}$$

and, since $k = \ell$ for $\omega = \omega_c$, the phase velocity takes the same value, being given by

$$\omega/k = -\tfrac{1}{2}\,\beta\mu_1^2 \,. \tag{23}$$

Since these waves attain unit steepness when the root-mean-square fluid velocity U reaches the same value as the phase velocity, comparison of (23) with (14) shows that for $U = U_c$ the resulting wave is a first mode baroclinic Rossby wave having $k = \mu_1^{-1}$. These waves have a special significance in the overall dynamics since, for $U = U_c$, $k_a = k_\beta$,and so turbulence with this characteristic energy results in waves which have the least tendency to finally evolve into a state of barotrophy — compared with $U > U_c$, in which the turbulence becomes barotropic before reaching the transition wavenumber to waves. Moreover, since bottom topography has the effect of 'detuning' the cascade and inhibiting the final state of barotrophy [16], it will therefore act as a preferential selection process, having greatest effect on the dynamics for which $U = U_c$, and thus the first mode baroclinic waves will retain their structure longer than any other.

Moreover, the nature of the dispersion relation for baroclinic Rossby waves changes dramatically at wavenumber $k = k_a$; for $k > k_a$ the dispersion relation has the same form $\omega \propto k^{-1}$ as barotropic Rossby waves, but for $k < k_a$ it has the same form $\omega \propto k$ as turbulence. Thus, in the particular case of $U = U_c$ where the baroclinic structure is retained for a long time, there is no conflict between the turbulence dispersion relation and wave dispersion relation. Hence in this particular case the anisotrophy mentioned previously will not occur and these baroclinic structures would tend to form near-circular eddies.

The net outcome therefore in a field of freely evolving turbulence is that the structures with the longest lifetime would be expected to be near-circular eddies having a characteristic horizontal wavenumber $k = \mu_1^{-1}$, and a vertical dependence characteristic of a first baroclinic mode Rossby wave.

This can be seen even more clearly by taking a generalized form of solution to the linearized wave equation for a continuously stratified ocean. Implicit in the analysis up to this point has been a solution in which the variations of ψ are harmonic in the horizontal. A more general form

$$\psi = J_0(kr) \tag{24}$$

in the horizontal, where $r = [(x{-}ct)^2 + y^2]^{1/2}$, is also a solution provided c takes the value of the westward drift given in (17). Thus all of the relationships previously derived are valid in this more general case.

Furthermore, the result derived in (23) is even more significant for the cascade resulting from baroclinic instability. This can be seen by noting that not only is the phase velocity given by (23) associated with the highest frequency possible for any baroclinic Rossby wave, but it is also the maximum value of phase velocity attainable by any baroclinic Rossby wave at any wavenumber (as can be derived by differentiation of ω/k with respect to k, where ω is given by (15)).

However, the steepness of a baroclinic Rossby wave is U/c_p, where c_p is the phase velocity. Hence as a baroclinic instability grows, the wave that can contain the highest energy (greatest U) before it becomes unstable (at a wave steepness of unity, for then $\varepsilon = 1$) is just that for which c_p is a maximum. Thus the baroclinic instability cascade which results in the most energetic eddies is the one involving the first baroclinic mode. Moreover, for reasoning similar to that for the turbulent cascade, these energetic eddies have the longest lifetimes.

The end result therefore for an ocean which contains turbulence, or long baroclinic waves, or both, is that the longest lifetime structures which evolve are near-circular mesoscale eddies having a horizontal wavenumber μ_1^{-1}, a first baroclinic mode structure in the vertical, and for which the linear and nonlinear effects are of equal magnitude.

The uniqueness of these eddies is perhaps epitomized by the realization that they must have a nondimensional number N_E associated with them which is invariant for all such eddies. This number, defined as

$$N_E = D_0/R^2 f^2 \tag{25}$$

is the ratio of the dynamic height of the eddy to the square of the product of the eddy radius and Coriolis parameter, where the dynamic height D_o is defined in the usual way as $g\eta_o/10$; η_o being the perturbation of the sea surface caused by the eddy. That this must be invariant can be deduced from the fact that since the vertically averaged energy is known ($U = U_c$) and the vertical structure is known (first baroclinic mode) then η_o can be determined provided the horizontal distribution is known. But the horizontal wavenumber is also known (μ_1^{-1}) and, taking the general form of an eddy given in (24), then η_o is known. Finally, the radius R is chosen so that kR is the first non-zero root of $J_1(kr)$, for here $\partial\psi/\partial r$ is zero, i.e., the horizontal velocity is zero, which defines the edge of the eddy.

Preliminary (approximate) evaluations of the eddy number indicate that for an arbitrary stratification N , its magnitude is independent of N and is given by

$$N_E \sim 6.8 \times 10^{-3} \tag{26}$$

which is in excellent agreement with the average for 52 quasi-circular eddies from all oceans reported in [17].

The real test of these results is whether actual ocean eddies exhibit the predicted properties. Now, since the major Western boundary currents of the world's oceans, e.g., the Gulf Stream [18], the Kuroshio [19], and the East Australian Current [20], all have long-lived eddies associated with them, they provide such a test; perhaps the best being the East Australian Current where the eddies form intermittently [21] and are, therefore, likely to have even longer lifetimes, since spatial isolation in an eddy field also acts to prevent barotrophy [5].

Shown in Fig.1 is one such eddy associated with the East Australian Current and described in [21]. Plotted is the position of the isothermal core which characterizes eddies formed from this particular current. This singular feature enables the drift velocities of such eddies to be calculated very accurately, in this case being 0.84 kilometers per day to the west. The predicted westward drift, based on the Rossby deformation radius and computed from (22), was found to be 0.85 kilometers per day, in excellent agreement with the observations.

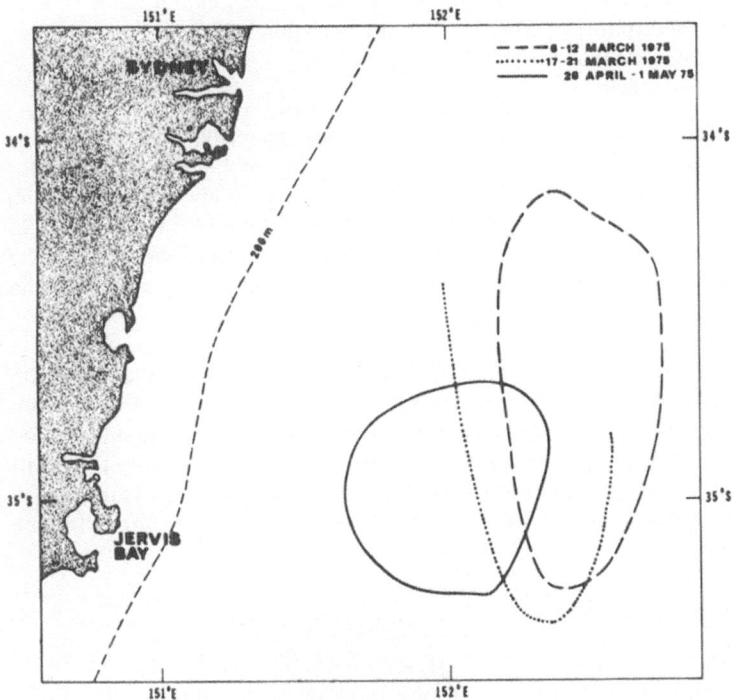

Fig.1   Westward drift of an eddy associated with the East Australian Current.

6.   Ocean Acoustics

The existence of unique eddies in the ocean offers the possibility of a deterministic evaluation of the effects of these inhomogeneities on the propagation of underwater sound. In order to elucidate such effects, an eddy was chosen which accurately matched a Bessel function structure in the horizontal and a first baroclinic mode structure in the vertical. This eddy, which is described in some detail in [22], has the vertical distribution of temperature shown in Fig.2a and the corresponding vertical distribution of sound speed shown in Fig.2b. It is a major perturbation to the normal horizontal stratification of the ocean; added to which the isothermal core results in closed contours of sound speed, thus forming a three-dimensional acoustic lens in the ocean.

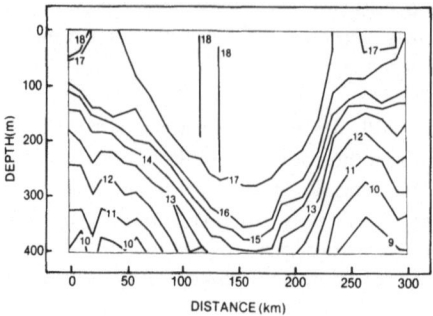

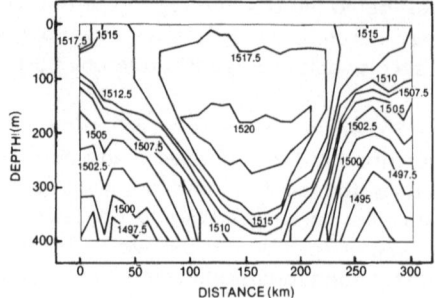

Fig. 2a Vertical temperature distribution (°C) across the eddy described in [22]. The lines marked 18°C denote the isothermal nature of the eddy core.

Fig. 2b Corresponding sound speed structure (ms⁻¹).

Acoustic propagation through this structure shows distinctly different results depending on the direction of propagation. Shown in Fig. 3a are the acoustic ray paths for a source situated at a depth of 250 meters in the center of the eddy and propagating in a direction radially outwards. This ray pattern is distinctly different from that of Fig. 3b in which the only change has been a reversal of the direction of propagation, in the sense that in Fig. 3b the source is situated at a depth of 250 meters at the edge of the eddy and is propagating radially inwards.

These ray patterns have been quantified in Fig. 3c, where the propagation loss with range has been calculated for a receiver near the surface of the ocean. The vast difference between the radially outward propagation (large dashes) and radially inward propagation (solid line) is immediately obvious. Even the most fundamental acoustic feature of the ocean, the convergence zone (seen here between 50 and 70 km), which results from the basic stratification of the ocean, is non-existent near the surface for one propagation direction.

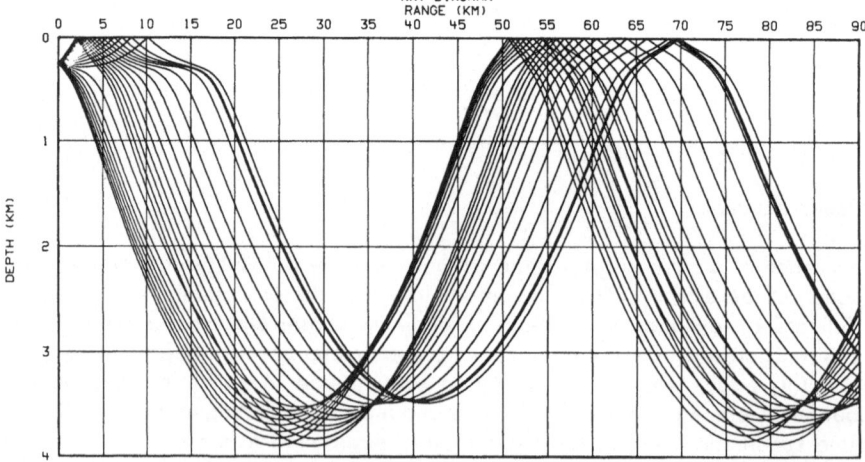

Fig. 3a Ray paths for acoustic propagation in a vertical plane across a radial section of the eddy in fig. 2. The source is at a depth of 250 meters in the center of the eddy.

304

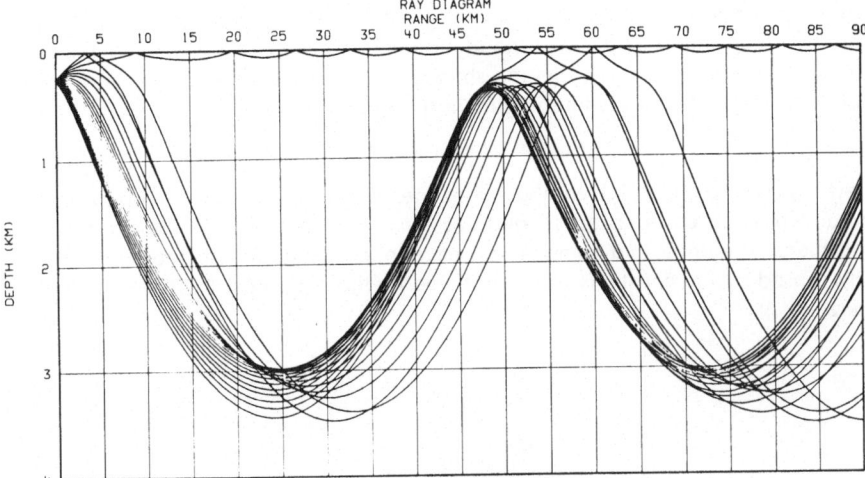

Fig.3b Ray paths for acoustic propagation in a vertical plane across the same radial section as (3a), but for which the propagation direction has been reversed. The source is now at a depth of 250 meters at the edge of the eddy.

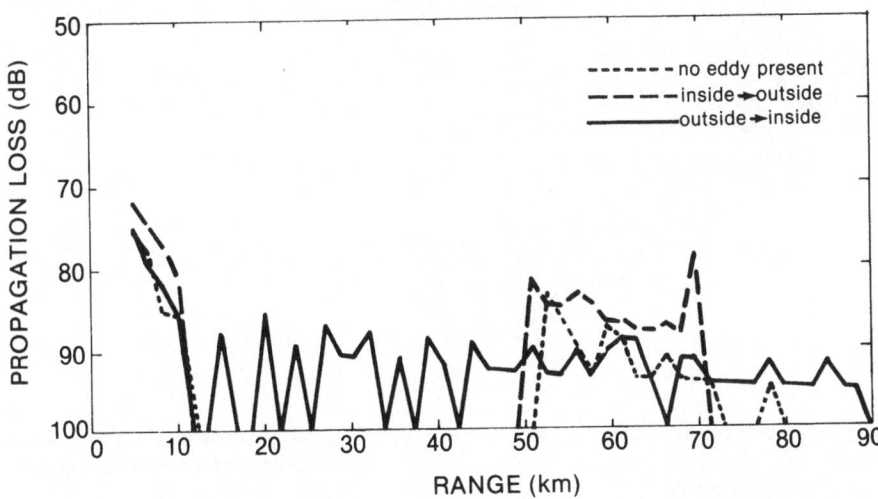

Fig.3c Propagation loss curves for the two cases shown in (3a) and (3b) together with the case for no eddy present.

For comparison, the propagation loss curve corresponding to the unperturbed ocean outside the eddy is shown (small dashes). It is different again, thus demonstrating that not only do these inhomogeneities markedly alter the acoustic propagation characteristics of the ocean, but they also make these characteristics highly sensitive to the direction of propagation.

Apart from modifying the acoustic propagation in a vertical plane, these inhomogeneities also alter the propagation in the horizontal. To evaluate these effects, an eddy was modeled in the horizontal plane simply as a circular sound speed contrast.

305

For a cold core eddy with a sound speed contrast of 30 ms^{-1}, the horizontal acoustic ray paths are shown in Fig. 4a. The sound speed inhomogeneity causes focusing of the near tangential rays, producing a caustic and shadow zone on each side of the eddy, with corresponding large changes in intensity in these regions.

A quantification of this situation is given in Fig. 4b. At the top, both the direct and refracted rays are plotted, which show that the azimuthal deviation caused by the horizontal refraction can exceed 10 degrees for some azimuths. At the bottom, the acoustic intensity relative to a homogeneous ocean is plotted, and the large increase in intensity caused by the caustic is seen, followed by the extremely low intensity in the shadow zone.

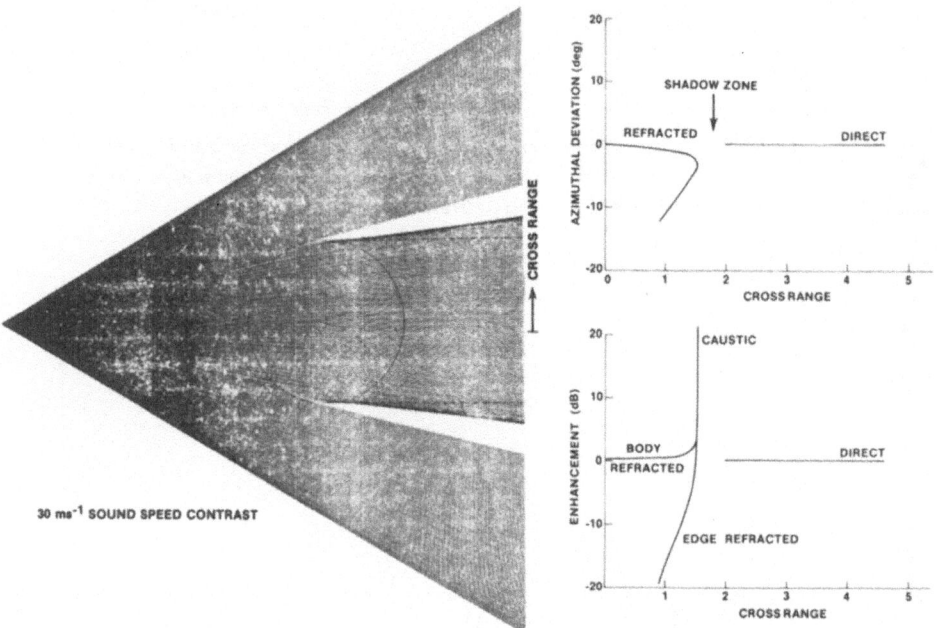

Fig. 4a   Ray paths for acoustic propagation in a horizontal plane through a cold core eddy.

Fig. 4b   Corresponding azimuthal deviation and intensity enhancement caused by the eddy.

These acoustic effects need to be further refined, but they do demonstrate that the perturbations to the acoustic field in the ocean caused by these mesoscale inhomogeneities can indeed be quantified in a largely deterministic manner.

7. References

[1] R. Thompson, Deep-Sea Res. **18**, 1 (1971).

[2] A. N. Kolmogoroff, C. R. Acad. Sci. U. R. S. S. **32**, 16 (1941).

[3] G. I. Taylor, Scientific Papers (ed. G. K. Batchelor), vol 2, 69 (Cambridge University Press, 1960).

[4] A. Roshko, AIAA Journal **14**, 1349 (1976).

[5] P. B. Rhines, The Sea (ed. E. D. Goldberg), vol 6, 189 (Wiley-Interscience, 1977).

[6] J. C. McWilliams, J. Phys. Oceanogr. **6**, 810 (1976).

[7] N. P. Fofonoff, The Sea (ed. M. N. Hill), vol 1, 323 (Wiley-Interscience, 1962).

[8] G. K. Batchelor, Homogeneous Turbulence, 186 (Cambridge University Press, 1953).

[9] G. K. Batchelor, Phys. Fluids (Suppl. II), **12**, 233 (1969).

[10] G. W. Platzman, Quart. J. R. Met. Soc. **94**, 225 (1968).

[11] P. B. Rhines, J. Fluid Mech. **69**, 417 (1975).

[12] G. Veronis and H. Stommel, J. Mar. Res. **15**, 43 (1956).

[13] M. J. Lighthill, Phil. Trans. A **265**, 45 (1969).

[14] A. E. Gill, J. S. A. Green and A. J. Simmons, Deep-Sea Res. **21**, 499 (1974).

[15] M. S. Longuet-Higgins, Proc. R. Soc. A. **284**, 40 (1965).

[16] W. B. Owens, J. Phys. Oceanogr. **9**, 337 (1979).

[17] J. C. Andrews, PhD Thesis, Flinders University of South Australia, pp 163 (1976).

[18] F. C. Fuglister, Studies in Physical Oceanography (ed. A. L. Gordon), vol 1, 137 (1972).

[19] K. Kitano, J. Phys. Oceanogr. **5**, 245 (1975).

[20] F. M. Boland and B. V. Hamon, Deep-Sea Res. **17**, 777 (1970).

[21] C. S. Nilsson, J. C. Andrews and P. Scully-Power, J. Phys. Oceanogr. **7**, 659 (1977).

[22] J. C. Andrews and P. Scully-Power, J. Phys. Oceanogr. **6**, 756 (1976).

On the Influence of Stochastic Sound Speed Variations on Acoustic Transmission Loss in Shallow Water

H.G. Schneider

Forschungsanstalt der Bundeswehr für Wasserschall- und Geophysik
Klausdorfer Weg 2-24
D-2300 Kiel, Fed. Rep. of Germany

1. Introduction

This study continues earlier ones [1], [2] in which the loss of acoustic
energy from a sound channel caused by stochastic sound speed variations
was successfully modeled. The excess loss relatively to a non varying medium was then found to depend on the amount of variability as well as on
the loss outside the channel e. g. the bottom reflectivity.

In extension of [1] and [2] we study here conditions in which the acoustic transmission loss (TL) in a stochasticly varying medium may also be
equal or less than in the corresponding invariant medium. A common explanation for these changes in TL is found in the development of the angular energy distribution (AED) with range. In a stationary stochastic environment the
distribution of energy over the propagation angles tends to become stationary for sufficiently large distances from the source, but the width and
center of gravity of the long range AED in a stochasticly varying medium may
differ considerably from the long range AED caused by the corresponding invariant sound speed layering and it may differ also from the AED's at very
short range, which are necessarily identical for a stochastic and deterministic environment.

The model study further shows that in an stochasticly varying medium different AED's at short range due to different initial conditions may converge to only one AED at large ranges. However different types of stochastic
variability of the medium may result in different stationary long range
AED's.

2. Description of the Acoustic Transmission Loss Experiment

In Fig. 1 sound speed profiles are displayed which were measured before
and after the acoustic experiment. Other profiles measured along the acoustic transmission track exhibit also a sharp thermocline with variable depth,
variable sound speed gradient and a varying difference between maximum and
minimum sound speed. More measurements of the same type can be found in [3],
where the variation of the thermocline in depth is several meters with horizontal correlation lengths from some 100 m to a few kilometers. Relative
to the water depth of 50 to 100 m these variations cannot be termed "microstructure" as seems to be appropriate for a deep water environment. On the
other hand for acoustic transmission ranges of about 30 NM these profile
variations can be assumed to be stochastic and mostly also stationary.

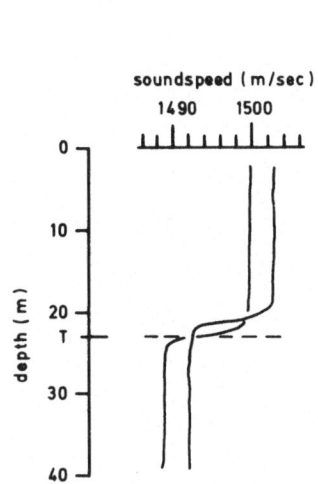

soundspeed (m/sec)

Fig.1 Soundspeed profiles before and after acoustic measurement

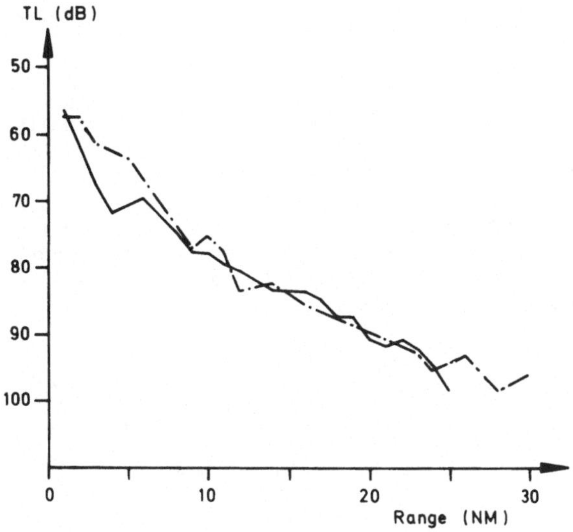

Fig.2 Measured TL at 4 kHz, 2 runs

Indicated in Fig. 1 are also the nominal transmitting and receiving depth of 23 m. The ocean bottom is to a good approximation flat with an average water depth of 40 m. The top layer of the sea floor consists quite uniformly of sand. Recorded wave height during the acoustic experiment was $H_{1/3}$ = 1,9 m.

The acoustic experiment was performed with hydrophones suspended from the anchored research vessel "WFS PLANET" and an aircraft dropping charges at intervals of 1 NM. Fig. 2 gives the measured transmission loss for one third octave band with a center frequency of 4 kHz. The two curves correspond to two tracks with different bearings and exhibit no peculiarities.

3. Model Computations

3.1 Description of the Model

The following computations are based on a stochastic ray tracing program which has been described in [1] and [4] and will therefore only be summarized. Since the TL data are in third octave bands the energy associated with a ray is added incoherently into vertical receiver windows. The loss of energy at the sea floor is entered via plane wave reflection coefficients [4] and no bottom roughness is assumed.

To incorporate stochastic variations of the environment into such a model we calculate beforehand according to wave theory the probability density (pd) for the change in the parameters which a plane wave will experience due to the stochastic process. This pd is then implemented into the program. For stochastic sound speed variations we first trace a ray through a deterministic layer from r_0 to r_1 and then compute the stochastic corrections to the ray parameters according to the pd and the pathlength in the layer by a random number generator.

This method itself is independent of the special probability density for the stochastic changes in the ray parameters, which must be selected for the special problem under investigation. In our case the acoustic wavelength is much smaller than any of the characteristic lengths of the medium variability, therefore we may use a frequency independent probability distribution and moreover we are only interested in the angular change $\Delta\phi$ of a ray caused by the stochastic process. The special probability distribution chosen is basically taken from [5] and reads with some modifications:

$$p(\Delta\phi) = \frac{1}{\sqrt{2\pi}\ \sigma_{\Delta\phi}}\ exp\left[-\frac{\Delta\phi^2}{2\sigma_{\Delta\phi}^2}\right] \tag{1}$$

where: $\Delta\phi$ angular change

$\sigma_{\Delta\phi} = \sqrt{2\cdot D\cdot S}$

S =   pathlength of ray in stochastic layer

D =   diffusion constant

     = $D_0 + D_g\frac{d}{gdz}\ C\ (z)$

$D_0,\ D_g \sim (\frac{\Delta C}{C_0})^2/a_0$

a_0 =   correlation length

The diffusion constants D_0 and D_g are measures of the sound speed variability.

All input parameters were determined from the experimental conditions. The bottom reflection coefficient was taken from [4] and the medium attenuation is according to Thorp's formula [3], [2]. The sound speed profile however could not be uniquely determined. Therefore three configurations were considered simultaneously which are shown in Fig.3. The transmitter depth is indicated by T and the horizontal bars define the receiver window centered

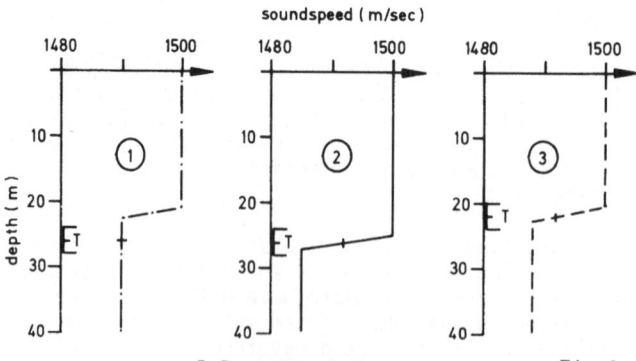

T = Transmitter depth
[= Receiver window

Fig.3 Three model input configurations

310

at T. All following figures display curves for these three configurations simultaneously and were computed for the third octave band centered at 4 kHz.

3.2   Results for a Deterministic Medium

A first computation was done for the three sound speed configurations with a deterministic layering (D = o) which gave the TL curves of Fig. 4. Characteristically for a deterministic model the large differences in TL stem from small variations of transmitter and receiver depths relative to the thermocline, which are not predictable. The measured TL curves lie in between those of configuration 1 and 3. Without doubt a sound speed profile can be found with which the measured TL can be fitted, but this type of fitting is rather undesirable.

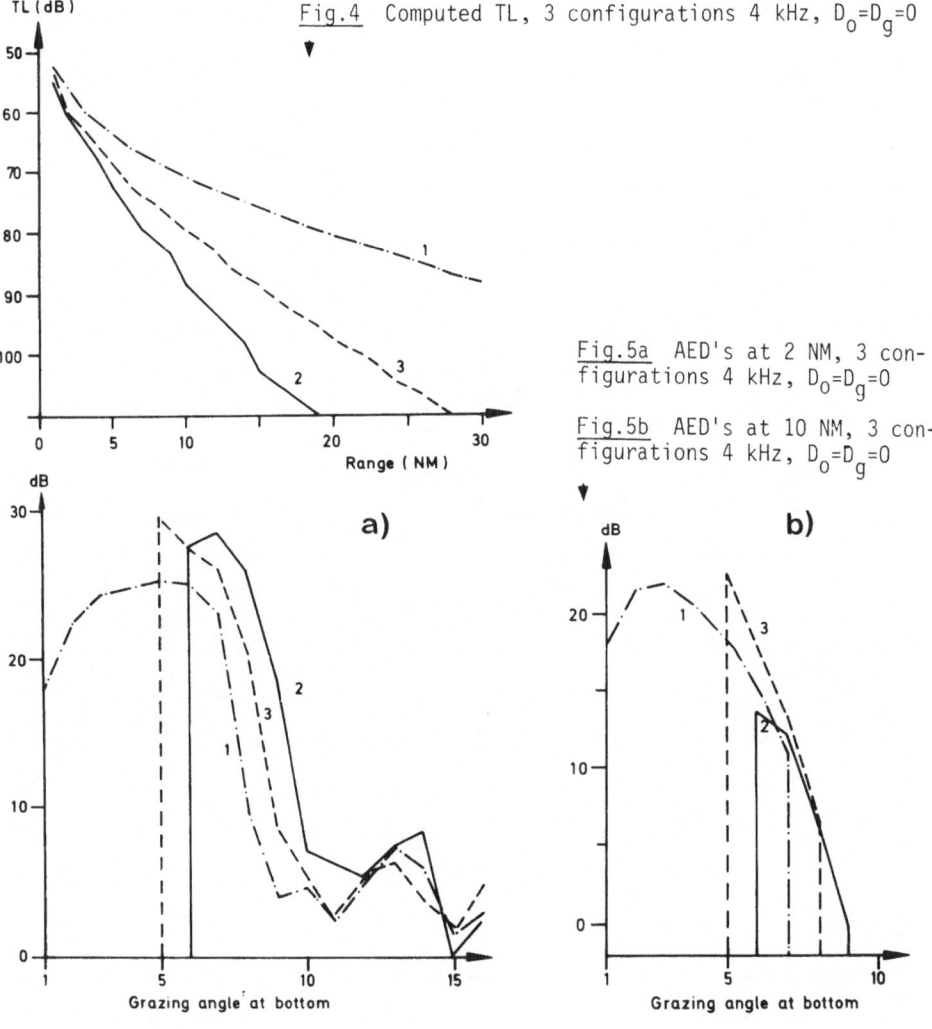

Fig.4   Computed TL, 3 configurations 4 kHz, $D_o=D_g=0$

Fig.5a   AED's at 2 NM, 3 configurations 4 kHz, $D_o=D_g=0$

Fig.5b   AED's at 10 NM, 3 configurations 4 kHz, $D_o=D_g=0$

a)

b)

311

The large differences in the TL curves can be explained by the angular energy distributions (AED's) at the sea bottom which are given in Fig. 5a, b for ranges of 2 NM and 10 NM. Because of the higher loss at the bottom occurring with increasing grazing angle the energy propagation is dominated at longer ranges by rays with the smallest grazing angle possible and the AED's become narrower with increasing range, where the minimum grazing angle is a function of the initial conditions.

3.3 Results for a Stochastic Medium

3.3.1 Thermocline Variability

The computations were redone with the diffusion coefficients
$$D_o = 0, \quad D_g = 10^{-6}m^{-1}$$
which is quite similar to those in [1] and correspond to a stochasticly varying thermocline. The resulting TL curves are shown in Fig. 6. In comparison to Fig. 4 it is noteworthy that: (i) all three curves are closer together

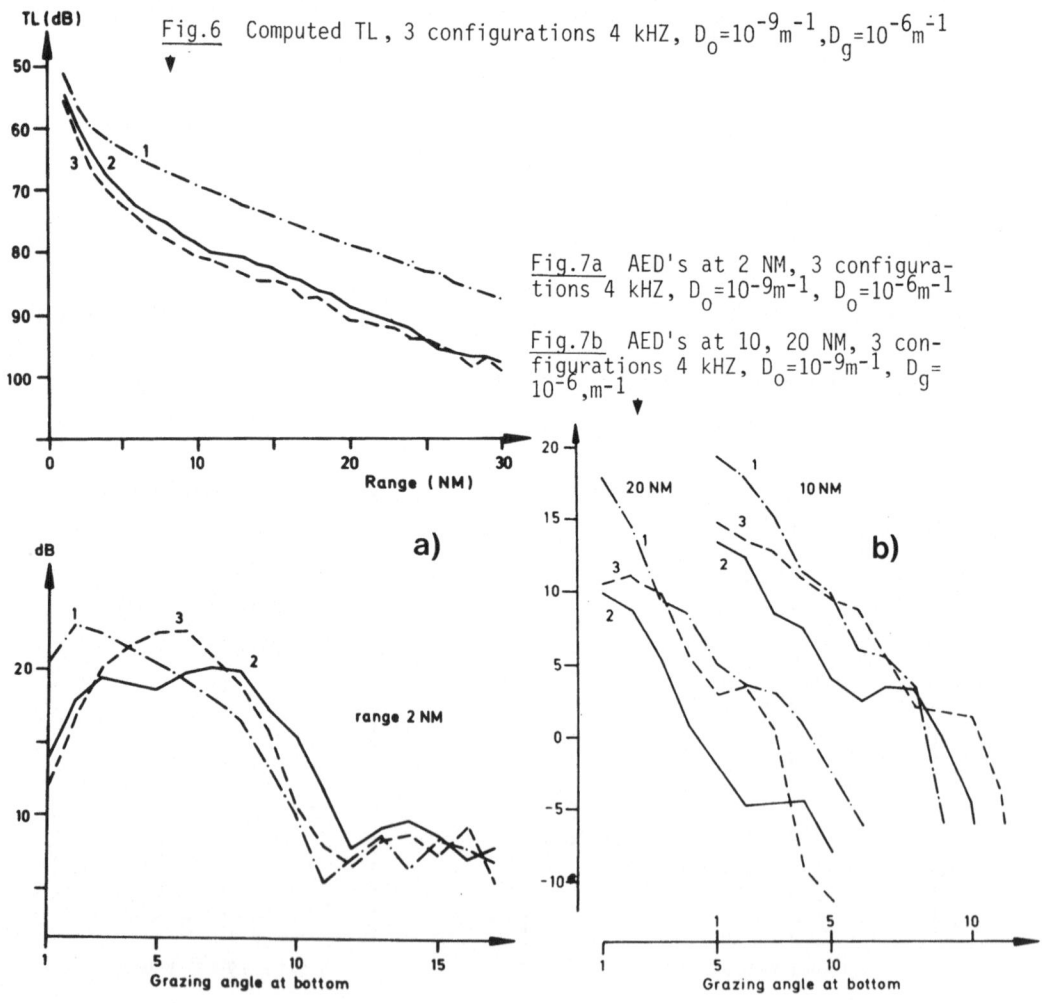

Fig.6 Computed TL , 3 configurations 4 kHZ, $D_o=10^{-9}m^{-1}$, $D_g=10^{-6}m^{-1}$

Fig.7a AED's at 2 NM, 3 configurations 4 kHZ, $D_o=10^{-9}m^{-1}$, $D_o=10^{-6}m^{-1}$

Fig.7b AED's at 10, 20 NM, 3 configurations 4 kHZ, $D_o=10^{-9}m^{-1}$, $D_g= 10^{-6},m^{-1}$

and the TL of model configuration 2 and 3 are practically undistinguishable and that (ii) the TL curves run parallel for ranges larger than approximately 7 NM.

This can be explained by the corresponding AED's given in Fig. 7a, b. At a distance of 2 NM from the source a considerable amount of energy has already been scattered into the small grazing angle region which has been empty for configurations 2 and 3 in the deterministic medium. For larger ranges the AED's are no longer significantly different which explains the equal slope of the TL curves. In addition the shape of the AED's does not change from 10 NM which means they have become stationary. This stationarity is an important result and should allow one to separate long range propagation characteristics from source and receiver geometry.

3.3.2 Overall Sound Speed Variability

The computation was repeated with the diffusion constants

$$D_o = D_g = 5 \cdot 10^{-7} \, m^{-1}$$

which assume also a sound speed variability in the isospeed layers. According to eq. 1 we expect a larger loss than in the previous case. Fig. 8 displays the corresponding TL curves, which show the same general behaviour as those in Fig. 7.

The AED's at 2 and 10 NM (Fig. 9) suggest that a stationarity is almost reached at only 2 NM. In comparing the AED's at 10 NM for the two stochastic parameter sets we notice a significant difference: while the AED at 10 NM of Fig. 7b is markedly peaked at low grazing angles, the one in Fig. 9 is considerably broader which is due to the larger stochastic influence.

The measured TL is approximated within roughly 5 dB by all stochastic TL curves shown here. Close agreement is reached for $D_o = o$, $D_g = 10^{-6} m^{-1}$ with

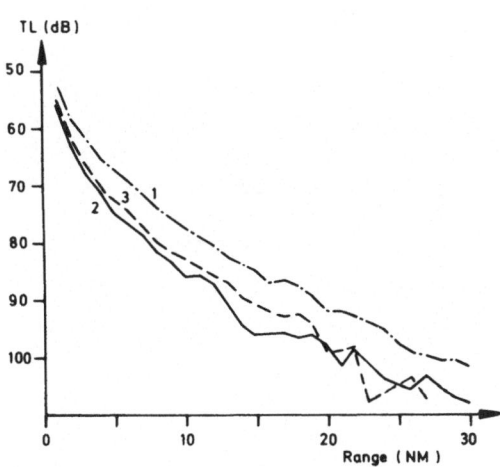

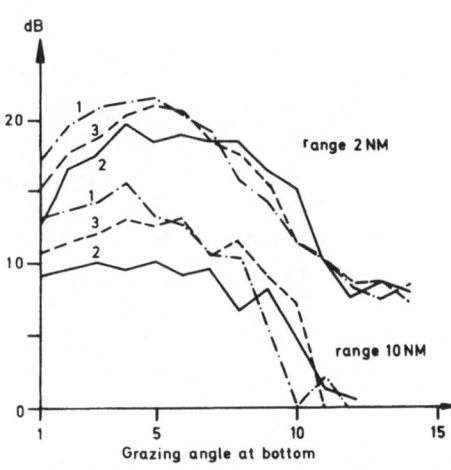

Fig.8 Computed TL, 3 configurations 4 kHz, $D_o=D_g=5 \cdot 10^{-7} m^{-1}$

Fig.9 AED's at 2,10 NM; 3 configurations 4 kHz, $D_o=D_g=5 \cdot 10^{-7} m^{-1}$

313

sound speed configurations 2 and 3 and for $D_o = D_g = 5 \cdot 10^{-7} m^{-1}$ with configuration No. 1.

However this agreement is of no significance if we cannot limit the arbitrariness of the choice of environmental parameters to achieve a fit for the TL data. In the cases considered here the measurement of the AED at only one or two ranges could at least lower this arbitrariness and avoid questionable fittings.

4. Conclusion

We have shown in a model study that

  i the angular energy distribution (AED) may in a stationary stochastic environment become stationary at sufficiently large ranges from the source,

  ii very different AED's at the acoustic source may in a stochastic environment converge to only one long range AED,

  iii different types of stochastic sound speed variability may result in different long range AED's,

  iv a larger influence of the stochastic variability on the acoustic transmission paths may cause a faster convergence towards the stationary long range AED.

Therefore we expect for the acoustic energy transmission that

  i in case of unfavourable initial conditions, which would give a high loss in a deterministic environment, the stochastic variability may cause less loss,

  ii in case of optimum initial conditions for a deterministic environment the stochastic sound speed variability may cause a larger loss.

It remains to be seen whether measured AED's can be used to estimate the variability of the sound speed layering.

References

1. H. G. Schneider, J. Acoust. Soc. Am., Vol. 62, No. 4, pp. 871-877 (1977)

2. R. H. Mellen, H. G. Schneider, J. Acoust. Soc. Am., Vol. 62, No. 4, pp. 1038 - 1041 (1977)

3. K. H. Keunecke, L. Magaard, Mémoires Société Royale des Sciences de Liège, 6e Série, tome VII, pp. 147 - 166 (1975)

4. H. G. Schneider, Acustica, Vol. 35, No. 1, pp. 18 - 25 (1976)

5. L. A. Chernov "Wave propagation in a random medium" McGraw-Hill, 1960

The Inverse Backscattering Problem – a Different Approach

S. Leeman and P. Vaughan

Department of Medical Physics
Royal Postgraduate Medical School and Hammersmith Hospital
London, W12 0HS, United Kingdom

1. Introduction

Many investigative procedures utilise the pulse-echo method for proving the acoustical structure of an unknown medium. While the technique is invaluable for providing an idea of the presence and arrangement of backscattering structures, it cannot be regarded as a quantitative one. This is because it does not allow the evaluation of that medium property which actually generates the backscattering, viz. the characteristic acoustic impedance. There is some interest, therefore, in developing methods for the reconstruction of the impedance profile itself, directly from the measurements of the backscattered echoes. Only backscattering is considered here, since pulse-echo methods are widely used; also, attention is focussed on a simple layered, lossy medium only. The extension from this one-dimensional case may, in fact, be made, by the introduction of the "effective" impedance [1]. The method proposed here is an elaboration of the impediography method, as developed by JONES and others [2].

2. First-Order Reflections

Consider a medium filling the half-space, $x \geqslant 0$, with impedance $Z(x)$, and (linear) absorption, α. Pulse-echo methods are described most usefully in the time domain, and, to include the case that the ultrasound wave velocity in the medium is non-uniform, we introduce the "travel-time" variable,

$$\varepsilon(x) = \int_0^x \frac{dx'}{c(x')}$$

where $c(x')$ is the appropriate velocity at position x' in the medium. Accepting, for the moment, that the impulse response of the medium can always be retrieved from the backscattered echoes by a deconvolution process, the essential problem is thus the relationship between the temporal behaviour of the impulse response, $h(t)$, and the impedance profile, $Z(\varepsilon)$. This is very easily obtained by the method of the Duhamel integral [3], a fact which seems to be overlooked in the rather tortuous calculations of other investigators [2]. This method has as its starting point the concept of the response to a unit input - in this case, equivalent to the reflection from a unit impedance step. For such a step, located at ε', and rising from impedance 0 to $Z(\varepsilon')$, the impulse response is:

$$e^{-2\alpha_o \varepsilon'} \delta(t - 2\varepsilon')$$

315

where α_o is the amplitude absorption coefficient, per unit travel time. This is now normalised to "unit input", on division by $Z(\varepsilon')$, and the impulse response of the general medium may be written down immediately:

$$h(t) = \int_0^+ \frac{dZ(\varepsilon')}{d\varepsilon'} \frac{1}{Z(\varepsilon')} e^{-2\alpha_o\varepsilon'} \delta(t - 2\varepsilon') d\varepsilon' \tag{1}$$

$$= \frac{1}{2} \frac{d \ln Z(\varepsilon)}{d\varepsilon} \Bigg|_{\varepsilon = t/2} e^{-\alpha_o t}$$

Thus, provided α_o is known,

$$Z(\varepsilon) = Z(0) \exp\{2 \int_0^\varepsilon h(2t) e^{2\alpha_o t} dt\}$$

3. Higher Order Reflections

The above treatment is clearly valid for weak backscattering only, but may be extended to include multiple reflections as well. This may again be done by the complicated limit treatments of other methods [2], or even by resorting to the advanced Gelfand-Levitan inversion method [4], but we present here a physically appealing approach which will enable higher order terms to be written down immediately, even for quite complex media.

We consider a first-order reflection occurring at ε. The impulse assumed incident on the medium is written δ_o, and, on arrival at ε has the form δ_ε, symbolically written as

$$\delta_\varepsilon = P(\varepsilon, 0) \delta_o$$

Here, $P(\varepsilon'', \varepsilon')$ is an <u>operator</u> (the "propagator") describing the propagation of an impulse from $\varepsilon = \varepsilon'$ to $\varepsilon = \varepsilon''$. Except for the very simplest media, P has a complicated form and δ_ε will be a time-stretched function.

The reflected pulse generated at ε is denoted $\delta_{R\varepsilon}$, and is symbolically written by introducing the reflection operator, R_ε (the "reflector"),

$$\delta_{R\varepsilon} \equiv R_\varepsilon \delta_\varepsilon = R_\varepsilon P(\varepsilon, 0) \delta_o$$

The pulse arriving back at the origin, from the single reflection at ε, is thus

$$h_\varepsilon(t) = P(0, \varepsilon) R_\varepsilon P(\varepsilon, 0) \delta_o$$

For the general impedance profile, therefore, the first-order impulse response is

$$h(t) = \int_0^+ P(0, \varepsilon) R_\varepsilon P(\varepsilon, 0) \delta_o d\varepsilon$$

316

This may be compared with (I), to give the equivalence

$$\delta_o = \delta(t)$$

$$P(\varepsilon,0)\, \delta_o = \delta(t - \varepsilon)\, e^{-\alpha_o \varepsilon} \qquad (2)$$

$$R_\varepsilon = \frac{d \ln Z(\varepsilon)}{d\varepsilon} \qquad (3)$$

Eq. (2) is valid only for the simple lossy medium discussed here; in general, the propagator is a complicated integral operator, whose evaluation may be possible only to some approximation. Eq. (3) is generally valid, for impedance-generated reflections. ·Also, it follows, quite generally, from reciprocity, that

$$P(\varepsilon'',\varepsilon') = P(\varepsilon',\varepsilon'')$$

The propagator-reflector method is a powerful one, allowing the general Nth-order reflection component of the impulse-response, $h_N(t)$, to be written down (I). The impulse response is then expressed as the sum

$$h(t) = \sum_{N=1}^{\infty} h_N(t) \qquad (4)$$

When the scattering is strong, this series may converge only slowly, and another formulation of the problem, in terms of an integral equation to which (4) is the iterative solution, may be devised.

4. The Role of Absorption

In practice, media show strong absorption (even for the longitudinal wave case treated here), which will, in addition, exhibit dispersion. Causality requires that dispersive absorption is associated with velocity dispersion, and this undoubtedly complicates the evaluation of the propagator. The following points now merit consideration: if the full form of the propagator is used, the impulse response should be evaluated by deconvolution with respect to the incident wave form. Secondly, in dispersive media, the concept of "velocity" needs some elaboration. We contend that the signal velocity should be utilised in the definition of the travel-time. Thirdly, in the presence of absorption fluctuations, backscattering may be generated which does not originate from the impedance variations.

The last point does not appear to have been previously noted, and it is worth looking into this in more detail. Consider a plane interface located at $x = 0$, between two media of different absorption, but of the same impedance. Let the incident (i) and reflected (r) waves (if present) in medium I ($x \leqslant 0$) be denoted as the velocity potentials

$$\Phi_i = A_i\, e^{-\alpha_I x}\, e^{iw(t - x/c_I)}$$

$$\Phi_r = A_r \, e^{-\alpha_1 x} \, e^{iw(t + x/c_1)}$$

where w denotes the circular frequency and c the velocity. The transmitted wave in medium 2 is

$$\Phi_t = A_t \, e^{-\alpha_2 x} \, e^{iw(t - x/c_2)}$$

Applying the usual boundary conditions (continuity of pressure and particle velocity) we find that the reflected wave does exist, with the amplitude

$$A_r = A_i \left\{ \frac{\alpha_1 c_1 - \alpha_2 c_2}{\alpha_2 c_2 - \alpha_1 c_1 + 2iw} \right\}$$

The reflected wave thus suffers a phase change, and the reflection coefficient is dispersive itself - i.e. there would be a change in pulse shape on reflection.

5. Conclusions

It is possible to reconstruct the impedance profile from the backscattered echoes in simple cases. The propagator-reflector method prescribed here is physically simple, and allows higher-order reflection terms to be written down without much difficulty. There is the question, however, of whether the inversion is unique when higher-order terms are taken into account (2). More problematic, is the influence of dispersive absorption, which will complicate the propagator, and introduce reflections, if it is varying throughout the medium. This last circumstance must raise doubts as to whether impedance reconstructions, from single pulse-echo sequences alone, have any validity.

References

1. Leeman, S. in "Acoustical Imaging", Vol. 8, ed. A. F. Metherell, Plenum Press, 1979.

2. Jones, J. P. in "Recent Advances in Ultrasound and Biomedicine", Vol. I, ed. D. N. White, Research Studies Press, 1977.

3. Magnus, K. "Vibrations", Blackie, 1965.

4. Lefebvre, J. P. in "Lectures on the Applied Inverse Problem", Springer, in press.

Index of Contributors

Apfel, R.E.   79

Bjørnø, L.   261
Bovis, A.   23
Butt, R.   225

Chahine, G.L.   23
Cramer, E.   54
Crowther, P.A.   194
Crum, L.A.   84
Curtis, G.D.   237

Desaubies, Y.J.F.   281

Ebeling, K.J.   35
Evans, A.   90

Fedoseeva, T.N.   177
Fenlon, F.H.   141
Fridman, F.E.   177

Gimenez, G.   101
Goby, F.   101
Goldberg, V.N.   177
Graham, E.   108

Haussmann, G.   219
Hedges, M.   108
Hentschel, W.   47
Hinsch, K.   225

Kedrinskii, V.K.   119,170
Kobelev, Yu.A.   151
Kuzavov, V.T.   119

Lastman, G.J.   72
Lauterborn, W.   3,42,47
Leeman, S.   108,315
Løvik, A.   211

Malykh, N.V.   164
Medwin, H.   187
Mellen, R.H.   272
Mørch, K.A.   95

Nakoryakov, V.E.   157

Ogorodnikov, I.A.   164
Ostrovsky, L.A.   151

Parvulescu, A.   237
Pokusaev, B.G.   157
Prosperetti, A.   13

Rath, H.J.   64

Schippers, Ir.P.   205
Schneider, H.G.   308
Scully-Power, P.   294
Shreiber, I.R.   157
Sutin, A.M.   •151
Sutton, P.   108

Teslenko, V.S.   30
Tilmann, P.M.   113
Timm, R.   42·

Vaughan, P.W.   108,315

Weitendorf, E.-A.   230
Wentzell, R.A.   72
Wijngaarden, L.van   127
Wonn, J.W.   141

Zarnitsina, I.G.   177

Ocean Acoustics

Editor: J. A. DeSanto

1979. 109 figures, 5 tables. XI, 285 pages
(Topics in Current Physics, Volume 8)
ISBN 3-540-09148-3

Contents:
J. A. DeSanto: Introduction. – *J. A. DeSanto:*
Theoretical Methods in Ocean Acoustics. –
F. R. DiNapoli, R. L. Deavenport: Numerical
Models of Underwater Acoustic Propaga-
tion. – *J. G. Zornig:* Physical Modeling of
Underwater Acoustics. – *J. P. Dugan:* Oceano-
graphy in Underwater Acoustics. – *N. Blei-
stein, J. K. Cohen:* Inverse Methods for Reflec-
tor Mapping and Sound Speed Profiling. –
R. P. Porter: Acoustic Probing of Space-Time
Scales in the Ocean. – Subject Index.

The purpose of the book is to discuss the
correlation between oceanic variability and
the propagation of acoustic signals in the
ocean. This correlation is presented using the
results from acoustic experiments in the
ocean, theoretical models, and computational
and controlled-tank experimental simulation
methods. Experimental results reveal that
ocean dynamics such as internal gravity waves
and circulations induce amplitude and phase
fluctuations in acoustic signals. Theoretical
and computational methods are developed to
quantify this interrelationship. Water-tanks
are useful in isolating critical ocean parameters, and because their controlled measurements are repeatable. This book presents
both an introduction to and the current status
of these various areas of research in ocean
acoustics.

Y. I. Ostrovsky, M. M. Butusov,
G. V. Ostrovskaya

Interferometry by Holography

1980. 184 figures, 4 tables. Approx. 280 pages
(Springer Series in Optical Sciences,
Volume 20)
ISBN 3-540-09886-0

Contents:
General Principles: Interference of Light.
Optical Interferometry. Holography. Holo-
graphic Interferometry. – Experimental
Techniques: Light Sources. Hologram Recor-
ding Materials. Setups. Experimental
Aspects. – Investigation of Transparent Phase
Inhomogeneities: Features of Holographic
Interferometry of Transparent Objects. Sensi-
tivity of Holographic Interferometry and
Methods of Changing It. Holographic
Diagnostics of Plasma. Use of Holographic
Interferometry in Gas-Dynamic Investiga-
tions. – Investigation of Displacements and
Relief: The Process of Interference-Pattern
Formation in Holography Methods of Inter-
preting Holographic Interferograms when
Displacements are Studied. Investigation of
Surface Relief. Flaw Detection by Holo-
graphic Interferometry. – Holographic
Studies of Vibrations: Influence of Object Dis-
placement on the Brightness of Reconstruc-
ted Imeage – The Powell-Stetson Method.
The Stroboholographic Method. Phase
Modulation of the Reference Beam. Deter-
mining the Phases of Vibrations of an Object.

Positrons in Solids

Editor: P. Hautojärvi

1979. 66 figures, 25 tables. XIII, 255 pages
(Topics in Current Physics, Volume 12)
ISBN 3-540-09271-4

Contents:
P. Hautojärvi, A. Vehanen: Introduction to
Positron Annihilation. – *P. E. Mijnarends:*
Electron Momentum Densities in Metals and
Alloys. – *R. N. West:* Positron Studies of
Lattice Defects in Metals. – *R. M. Nieminen,
M. J. Manninen:* Positrons in Imperfect Solids:
Theory. – *A. Dupasquier:* Positrons in Ionic
Solids.

Springer-Verlag
Berlin
Heidelberg
New York

Electrets

Editor: G. M. Sessler
1980. 205 figures, 24 tables. XII, 404 pages
(Topics in Applied Physics, Volume 33)
ISBN 3-540-09570-5

Contents:
G. M. Sessler: Introduction. – *G. M. Sessler:*
Physical Principles of Electrets. – *J. van Turn-
hout:* Thermally Stimulated Discharge of
Electrets. – *B. Gross:* Radiation-Induced
Charge Storage and Polarization Effects. –
M. G. Broadhurst, G. T. Davis: Piezo- and
Pyroelectric Properties. – *S. Mascarenhas:*
Bioelectrets: Electrets in Biomaterials and
Biopolymers. – *G. M. Sessler, J. E. West:*
Applications.

H. Haken

Synergetics

An Introduction
Nonequilibrium Phase Transitions and Self-
Organization in Physics, Chemistry and
Biology
Springer Series in Synergetics
2nd enlarged edition. 1978. 152 figures,
4 tables. XII, 355 pages
ISBN 3-540-08866-0

Contents:
Goal. – Probability. – Information. –
Chance. – Necessity. – Chance and Necessity. – Self-Organization. – Physical
Systems. – Chemical and Biochemical
Systems. – Applications to Biology. – Socio-
logy: A Stochastic Model for the Formation
of Public Opinion. – Chaos. – Some Historical Remarks and Outlook.

B. Saleh

Photoelectron Statistics

With Applications to Spectroscopy and
Optical Communication
1978. 85 figures, 8 tables. XV, 441 pages
(Springer Series in Optical Sciences,
Volume 6)
ISBN 3-540-08295-6

Contents:
Tools From Mathematical Statistics:
Statistical Description of Random Variables
and Stochastic Processes. Point Processes. –
Theory: The Optical Field: A Stochastic
Vector Field or, Classical Theory of Optical
Coherence. Photoelectron Events: A Doubly
Stochastic Poisson Process or Theory of
Photoelectron Statistics. – Applications:
Applications to Optical Communication.
Applications to Spectroscopy.

P. S. Theocaris, E. E. Gdoutos

Matrix Theory
of Photoelasticity

1979. 93 figures, 6 tables. XIII, 352 pages
(Springer Series in Optical Sciences,
Volume 11)
OSBN 3-540-08899-7

Contents:
Introduction. – Electromagnetic Theory of
Light. – Description of Polarized Light. –
Passage of Polarized Light Through Optical
Elements. – Measurement of Elliptically
Polarized Light. – The Photoelastic Pheno-
menon. – Two-Dimensional Photoelasti-
city. – Three-Dimensional Photoelasticity. –
Scattered-Light Photoelasticity. – Interfero-
metric Photoelasticity. – Holographic Photo-
elasticity. – The Method of Birefringent
Coatings. – Graphical and Numerical
Methods in Polarization Optics, Based on the
Poincaré Sphere and the Jones Calculus.

Springer-Verlag
Berlin
Heidelberg
New York